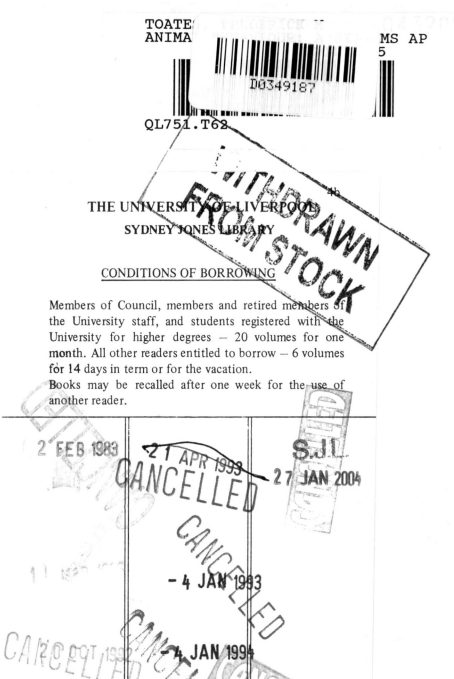

THE UNIVERSITY OF LIVERPOOL
SYDNEY JONES LIBRARY

CONDITIONS OF BORROWING

Members of Council, members and retired members of
the University staff, and students registered with the
University for higher degrees — 20 volumes for one
month. All other readers entitled to borrow — 6 volumes
for 14 days in term or for the vacation.
Books may be recalled after one week for the use of
another reader.

Animal Behaviour—
A Systems Approach

Animal Behaviour—
A Systems Approach

Frederick M. Toates
Biology Department, The Open University

JOHN WILEY & SONS
Chichester · New York · Brisbane · Toronto

British Library Cataloguing in Publication Data:

Toates, Frederick M.
  Animal behaviour.
  1. Animals, Habits and behavior of
  I. Title
  591.5   QL751   79-41485

ISBN 0 471 27724 X
ISBN 0 471 27723 1 Pbk

Photosetting by Thomson Press (India) Limited and
printed in the United States of America

Contents

v

Preface

I had several kinds of student in mind when writing this book. Like most authors, I had a series of imaginary readers standing over my shoulder for most of the time. Primarily, it was designed for 2nd- and 3rd-year undergraduate and postgraduate courses in psychology and zoology, but I would hope also to interest physiology and medical students.

It is usual for an author to attempt to justify a new book by pointing out where similar books already available in the area are either deficient, out of date, or difficult to understand. In this respect my task is particularly easy since there are no introductory books available on a modelling or systems approach to motivation and animal behaviour. The idea that underlies writing this book is that a massive and harmful gap exists in the literature. If I have filled just a small bit of this gap my time will not have been wasted.

Largely by means of teaching and meeting numerous students who were taking two sets of courses (on the one hand animal behaviour/ethology, and on the other something variously called physiological psychology, psychobiology, biopsychology, or motivation and emotion), I have become convinced that something fundamental and vital is lacking in the way the subject is presented in the literature. This is particularly so in the physiologically based literature. In many cases the textbooks themselves are excellent in the areas that they cover. Nonetheless, the overall picture painted is usually not that of a live organism containing subtle control systems that enable it to interact with its environment. Rather, in many cases one gets a detailed list of the physiological components. It is as if one were to be shown, in minute detail, a gearbox, accelerator, steering wheel, and brake, but left to speculate about the energy yield of petrol, pollution, starting and driving the car, obeying traffic lights, and the rules of priority in traffic.

The student who wishes to know about the animal's interaction with its environment needs to consult the ethology literature, but here of course much of what is vital about the organism's physiology gets left out. The interface between natural behaviour, its physiological bases, and consequences, appears to be largely missing from the literature. Interestingly, it is precisely those few areas where the truly comparative *and* physiological aspects of *behaviour* are pursued that arouse the most interest from students. This approach is exemplified by Beach, Bolles, Garcia, Rozin, and Seligman amongst others.

My misgivings about the subject area of physiological psychology were very strongly reinforced when I saw the type of answer given in examinations, since I assumed that this reflected what the student considered essential. It was in many cases comparable to the way I remembered answering questions in history and geography, i.e. by giving the names and dates of the kings and queens of England, or perhaps more closely comparable, the location of capes and bays in Africa. The student often attached great importance to the names of brain 'centres', fibre tracts, and chemical transmitters. But what seemed to be missing was a feel for working *processes* serving functions and allowing commerce with the animal's environment. In the life of the animal (or bit of an animal) that emerged there was little sense of decision-making, competition, or challenge. Although the student often knew the names of the brain regions that could be lesioned or stimulated to obtain, say, polydipsia, the actual economics of water supply and turnover, osmotic gradients, kidney function, etc. were topics seemingly of less importance. Little sense of complex interaction and competition with other activities was obtained from reading the standard textbooks.

Now let me hasten to add that, in using these examples, in no sense am I criticizing the students or their lecturers. On the contrary it is largely through discussions with them that I have been able to formulate my position.

In writing this book I have taken a modelling or systems approach because I believe it to be absolutely essential to understanding the processes by which animals behave and distribute their time between competing activities. I can see no alternative approach to understanding processes of such intimidating complexity as, say, hunger and energy balance. Knowing what to call this study caused me a few problems. I originally wanted to put 'Motivation' in the title, but by 'motivation' some readers might expect a standard psychology text, involving such topics as achievement motivation, cognitive, dissonance etc., but little consideration of ethology.

Quite deliberately, little or no attention is paid to brain mechanisms and the neurophysiology of behaviour. I realize that this will lose me some potential support. I make no apologies though, and there are several reasons I can give to justify this approach. First, at the level of explanation adopted here it is most appropriate to take a 'black-box' approach to the brain. I am concerned with overall performance characteristics and not with the precise embodiment of the machinery. Secondly, there already exists a vast literature on the neurophysiology of behaviour, written by people much more competent than I am. The student should turn to this literature for the more neurophysiological side of the story.

Ultimately a synthesis will be needed, and where it seems appropriate I point the reader towards neurophysiological findings. However, it is my belief that there is something profoundly wrong with the subject-matter of the neurophysiology of behaviour. There has been such a rapid growth in the volume of literature in this area that it threatens to go critical at any moment. I feel that it badly needs a better theoretical base and organizing structure, and I am convinced that if people from this background became interested in systems then they would find such a base.

To many students the mere suggestion of a modelling or systems approach causes immediate fear. On numerous occasions I have encountered the argument 'I can see it's interesting, but my mathematics is not good enough for that'. Unfortunately, and quite wrongly, the systems (wo)man is often seen as being brilliant at mathematics but somewhat eccentric. They are thought to survive by entertaining a very narrow circle of academic friends with esoteric mathematics that are guaranteed to be incomprehensible to lesser mortals. I have a number of comments to make about this. First, whereas the more conventional investigators of motivation are to be numbered throughout the world in at least four figures, one could count on the fingers of one hand (or at most two) the number of people actively pursuing a systems approach to motivation. Therefore at worst we can only waste relatively few academic person-hours.

Fear of mathematics arises largely because in the past the approach to the subject has paid too much attention to getting answers and not enough to illuminating the fundamental principles involved. However, this is not the place to pursue this favourite hobby-horse of mine, since I employ almost no mathematics for this introductory text. Rather than being difficult, a systems approach, provided it is presented simply, clearly, and with a total absence of jargon and unanchored constructs, can uniquely and vividly illuminate how biological systems work.

As I will show in the chapters that follow, we are now in a position to construct formal but tentative models of hunger, thirst, sex, and behavioural temperature control. Although, as yet, we can't do the same for fear, aggression, exploration, or sleep, we can at least ask the appropriate questions. I am sure that by adopting a *philosophy* of modelling and systems we will illuminate how such behaviours are organized.

Therefore I do not feel I am cheating in the title of the book. I hope that the student is not disappointed in what follows. Please write to me if anything is unclear, or if I can be of help.

Finally, I would like to acknowledge those who have helped me in this project, particularly Dr J. Archer, Professor R. C. Bolles, Dr D. A. Booth, Dr N. Chalmers, Dr T. Halliday, Dr J. Horne, Dr A. Houston, Dr J. Lambley, Dr S. Murphy and Dr J. Thomas who carefully read the initial version of one or more individual chapters.

Milton Keynes,
June 1979   F. M. Toates

CHAPTER 1

Introduction

1.1 MOTIVATION AND CAUSALITY

This book is primarily a study of motivation. To motivate means to move, and is concerned with what 'moves' an animal to act in a certain way. Our interest concerns behaviours having rather clear goals and characterizable in purposeful terms, such as feeding, drinking, copulating, and fleeing. We will consider the causal factors which instigate such behaviour and examine what terminates it. Sleeping, avoidance, exploration, fighting, and temperature-regulation are also examined; in other words most of the major activities which occupy the time of the animal.

The present investigation works on the basis that a set of *causal factors* determine the animal's behaviour. The philosopher and mathematician Laplace once argued that if we knew the state of every object in the universe and every force acting at a point in time we could, at least in theory, predict the subsequent outcome of all things for all time. The tacit assumption behind motivation studies of the kind pursued here is more modest than this, but is in the same philosophical tradition. It is that an animal's behaviour is definable in terms of its organismic structure and the stimuli that impinge upon its sense organs, and therefore we should be able to identify some of the important causal factors that determine its behaviour.

What is sometimes called the Harvard Law of Animal Behaviour (Dubos, 1971), must be one to which every observer of animals has at some time felt sympathy:

'Under precisely controlled conditions an animal does as he damn pleases.'

However, to view each rat as a unique existential being, shaping its own destiny in a manner closed to scientific investigation, may prove to be a premature surrender. The justification for writing books of this kind is that motivation is valid material for scientific study. Such study consists of looking for general rules of behaviour that enable predictions to be made.

Causality means that a particular behaviour does not arise spontaneously but is attributable to antecedent conditions; without these, the behaviour would not have occurred. Causality is a concept which through several hundred years has caused headaches to philosophers, but we will not pursue the philosophical angle here. It is interesting to note that many people strenuously deny that their

1

behaviour is caused by antecedent influences; even if one knew all of their past history, it is claimed, this still would not enable one to predict their behaviour with any certainty. We just as strongly deny spontaneity outside of ourselves. Mysterious cases of where physical objects move but no antecedent event could be identified are, depending upon the particular observer, either attributed to hidden electromagnetic fields or to poltergeists. Also, as Skinner argues in *Beyond Freedom and Dignity*, those behaviours of which we are not so proud (e.g., a sudden loss of temper) are usually attributed to external antecedent influences.

Bolles (1975) uses the expression 'traditional rationalism' to describe the kind of explanation for behaviour that most people believe, and virtually all employ in their day-to-day social commerce. Thus we attribute behaviour to events in the mind of the other person; only he or she is aware of the cause of their own behaviour and by implication responsible for it. But this is of little help since it only serves to raise the question of what caused these mental events.

Mechanism is a word which implies a physical cause for an event. The faith of the mechanist is supported by the regularity of events in the outside world, and, within psychology, mechanism means extension of these principles to behaviour. It means that particular physical causes lie behind an animal's behaviour; causes which, it is argued, will ultimately be revealed. Rejection of explanations in terms of mechanism can perhaps only be defined in the negative; that behaviour does not have identifiable physical precursors. Again we must point out that almost any word used in these discussions: cause, effect, mechanism, physical, etc., can form the subject of interminable philosophical speculation.

A lucid account of the way in which one experimental psychologist views the evolution of the concept of motivation was given by Bolles (1975). Thus, sometimes we observe behaviour which is the immediate consequence of a stimulus impinging upon the organism. An example might be the withdrawal of a foot in response to touching a hot object. The behaviour is simple, predictable, and immediate, while the intellectual energy involved in picturing what is happening is not much more than that demanded for understanding, say, an automatic ticket-actuated barrier. However, it is more often the case that a simple cause cannot be identified. It is usually not observable just before the behaviour concerned was displayed. One is therefore forced to formulate explanations in terms of states within the organism, its past history of stimulation and learned associations. These form the minimum information components which are needed to account for the animal's behaviour. But, as Bolles reminds us, these seem remote and not easily understood. By contrast, if we can postulate 'a single active internal agency', in other words either a drive or motivation, this seems to make the situation rather simpler. It appears to be a necessary intellectual tool to have available in a discussion of the behavioural spectrum of animals. The fact that drive cannot be observed directly and is dependent upon such things as hormone level and learning is perhaps no argument against it having a provisional usefulness.

Bolles quotes Skinner (1938), who brings the subject into relief. Since a rat sometimes responds, but does not always respond, to the presence of food by eating, we attribute the variability to an internal state: hunger. Where variability does not exist then there is no necessity to postulate an internal state. The probability of a rat flexing its leg in response to shock is very high, so a 'flexing drive' is redundant.

Whether or not a drive concept is used, or even considered important, depends upon the purpose of the investigation. Where behaviour is being examined, and is clearly dependent upon events within the animal which are poorly understood, then it may prove inescapable to use a word such as drive to indicate the tendency of the animal to perform in a certain way. At other times it may be either misleading, or leading to an intellectual cul-de-sac, to attribute behaviour to a drive, when the only evidence for a drive is the behaviour we are observing (Beach, 1970). As more and more evidence about the neural, hormonal, and experiential determinants of behaviour becomes available, it would appear that theoretical drive constructs will suffer natural extinction. The present study is concerned primarily with the observable and quantifiable antecedent causes of behaviour and has little or nothing to say about the drive concept.

When looking at the causal factors which determine behaviour we work on the basis that the nervous system is the final determinant. Although it may be said that a rat drinks because it is 24 h water deprived, on closer scrutiny what is meant is that water deprivation changes the state of the animal's body-fluids, which in turn causes a detectable signal to arise in the nervous system. We are not in a position to look at this neural activity and the way in which it acts to generate the motor action of drinking. However, the physiological events which give rise to behaviour, such as body water, body energy, and levels of sex hormones in the blood can be analysed. From such information the important physicochemical events that influence the animal's behaviour via the nervous system can be revealed. Factors external to the animal, after impinging upon its sense organs, influence its subsequent behaviour. For example, the size of a meal is determined conjointly by the energy state of the animal and by the sensory qualities of the diet and its past associations.

1.2 MODELS OF BEHAVIOUR

A particular view will be advocated in the pages which follow: that in order to understand the causes of the animal's behaviour we should attempt to model the essential processes in formal terms. What is meant by a model? A model bears a strong resemblance to something else in certain important respects, but is none the less different in its actual embodiment. A model train is very different from a real train but shows some common features. Some of the real train's performance characteristics can be demonstrated by the use of a model train. In the immediate context of this discussion, by a model we mean a representation which captures the important processes underlying behaviour. If our understanding of the component processes of drinking is embodied into a model and this model then exhibits realistic 'drinking' we have done as much as we possibly can to construct an unambiguous theory of thirst. One rapidly learns to distrust theories that are simply written in words. What the author intended is often not the same as the reader's interpretation. Models that take the form of a diagram showing the important processes are much better, and better still is computer simulation. We write into a computer simulation our understanding of the component parts. The computer then predicts the outcome of our assumptions and this is compared with the behaviour of the animal. In this way a truly 'working' model of behaviour is built. This study proceeds no further than presentation of formal diagrams, but we will certainly have our eye on the possibilities of simulation.

Models are probably never final, in the sense that they cannot be improved. Invariably they show only some important aspects of behaviour and omit or simplify others. The history of modelling is one of evolution towards models which apply to an increasingly broad range of data. A model is built, tested, and compared with the real system. Disparity between the model and the biological system suggests improvements in the model. These are incorporated and it is retested. By this iterative process better models evolve. Models are realistic only in that their construction *and* performance have important features in *common* with the organism, not merely that they can copy its behaviour, since that is fairly easy to achieve.

1.3 LEVELS OF EXPLANATION

We will be concerned with levels of explanation other than the causal kind. Ethologists speak of *functional* explanations. Functional explanations refers to how behaviour increases the survival and reproduction chances of an animal. A functional explanation of courtship is that it increases reproductive success. Functional explanations are very difficult to test. How is one to measure, in the natural habitat the number of offspring an animal has, and follow reproductive success over several generations? Functional explanations are therefore usually somewhat speculative. *Fitness* is a term commonly used in the ethology literature. An animal contributes to its fitness by engaging in activities that increase the chances of its genes being perpetuated. To avoid poisoned food contributes to an animal's fitness because it increases its own survival chances, and hence its chances of gene perpetuation through reproduction. The *Coolidge effect* (see later) refers to arousal of sexual behaviour by a change of partner. This provides difficulties of explanation at a causal level, but a functional explanation is clear: it increases the chances of successful reproduction.

One way of viewing the situation is to imagine that we were designing the animal so as to maximize its fitness. What design criterion would we employ? The answer to this question should be compared with how the living system functions. To speak of evolutionary design is not to appeal to teleology or mysterious forces, but rather to consider that evolution tests each design according to its fitness. The best 'designs' survive.

We imagine that no individual animal actually knows that mating leads to gene perpetuation in quite the same way that we do! If we are to believe the writings of some early Pacific explorers the 9-month gestation period is such that even in some human cultures the connection between sexual behaviour and repro-duction was not fully appreciated. Functional explanations tell us how mating increases gene perpetuation chances, an aspect of the explanation being that an animal with a low sexual drive would have little chance of gene perpetuation. But the functional explanation is of no value in directly explaining the sexual drive of an individual animal. Here our explanation must be in terms of hormone levels, incoming sensory information from a partner, executive brain regions control-ling mounting, etc. This is the *causal* analysis, and we will need to refer to both kinds of explanation, hopefully avoiding the common error of indiscriminately mixing levels. With care the causal and functional levels of explanation can usefully be studied in parallel. If the adaptive significance of an aspect of behaviour is known then the characteristics to look for at the causal level may be hinted at.

Fitness and the functional explanation are intimately connected with *ecolog-ical* information. For example, when it is deprived of food in the laboratory, the Mongolian gerbil drinks relatively enormous quantities of water, and as yet no entirely convincing explanation is available. When we consider the environment which has shaped the behaviour mechanism of this species it is clear that the situation of an abundance of free water in the absence of food is one that is hardly likely to have been encounted in the arid desert regions of Asia. Therefore it would seem inappropriate to ask what might be the adaptive significance of such behaviour. To take another case, rats cut down on their intake of food when they are deprived of water. A shortage of water in the presence of food may be a not uncommon environmental situation when viewed in terms of the rat's evolution. Reduction in feeding under these circumstances can be seen to be beneficial to the survivial of the individual animal, unlike excessive drinking during starvation in the gerbil, which hastens death. Thus ecology can both provide useful hints on the nature of causal mechanisms, quite apart from being worthy and interesting subject matter in its own right.

1.4 ETHOLOGY AND PSYCHOLOGY

Traditionally, experimental psychologists have not paid a great deal of attention to ecological and evolutionary arguments. Evidence has been obtained from rather few species in a restricted set of environmental circumstances. The white rat has been the favourite subject. Some (e.g., Rowland, 1977) have even gone so far as to ask, 'is the white rat a red-herring?' Others have answered this in the affirmative. I do not believe that psychologists are wasting their time looking at rats in Skinner-boxes and small cages. Any organism in any environment must reveal the reaction of that species to that environment and these are surely valid data. However, more useful information could probably be obtained if a slightly more complicated and natural environment were to be used. By comparing rodents of several ecological backgrounds we may be able to illuminate our understanding better than by concentrating on just one species. By testing a different species the aspect of behaviour in which we are interested may be exaggerated because of the animal's ecology, and therefore easier to manipulate. 'Safe' assumptions may be questioned by interspecific comparisons. That is in no way a declaration of support for the anti-rat brigades.

I believe that there is a danger of missing important features of the rat's behaviour by not taking a more ethological view of its behaviour. The models that are derived from simple environments may be perfectly valid and a necessary step in developing more comprehensive models. However if we dwell too long with them they may prevent an understanding of some crucial processes underlying the animal's behaviour. For example, the small laboratory cage allows no insight into how the rat may take distance to food and water into account in timing ingestion. Therefore the models that are discussed later include factors which are appropriate to the animal's natural habitat and indicate how the animal would divide its time between several activities.

Behaviour may at times be satisfactorily understood in terms of simple physiological factors; for example the water-deprived animal on being given water appears to be largely under the dictates of the body-fluid deficit. Models of behaviour should obviously be capable of accounting for this; but they must not blind us to the fact that animals at times drink when there is no obvious fluid deficit present, and the model should be sufficiently flexible to include such aspects of behaviour. We must consider the 'decision rules' which the animal employs in determining its choice of behaviour. Given both food and water the animal which has previously been deprived of food and water will need to make some decisions about its ingestive activity. Simply observing its minute-by-minute state of body-fluids and energy, even if we could perform this measurement, would not necessarily yield a convincing explanation of its timing. In the next chapter we consider the physiological aspect of motivation using the concept of homeostasis as an integrating tool. In Chapter 3 we discuss the timing of behaviour: how animals divide their time between competing demands. We then consider each motivational system in turn.

CHAPTER 2

Physiological regulation, control theory, and model building

2.1 HOMEOSTASIS

2.1.1 Introduction

A subject which forms an important part of this study is *homeostasis*. Although we will use it to introduce the discussion and it appears throughout the study, homeostasis will not be raised to the status of a panacea that can explain all our observations. We will show where homeostasis can illuminate aspects of behaviour, but also where additional but closely allied theoretical concepts are needed.

There are a number of situations where a living organism appears frail and vulnerable. Lack of oxygen in the brain for a few seconds leads to death. For the animal to stay alive body temperature can depart only a little from its normal value. Yet, what is so impressive is just how well the body maintains itself in the face of hostile environments. In the case of humans and many other species, body temperature does not depart far from 37°C despite enormous changes in environmental temperature. The body has evolved the ability to make adjustments such that its internal environment is relatively stable. In the case of fluid balance, although we constantly lose water by evaporation, urination, etc., we fairly accurately replace it. This stability of the body's interior was termed *homeostasis* by W. B. Cannon in his book *The Wisdom of the Body*, first published in 1932. Homeostasis means that not only are bodily conditions relatively constant but, within bounds, whenever something threatens to shift conditions away from equilibrium corrective action is automatically taken so as to return to equilibrium. Homeostasis is one of the absolute fundamentals of physiology.

Each system has its effectors. For example, relative constancy of body-fluids is crucial to survival; drinking and water excretion being the instruments of action. When body-temperature falls or is in danger of falling we start to shiver. This generates heat and brings temperature back to normal. If temperature rises above normal then we sweat. This causes heat to be lost from the body at an increased rate until temperature comes down to normal.

At this point it is appropriate to introduce an analogy or model. We can take a system in engineering, and from examining how it works illustrate some features

7

that it has in common with a biological system. It is important that we are prepared to modify our model. Analogies are useful up to a point but we should constantly challenge our analogy or model in the belief that it may give rise to a better one. The models which follow have undergone this evolutionary development.

We will consider the thermostatically controlled system which heats a room, a favourite example in this area. The occupant has set the temperature on the dial to, say, 20°C, and at the time of observation room temperature is indeed 20°C. The temperature outside is 0°C. The occupant opens a window for a few minutes and some heat escapes, taking the temperature down a few degrees. The temperature detector in the room perceives this, and the heating comes on automatically. Heat is generated until the temperature returns to 20°C, and then the heating is automatically switched off. We would explain what has happened in the following terms. The system is constructed to compare the actual temperature of the room with the temperature set on the dial. When actual temperature falls below set temperature then heating is switched on. When temperature returns to the set value heating is switched off. Suppose the occupant decides the room is not warm enough and moves the setting from 20 to 22°C. Heating will come on for long enough to bring temperature to 22°C. Perhaps half an hour later temperature will fall sufficiently to switch the heating on again.

We can now introduce some control systems terminology. The temperature which the occupant sets on the dial is the *set-point* of the system. It may also be called the *input* to the system. 'Input' is not to be confused with the energy input of the heater, though beginners usually do make that mistake. We are concerned with the information input on the temperature dial, not the energy input. The temperature of the room is the *output* or *actual value* of the control system. If everything is working satisfactorily and we don't make unreasonable demands then the output will closely match the input. But if we impose excessive demands this match may be unobtainable. If it is very cold and we ask for 26°C then even with heating on for 24 h a day the room may never reach this value. In German, set-point is known as *Sollwert*, 'should-value' and output is known as *Ist-wert*, 'is value'. This perhaps conveys the meaning more vividly than the English equivalents, drawing the useful distinction between command and response.

Engineering science can often be extremely helpful in illuminating how living systems work. However, it can also be misleading, sometimes to an alarming degree, unless we are clear as to how far we can take any particular analogy. In the case of a set-point in engineering, a human set it. An engineer designed it and a room occupant adjusts it. Now clearly a rather different set of circumstances lie behind any biological control system. Even so, it may be the case that features can be identified which are in some sense analogous to those of the engineering system. Often, though, the biologist tends to infer the existence of components in the biological domain simply because the behaviour of the biological system suggests that they might be present. It is precisely here that the kind of model that may prove to be more harm than help can start its gestation. I believe, though, that often even if a model ultimately proves to be highly inadequate it is better than no model at all. At least it provides a means for critical discussion. We will return to this subject after we consider a biological system.

To take a particular example, body-water content is maintained at a fairly constant level. A hormone, known as anti-diuretic hormone (ADH), controls the rate at which the body excretes water. As its name implies, the higher the concentration of ADH in the blood the lower is the excretion rate of water. If an organism in normal fluid balance is loaded with water the secretion of ADH is temporarily halted, and the body produces urine at a high rate. When the excess has been lost ADH secretion rate rises back to normal and consequently urine flow falls back to its normal value again. To consider the opposite disturbance, when the body has lost water through, for example, being exposed to heat, production of ADH is high and little urine is formed. Although, of course, this cannot provide water, it can at least ensure that only a minimum amount is lost from the body. In other words, when body-water content departs from normal the ADH–kidney system automatically takes action to minimize the effect of the disturbance, which seems to bear a strong similarity to the room temperature control system. This is indeed the case and the analogy has proven to be extremely valuable. There appear to be detectors of some aspects of body-fluid volume. A detector in the blood signals that volume has increased following a water load and causes ADH secretion to be inhibited. A detector of the size of the cellular compartment will also signal an excess and contribute to excretion. Conversely if the body were to lose water these same detectors would cause ADH secretion to be increased and would also cause the animal to ingest water if it were available.

The essential features which the room temperature and body-water systems have in common are as follows. The causal agents drinking and excretion are influenced by body-fluid volume. They are influenced such that if body-fluid volume is disturbed it tends to return to its original value. For a fixed setting on the dial, generation of heat bears the same relation to room temperature. For some purposes it may convenient to call the normal value of body-fluids the set-point, and this is the subject to which we now direct attention.

2.1.2 Set-points

Body-water volume is fairly constant from day to day. If it departs from normal, action is quickly taken to return it to normal. Why then should there be any harm if biologists use the expression 'set-point' in such a case? After all, the similarity to room temperature control would seem very striking. At first 'set-point' may be useful as a provisional term, or metaphor. But we may find ourselves stuck with the term after it outlives its usefulness. The point is that until we scrutinize the biological system we may find more than one different model that can appear to explain its behaviour.

Although total body-water remains relatively constant, and indeed a single set-point system yields predictions entirely compatible with the behaviour of this biological system, we believe control to be exerted by detectors of cellular and

10

extracellular components of body-water. If we were to insist upon a total body-fluid set-point just because it seemed to work in some situations we would miss the precise mode by which regulation is achieved. We might spend all of our time wondering how a single detector could measure total body-water when in reality that is not how control is achieved.

The situation may be even worse than this. It can be illustrated by a familiar example. In England the amount of water in the village pond usually stays reasonably constant throughout the year (except the freak drought of 1976). If it rains heavily the level rises a little, but soon water leaves the pond by ditches and level drops to normal. If, because of low rainfall, little water enters the pond the level may fall slightly. However, as a result of this, ditches either slow down or halt their rate of removal of water so the fall is checked. In other words, water level is fairly constant and disturbances are checked, but would we want to call the normal level a set-point value? I think not, and even as a convenient fiction or metaphor it might be deceptive to employ the expression. Consider another village pond, a man-made one with a full-time attendant. A tap controls flow out and the attendant's hand controls the tap. When the level rises above a certain

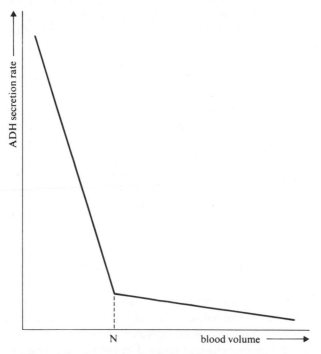

Figure 2.1 A suggested relationship between blood volume and anti-diuretic hormone (ADH) secretion rate. The normal value of blood volume is N. A fall in blood volume below this value sharply increases ADH secretion rate, whereas a rise has a less powerful effect in decreasing secretion rate

reading on a depth gauge the attendant opens the tap. When it falls below this level he closes the tap. Here we have what is, beyond dispute, a set-point controlled system in the traditional engineering sense, and yet, if all we do is observe water level and flow, its behaviour may be indistinguishable from the first pond we described. If we were simply to examine the performance characteristics of an ordinary village pond, we might start a fruitless search for a tap and human hand controlling it. In biology the moral is that a particular engineering model should never dominate our investigations. We must be cautious and consider each system on its merits. Our model should be based upon observables and suggest possible alternative mechanisms but should not entirely colour our interpretation of the data. Sometimes the biological system may possibly be explained in terms of set-points, at other times a simpler mode of operation better fits the evidence.

How about our water excretion model? At present we do not have a perfectly clear picture of how control is exerted. Suppose it is as shown in Figure 2.1. When blood volume falls below value N then ADH secretion rate rises sharply, and when it goes above N then secretion is soon inhibited. 'N' appears to take on the property of an engineering set-point, and some may wish to call it that. I would not do so, since the precise analogous component of a set-point in the engineering system is not present. I think it is analogous to the unattended village pond. We should simply describe the biological problem in the formal terms we have just used. There is no benefit to be served by calling N a set-point unless we can find some physiological signal which is directly analogous to its engineering equivalent. Engineering has served us well enough already in indicating that a displacement is self-correcting and showing us the essential performance characteristics.

2.1.3 Homeostasis and behaviour

So much for physiological homeostasis, but where do the particular interests of the experimental psychologist and ethologist lie? The organism is not self-sufficient, but only survives by its commerce with the environment. In response to water loss only behaviour can correct the physiological deficit. Feeding and drinking are aroused by the need for appropriate commodities, food and water. When body temperature is above normal the organism will take corrective physiological action, but behaviour is also involved: the rat spreads saliva on its skin, and both lizards and humans seek the shade. As students of behaviour we should be concerned with both physiological and behavioural aspects of regulation. At the time of water shortage the kidney and the mechanisms of thirst work in harmony to minimize loss and maximize gain of water. It is only through examination of this concerted effort that we can fully appreciate the result.

If a researcher deprives an animal of water for a period of time (s)he is working on the assumption that this will motivate the animal to drink when water is returned. Water is being lost from the body and, it is claimed, the resulting deficit sets up a state of thirst motivation or drive. When water is available the animal drinks, and so its behaviour forms part of a homeostatic system. If the animal has learned to press a lever in a Skinner-box for the reward of water it will exhibit this following water deprivation. According to a traditional school within psychology, motivation arises because of a physiological displacement, in this case a lower level of body-fluid than normal. When the animal ceases to drink or work for water reward, it is said to be sated (or satiated), and the displacement assumed to exist no more. This is one aspect of homeostasis applied to motivation, and it has played an important role in the development of psychology. It is closely associated with the American psychologist Clark Hull, who based much of his theorizing on the fact that organisms seek to reduce drive levels and learn responses that are associated with drive reduction.

It is worth distinguishing 'broad' and 'narrow' homeostatic interpretations of behaviour. According to the broad view the drinking behaviour of an animal has evolved to serve a regulatory system and so almost every act of water ingestion, except possibly the most bizarre, is seen as homeostatic. The narrow homeostatic view would refer to homeostatic drinking only as that caused by an observable deficit in body-fluids. The broad view stems from Cannon, who recognized that an organism may employ a variety of processes by which it achieves relative constancy (see discussion in Toates, 1979b). For example, it may learn to drink and so avoid dehydration. The narrow view arises because regulation can so obviously and simply be understood in terms of responding to a disturbance so as to eliminate the disturbance. We should recognize that, important as this disturbance-actuated process is, it is only one amongst several that may be employed in the interests of constancy.

An aspect of the narrow homeostatic interpretation of behaviour is as follows. If we observe an animal in the presence of water *ad lib* it will take, say, 15 drinks per 24 h. Some of these will be in association with meals, but let us just consider those drinks that occur in isolation. To explain such behaviour by means of narrow homeostatic interpretation one would argue that at the time of drinking the body-water deficit has built up to a magnitude such as to trigger drinking. The animal then drinks until the deficit is eliminated. This view of motivation sees an otherwise passive organism being goaded into action only when the relevant physiological quantity had departed significantly from normal. Further, it sees the animal as being goaded for a sufficient period of time to return the body to physiological quiescence. The broader view would not deny that deficits can cause drinks, but would not inevitably assume that every drink was caused by a deficit.

2.1.4 Closed and open loops, positive and negative feedback

We have already discussed *feedback*, though we did not call it that. Here we will familiarize ourselves with this expression, as well as another having the same meaning, *closed loop*. Our thermostatically controlled room exhibits feedback, the actual temperature of the room is fed back and is instrumental in determining the heating effort. By contrast, if we had no thermostat but simply switched the heat on full in, say, October and left it on until April, irrespective of the consequences, we would have what is known as an *open loop* system. In this case 'open loop' means that although room temperature is dependent upon the output of energy by the heater, the activity of the heater is *not* dependent upon room temperature. The path of causality is open rather than closed. In thermostatically controlled heating not only is the temperature of the room dependent upon the heater but the heater is dependent upon room temperature. This is called a closed loop system because the direction of causality forms a loop (open loop is something of a misnomer). Output (actual temperature) is fed back and compared with input (setting on the dial).

Other examples of closed loops which are simpler than room heating may be given. It is perhaps just because they are so simple that the student is reluctant to describe them in the same terms that apply to the more complex systems, but this is a reservation that should be overcome. There are closed loops between volume and 'flow out' in the case of the unattended pond. It is important that essential common characteristics should be identified. Consider the example shown in Figure 2.2(a) of a piece of material free to swing around a pivot at one end. Left

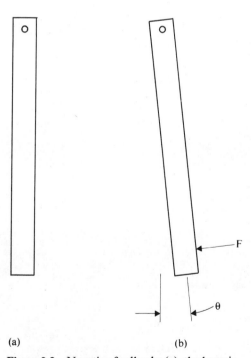

(a)   (b)

Figure 2.2 Negative feedback: (a) the lever is free to rotate around the pivot; (b) by displacing the lever through an angle θ, a force F is introduced that tends to restore the previous conditions

alone it will of course remain with its centre of gravity directly below the pivot. If we suddenly displace it with a flick of the finger to the position shown in Figure 2.2(b) it will automatically swing back to its original position. In introducing displacement θ we have also introduced a restoring force F, which returns it to equilibrium. Displacement introduces a force, and the force then affects the displacement. Force has the effect of eliminating any displacement and returning the system to its equilibrium position. Because of this self-elimination property the system is known as *negative* feedback. The system of fluid balance is also an example of negative feedback: deficits cause ingestion of an appropriate amount of water to *eliminate* the deficit.

Feedback systems can also show *positive* feedback. In Figure 2.3 we picture the difficult task of balancing a lever in the unstable position with the centre of gravity directly above the pivot. If we displace the lever then, as before, the displacement introduces a force that changes the position of the lever, but now, by contrast, the force takes the lever even further away from its original position. This is known as positive feedback, because displacement is not self-correcting

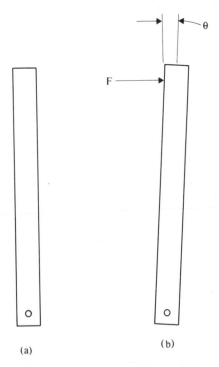

(a)   (b)

Figure 2.3 Positive feedback: (a) the lever is balanced vertically above the pivot, a very precarious position; (b) a displacement from this position, i.e. angle θ, introduces a force F that tends to take the lever even further away from its original position. Contrast this with Figure 2.2

but self-enlarging. Because we live in a finite world, positive feedback is usually short-lived. In this example, once the lever is set in motion it soon falls down to the stable position with its centre of gravity immediately below the pivot.

Short-lived examples of positive feedback are to be seen in biology. If the body loses a large amount of blood this diminishes the supply of blood to the heart itself, which weakens it. The heart therefore pumps less effectively, which in turn diminishes the amount of blood going to the heart still more, and so on. Death quickly follows, at which point of course the system, as such, is totally destroyed.

Positive feedback will probably be recognized by the reader as the more commonly expressed 'vicious circle'. It is not only in pathological states that biological systems exhibit positive feedback, but it is sometimes employed for specific purposes in the normal behaviour of an animal. Consider the effect of drinking on the tendency to continue drinking. In the long-term, then of course the effect is one of negative feedback; water serves to terminate further ingestion. However, in the case of the mouse, before this effect occurs the initial licks appear to potentiate the animal's drive as measured by lick intensity (Wiepkema, 1968).

2.1.5 Feedback, feedforward, and homeostasis

As we have seen, homeostasis and homeostatic behaviour both take slightly different shades of meaning in the hands of different authors. To some, homeostasis refers simply to the relative constancy of the various body functions over time, without much concern for the exact mechanisms by which constancy is achieved. Others are concerned with a variety of mechanisms actively involved in homeostasis, such as memory, learning, anticipation of future states, and negative feedback. To some, 'homeostatic' strictly means a particular type of mechanism in which a departure from equilibrium causes quantitatively appropriate corrective action to be taken, i.e., negative feedback. In the latter terms, homeostatic drinking is that in response to a water deficit somewhere.

It is clear that the body-fluid regulation system can be made to exhibit negative feedback, for example loss of blood, water-deprivation and salt injection are all drinking stimuli. However, it will be argued later that it is wrong to attribute the relative constancy of body-fluids exclusively to its negative feedback aspect; constancy may be better attained by other mechanisms working in concert with it. An example is as follows. When food arrives in the gut, unless sufficient water is already present there, it attracts water by an osmotic process. In so doing, it dehydrates the blood. If the animal were simply to behave according to negative feedback it would drink in response to this dehydration, and some species may indeed do this and no more. However, in the case of the rat it drinks so quickly after feeding that, given the speed of movement of water between blood and gut, it is clear that by then the blood has not had sufficient time to be dehydrated (Oatley and Toates, 1969). In other words, rats take action before the disturbance even occurs and in so doing they avoid it. Clearly this serves homeostasis in that body-fluid volume is well maintained, but this is not achieved by the organism simply reacting to a body-fluid displacement.

By analogy, if it were known that, in a thermostatically controlled room, departures from the temperature set-point would arise, then an additional mechanism would be needed for precise regulation. Suppose that we set 20°C on the dial, but find it necessary to open the window every 2 h to get fresh air. In the past this has caused the temperature to fall to 17°C before corrective action is able to bring it back to 20°C. To pre-empt the disturbance, next time we open the window we could simultaneously switch on an additional source of heat for a few minutes, and experiment with its intensity until opening the window causes almost no error in room temperature.

A 'trial and error' process may be at work in the evolution of the interaction from feeding to drinking. 'Evolution' could be in the experience of an individual animal. It learns that unless action is taken in advance, meals are associated with dehydration a little later. Alternatively, there could be an innate neural connection, in which case we literally mean evolution through the experience of the species. Either way the animal anticipates the deficit and drinks to avoid it, an example of *feedforward*. Feedback is response to a displacement, while feedforward is response in anticipation of a displacement that would otherwise occur.

Researchers in animal behaviour may have overemphasized the role of feedback mechanisms and may have given insufficient weight to feedforward. There is no doubt that feedback may be very widely demonstrated in physiological/behavioural systems, but it is accompanied in many cases by feedforward processes.

2.1.6 The detector in the system—a likely site of misunderstanding

After making explicit the mechanism of control we can still be misled in our line of argument unless every move is critically questioned. For instance, it may be argued that when the size of the body's cellular fluid compartment is reduced below normal, by salt injection, drinking is aroused and a quantity of fluid is ingested to return cellular size to normal. A cell in the brain appears to act as a microcosm, representing the fluid state of all the cells of the body. Such an assumption allows us to speak of a single entity, cellular water, provided that the stimulus at the detector cell is representative of the whole cellular compartment. Under some circumstances it may not be so. If minute quantities of hypertonic saline are injected into the brain then this causes local cellular dehydration and the animal ingests massive quantities of water, but the cells of the body as a whole may be hardly dehydrated at all. By analogy, it is possible to disrupt room temperature by holding a cigarette-lighter near to the thermostat. The detector would get hot and switch off all of the heating, while the occupants of the room would be cold. Conversely, if we were to pack ice cubes around the detector then it would pour out heat to a room having the correct temperature already. These are rather extreme examples, both in the biological and engineering cases, but we will have cause to meet more insidious cases of where we can be led astray by making comfortable models and/or blindly assuming homogeneity throughout the controlled quantity.

2.2 MODELS OF SYSTEMS

We have verbally described some systems that illustrate important principles of regulation, and the next step is to produce a diagram showing the relationship between the components. This is slightly more rigorous than verbal explanation since the opportunity for ambiguity and misunderstanding is less. Once our understanding is in the form of a diagram it is easy to translate to the more formal stage of computer modelling, but that takes us beyond the brief of this text. We will go no further than constructing a representation in *block-diagram* form. Such a diagram shows arrows, each of which represents a variable in the system. These signals are acted upon by the parameters of the system in order to yield other variables. An example is shown in Figure 2.4.

The net flow rate of water into a pond is the difference between the rates at which water enters and leaves. As Figure 2.4 shows, flow out is subtracted from flow in. The difference is net flow. An arrow represents each flow. The plus and minus sign at the summing junction indicate that flow out is subtracted from flow in. The direction of the arrows indicates that net flow depends upon the other two flows. A slightly different alternative representation which some authors employ is shown in Figure 2.4(b). Here a filled segment stands for the operation of subtraction and an unfilled segment stands for addition.

The volume of water in the pond depends upon net flow over all time that the pond has existed. Over all time, flow in must of course have been slightly larger on average than flow out for the pond to have any contents at all, but, over a period of time when the volume is constant, flow out must be equal and opposite to flow in. The relationship between net flow and volume of water in the pond is defined by *integration*.

We can illustrate the meaning of integration as follows. A bucket is empty at the start of observations. At any subsequent time the bucket's volume of water will be given by 'flow in' times the length of time over which the particular flow in occurs. If 'flow in' is 2 litres/min for 2 min, volume will be 4 litres after 2 min. If flow is 2 litres/min for 1 min and then 3 litres/min for 2 min volume is $(2 \times 1) + (3 \times 2)$. In other words, we add together each segment of flow, where the magnitude of volume for each segment is given by flow times the length of time over which that particular flow took place. This is the essence of integration.

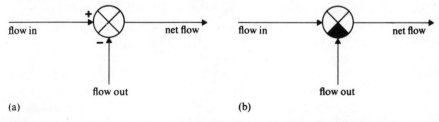

(a)　　　　　　　　　　　　　　　　　　　　　(b)

Figure 2.4 Block diagram representation of information. Net flow is obtained from the flow into a container minus the flow out. (a) and (b) are alternative ways of representing the same process of subtraction

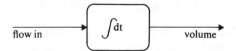

Figure 2.5 Block diagram showing the relationship between the volume of water in a container and the flow in. Since we assume that the container does not leak there is no flow out. By integrating flow, we obtain volume; i.e. we store the incoming flow

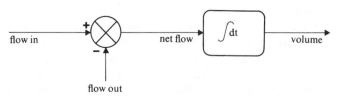

Figure 2.6 The relationship between volume and flow for a container that has both a flow in and flow out; e.g. a village pond

We will not be concerned with the mathematics of working out integrals; but it is necessary to be familiar with what is meant by the term, and it really is nothing more than taking each bit of flow and adding it together. We represent this as:

$$\text{volume} = \int \text{flow (dt)}$$

which simply reads that volume is the integral of flow with respect to time. Examples of integration are numerous. The number of people in a football stadium is the integral of the rate at which they enter through the turnstiles after opening. As Figure 2.5 shows, we can represent the process of integration in the form of a block diagram. Volume is represented by an arrow, as is flow. The operator which relates volume to flow, integration, is shown as the contents of the box.

For the village pond if net flow is integrated we obtain volume (see Figure 2.6). Not only is volume dependent upon net flow, but flow out depends upon the contents of the pond. A possible relationship between flow out and volume is shown in Figure 2.7. At a normal volume flow out is given by f_1 which equals the rate at which water enters. When volume falls below value V_1 flow out is essentially zero. Above V_1 flow out rises as a function of volume in the manner shown. Thus for any volume, we have defined flow out and this is precisely the meaning of the representation shown in Figure 2.8 where the whole system is assembled. Flow out is fed back and subtracted from flow in to give net flow. Net flow is integrated to give volume, and volume determines flow out according to the operation defined in Figure 2.7.

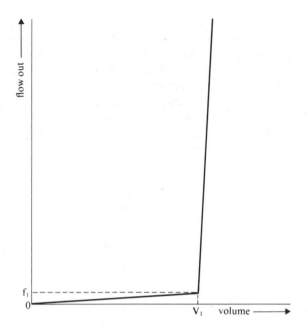

Figure 2.7 Suggested relationship between the flow of water leaving a pond and volume. V_1 is the normal volume, and a very sharp increase in flow occurs when this volume is exceeded

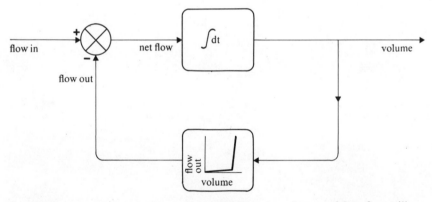

Figure 2.8 Block diagram of the relationship between volume and flow for a village pond

2.3 SUMMARY, CONCLUSIONS, AND DISCUSSION

Physiological functions such as body-fluid volume, temperature and electrolyte concentration vary only slightly over long periods of time, despite a hostile environment. Homeostasis refers to this aspect of the organism's performance and to the mechanisms that ensure stability. One such mechanism is negative feedback. If we disturb the physiological function then detection of the disturbance instigates corrective action so as to eliminate the disturbance.

To some investigators 'negative feedback' and 'homeostatic' are synonymous; the organism is said to be responding homeostatically when the stimulus is an internal displacement. This was termed the narrow interpretation, not in any pejorative sense but merely to distinguish it from other interpretations. The broad view considers a variety of processes contributing to homeostasis such as avoidance of disturbances by anticipation, memory, etc. The negative feedback mode is seductive in that it so obviously can contribute to constancy; the thermostat analogy is intuitively appealing. Other processes working in concert with negative feedback are perhaps more subtle.

The negative feedback aspect of behaviour may be clearly shown; for example, the post-deprivation drinking behaviour of an animal or its response to salt injection. Biological negative feedback systems do not necessarily involve anything directly corresponding to an engineering set-point. The 'passive' regulation of the village pond may sometimes turn out to provide a suitable analogy.

The evolution of our understanding of behaviour is in part reflected in a series of models each often more complex than its precursor (though sometimes simpler). Our arguments can perhaps be most unambiguously transmitted as formal diagrams and simulations.

CHAPTER 3

Decision-making and motivational systems

3.1 INTRODUCTION

In the last chapter we discussed the explanatory approach that a physiological psychologist might adopt towards a behavioural problem. In the case of thirst (s)he would generally try to isolate the variables of body-fluids and drinking from the influences of other activities. As far as possible competing demands would be minimized, the animal normally being housed in a small cage with adequate food and no other animals present. Body-fluids would then be disturbed by, say, salt injection, and the behavioural reaction of the animal measured. The animal would need to invest almost no energy in getting to water (when available), the spout being only a few centimetres from its mouth. It is presumed that following the disturbance the animal's sole concern is to regulate body-fluids, and the experimenter's sole concern is to measure body-fluid state and water ingested.

In one form or another this reflects the tradition of much behavioural investigation. It is necessary rigidly to control experimental conditions and make accurate measurements, and so this kind of uncontaminated laboratory experiment is obviously appropriate. The physiological basis of thirst is extremely complex to understand even with the help of rigidly controlled conditions. However, ethologists and, in recent years a number of psychologists, have shown interest in broader issues. They ask what does the animal do in a situation where it simultaneously has several competing demands to satisfy? What does it do in a complex environment where food and water are not readily available? In asking this kind of question we inevitably ignore some of the details of how it regulates any one physiological condition and try to see the overall picture. Both approaches are valuable; they ask fundamentally different questions and yield different answers. We will attempt here to initiate a synthesis.

The approach of this chapter is to some extent an adaptation of the work of David McFarland and associates (A. Houston, R. McCleery, R. Sibly, S. Larkin, and I. Lloyd) at the Animal Behaviour Research Group of Oxford University. I have deliberately simplified their position in some respects, and extended it in others so as to fit our present interests. Later I will attempt to broaden this work and integrate with physiologically based data.

3.2 DECISION RULES

Animals can be described as behaving according to a set of rules. Our task is to try to establish those rules that organize behaviour. Animals avoid the potential problem of ambivalence; stimuli appropriate for more than one behaviour may be present but the animal usually can only perform one behaviour at a time. Some species seem to have a very simple set of rules that govern their behaviour, and that ensure *singleness of action*. Consider, for example, the carnivorous gastropod *Pleurobranchaea* (Davis, 1979). Feeding takes very high priority in this species; it will even abandon its mate during copulation if offered food. As Davis argues, one must conclude from this priority that in its evolutionary history mates were more frequently encountered than food. However, there are circumstances where survival dictates that energy intake must be subordinate to other needs. During egg-laying there is a hormonal suppression of feeding which ensures against the animal eating its own eggs. Life does not appear to be terribly rich for this species; with a very simple set of commands and a hierarchy one could mimic its life-style. Using commands like 'IF' FOOD 'THEN' HALT MATING, a computer would have no difficulty in simulating its activities. With a set of hard and fast rules it would avoid ambivalence such as being unable to decide whether to feed or copulate.

When we come to more complex organisms many more factors must be taken into account. For instance food may vary in quality along dimensions of salt content, protein *v* carbohydrate, dry *v* moist, etc., that the animal must discriminate. It may be obtainable at fixed locations which means that the animal may acquire a 'cognitive map' of its environment.

It seems a daunting task to attempt to establish decision rules for the behaviour of an animal such as a rat or bird. However, we can try to identify some of the controlling factors and make intelligent guesses as to what the rules are.

3.3 FUNCTIONAL CONSIDERATIONS AND CAUSAL EXPLANATION

The rat maintained in a laboratory cage with *ad lib* water and food has minimal individual survival problems. Indeed, because of considering this environment students are frequently lulled into a false sense of security in their understanding of behaviour. In other words, the thermostat models the ingestive behaviour of the laboratory rat rather well. In the natural environment the problems confronting the animal in securing an optimal physiological condition may be much greater. The environment that has shaped the animal's behaviour tendencies through evolution is infinitely more complex than the laboratory cage.

Life in the wild is often a struggle. In order to stay alive the land-based animal must find food and water, yet in searching for these it uses energy and in a hot environment may dehydrate its body. Possibly it also exposes itself to predators. There is no obvious single strategy that maximizes survival chances for all species. Rather, it seems that they are 'programmed' by evolution to make responses that are often a compromise between conflicting requirements. Reasons could be given for an animal engaging in any one of several activities. Eating safeguards against future energy depletion; it makes sense frequently to seek food. A similar argument may apply to drinking. On the other hand sleep may serve a needed restorative function, or at least minimize risk from accidents and predation. Copulation would increase the chances of gene perpetuation.

For obvious reasons, an animal can usually only do one thing at a time (though I have heard of a rat that could eat and copulate simultaneously!). Furthermore engaging in one activity usually has certain harmful side-effects. If we consider only energy regulation it may make good sense for an animal to seek food very frequently. However, we must also consider the cost, in terms of predation, loss of water on exposure to the sun, consumption of energy, etc., involved in such meticulous attention to energy needs. When we do so, it may appear to be more sensible for the animal to suffer much bigger shifts in its energy state; i.e. to eat less often and eat more at each meal.

I may seem to be suggesting that animals are systems analysts, control engineers, and economists, and in one sense I am indeed doing that. We would expect that through evolution those strategies which have best served the animal's interests will survive at the expense of others. For example, thirsty rats reduce food intake because it has paid rats to do so in the past. To say it has contributed to their *fitness* is another way of expressing it. In the case of any individual rat, though, there must be a *causal* explanation of this behaviour, and that is our primary interest. The causal explanation might be in terms of a neural inhibition from the thirst controller to the feeding controller. Functional explanations give us vital clues about possible causal mechanisms. A more intellectually satisfying account can emerge by attention to both aspects; but I would argue that it is no stronger than a clue that we are given.

Optimality is a popular term at the moment and one which poses enormous problems for biologists; indeed, the problems are exceedingly complex even in engineering (McCleery, 1978). For the present discussion I will take the following view. I believe that it is never possible to say that an individual animal or species is perfectly adapted to a particular environment. That is if, by 'perfect', one means that given the physical structure of the organism an engineer could not, in theory at least, design better behavioural strategies than those exhibited. This possibility is particularly apparent when the environment is subject to rapid change. However, it seems reasonable to assume that after millions of years of evolution animals would approach optimality. Furthermore, one can produce numerous examples of where organisms make compromises between conflicting demands, when slavishly to follow any one would not be in the best interests of survival.

Animals must constantly make decisions. Monitored physiological states in the body provide sources of information to the brain. A physiological state influences the decision to take action directly appropriate to that particular physiological state. It may also influence decisions directly appropriate to other physiological states. For instance, as we have seen, the decision to feed is influenced by both energy and fluid states. The underlying theme of this book is that information on physiological state is used along with other sources of information in producing decisions.

If a man is hungry his physiological energy state will influence his decision of whether or not to visit the local Chinese take-away. If it is pouring with rain, it is late at night, and there is a considerable journey involved, then he may put off going until he is hungrier, or go to sleep hungry. But if it is light and warm, and the journey is not so far, then with the same energy state he may readily go and buy food. Now I am not suggesting that there is necessarily the same kind of decision-making processes in humans and rats. There may be, but we simply don't know. Viewed though from the outside the outcome is sometimes amazingly similar. Perhaps the word decision is too anthropomorphic, and should be put in inverted commas in the case of animals. I think this is unnecessary since it vividly conveys what we mean. All we are arguing is that, in this connection, the nervous system of an animal is in receipt of several sources of information, such as the appropriate physiological state, estimated distance to food, etc. On the basis of computations carried out on this information the animal seeks food or does not. We will employ the term *causal factor*. The set of causal factors is the set of factors that influence the animal's decision. Thus the perception of food, bodily energy state, fluid state, etc., are some of the causal factors for feeding and food seeking.

The discussion now turns to some simple examples of how animals make decisions and, in so doing, divide their time between various activities.

3.4   CUE STRENGTH AND PHYSIOLOGICAL STATE

Imagine a rat housed in a small metabolism cage in the laboratory. Water and food will, at most, be about 20 cm from its mouth. Typically it takes 15 drinks in 24 h (Oatley, 1971), and this would be explained by saying that on each occasion the causal factors have been sufficient to stimulate ingestion. If we take the traditional homeostatic view, we argue that body-fluids reach a certain minimum level that causes the animal to ingest. It takes enough water each time to restore body-fluids to optimum. We then translate this situation to the natural environment, where the rat must make a long journey to water. Does it still take 15 drinks in 24 h? Although the necessary observations are not available, extrapolating the available evidence suggests that it would take considerably fewer.

Would we then expect the rat to become seriously dehydrated in the natural environment? Not necessarily, since the intake in the laboratory may be in excess of minimum need. In the wild it could of course drink more at each of its fewer drinks, and that probably does occur to some extent. However, if the rat drank less frequently it might need to drink less overall. For one thing its kidneys might work more efficiently and lose less water. It might eat less than in the opulent conditions of the laboratory and hence need less water.

In other words I am arguing that environmental cues probably influence the decision of when to seek water. If water is far away I will call this low *cue strength* of water. In the laboratory cage, water has a high cue strength because it is near and easily obtainable (McFarland and Sibly, 1975).

What evidence is there for this view of drinking? We will pursue the details later, and some arguments are speculative. However, we do know that ingestion is partly under environmental control. Nicolaidis and Mather (private communication) found that simply supplying a box in which to hide, placed inside the laboratory cage, caused rats to take fewer drinks. During the light phase they would have been forced to leave a dark sanctuary for a light environment to get water. Diurnal drinks decreased. Larkin and McFarland (1978) found that birds alternate less frequently between feeding and drinking if a journey is involved in the transition.

I believe that if cue strength is low there must be a relatively high physiological signal (i.e. large dehydration in the case of drinking) to motivate water-seeking. Conversely, if cue strength is high then according to this view even if body-fluids have a relatively high volume the animal may still ingest water. With reference to the latter, in the natural habitat the animal may drink if, in the course of some other activity, it happens to pass water (i.e. high cue strength) even though at the time it does not appear to be dehydrated.

Laboratory studies suggest that cue strength and the associated physiological state determine how well-placed a particular motivation is when in competition with other motivations. The animal may be pulled to, say, food according to some complex function of physiological energy state and food cue strength signals.

To propose that an animal in the wild, such as a rat, takes distance to food and water into account in timing its ingestive excursions implies in one real sense that it *knows* how far distant they are located. In other words, I am arguing in the terms of Tolman (1948), that animals construct cognitive maps of their environment: they have a neural representation of the environment, which influences their decisions.

Cue strength is dependent upon distance to the commodity concerned. Under some conditions it may reflect the rate at which the commodity can be obtained, a term which Sibly (1975) calls incentive. Sibly demonstrated a trade-off between cue strength (incentive) and physiological state (hours of deprivation) in the Skinner-box. In this case a high cue strength meant a high rate of return for responding. Birds were trained to obtain either high or low rates of reward of food or water by pecking a key. They were then made both hungry and thirsty so there was some competition between the two activities. It was found that tendency to engage in an activity was dependent upon the animal's relevant physiological state (measured in terms of length of deprivation) and the appropriate cue strength.

We can construct what are known as motivational isoclines, one such for hunger being shown in Figure 3.1. Consider the bird at point $x_1 y_1$. Its energy deficit (x_1) is fairly small, implying a relatively short food-deprivation period. The rate of obtaining reward in the Skinner-box (cue strength) is high (y_1). The two component causal factors give a strength of feeding tendency that we label t_1. Consider a bird having a much larger energy deficit x_2 but a relatively low cue

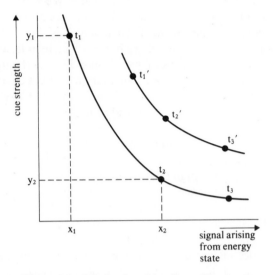

Figure 3.1 Motivational isoclines. The isocline
running through points $t_1, t_2 \ldots$ links points of
equal motivational tendency. Each point on the
isocline (e.g. t_1) arises from a combination of an
energy state (e.g. x_1) and a cue strength (e.g. y_1).
Another isocline links points of a higher motivational tendency, such as $t_1', t_2' \ldots$ Energy deficit
is in fact length of food deprivation, e.g.
$x_1 = 12$ h, $x_2 = 24$ h. Motivational tendency is
the product of cue strength and deficit. For
example, the product $x_1 . y_1$ equals $x_2 . y_2$ and
hence t_1 and t_2 are on the same motivational
isocline

strength y_2. This combination yields a feeding tendency of strength t_2 having the
same magnitude as t_1, and therefore placed on a common isocline.

Sibly argues that we can simply multiply cue strength by deficit to get tendency.
I believe that this is a useful first approximation, but I will add some
qualifications later. However, for the moment it is sufficient to note that there are
a series of combinations $t_1, t_2, t_3, \ldots$ all of equal magnitude. Another series
$t_1', t_2', t_3' \ldots$ etc., have a higher magnitude; the further from the origin, the larger
the tendency.

According to the evidence that Sibly presents, it is the relative strength of the
tendencies that determines which will dominate in a competition. Thus, in a
competition between hunger and thirst, the strength of feeding tendency will be
determined by a multiplicative relationship between deficit and the rate at which
the bird has been used to getting food rewards. If the corresponding product for
thirst is lower in magnitude then hunger will dominate.

3.5 COMPETITION AND TIME-SHARING
BETWEEN ACTIVITIES

3.5.1 Inhibition

There is assumed to be a set of causal factors for each possible behaviour. A single causal factor may influence more than one behaviour. For example, body-fluid state is clearly a causal factor for drinking. It is also a causal factor in feeding; animals stop eating if they are dehydrated. If a rat is given food *ad lib* but no water, and is observed to reduce its food consumption, then one might say that thirst inhibits feeding (Toates, 1979a). This is the kind of inhibition which interests physiologists and physiological psychologists. Let us call it *physiological inhibition*. There is also *behavioural inhibition*, and this has been studied by ethologists. The distinction is best illustrated by means of examples.

Hinde (1970, p. 396) introduces behavioural inhibition in the following way. A tit, in the process of feeding, will abandon this task and dash for cover if a predator appears. Since the causal factors for feeding (energy state and food cues) are still present they must be cancelled by inhibition arising from the now dominant motivation, escape. In Hinde's terms, behavioural inhibition is present when:

'. . . the causal factors otherwise adequate for the elicitation of two (or more) types of behaviour are present, and one of these is reduced in strength because of the causal factors for the other.'

Suppose we simultaneously water- and food-deprive an animal for 24 h, and then put it into a familiar environment where it has both food and water. Obviously it can't eat and drink simultaneously and must have mechanisms for ensuring that a fruitless tug-of-war does not result. In McFarland's (1974) terms, only one activity can occupy the *behavioural final common path* (see also Ludlow, 1976).

If an animal were deprived of food and water and put into an environment containing only water it would drink. If solely food were available it would eat. Either motivation is sufficiently strong to take command of behaviour. We put the animal in an environment having food and water. In the case of 24 h food–water deprivation it is known that rats eat first and drink secondly. The tendency to drink must be restrained by the processes underlying the act of eating, i.e. behavioural inhibition.

The distinction between behavioural and physiological inhibition is that in the former case the execution of the behavioural act of feeding or drinking causes simultaneous inhibition of potential rival activities (but only while feeding). In physiological inhibition, to take the familiar example, depleted body-fluids (not drinking, since no water is available) inhibits feeding until thirst is corrected.

3.5.2 Time-sharing

Consider an animal deprived of food and water which has started eating. After a period of eating it switches to drinking. What determines when this switch

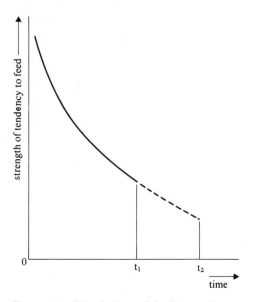

Figure 3.2 The decline of feeding tendency and a switch to drinking. At time zero a hungry and thirsty animal is allowed to feed and drink and, as the graph shows, it feeds and the strength of the feeding tendency falls as food is ingested. At time t_1 it switches to drinking. With a lower level of thirst and the same conditions of hunger it repeats this same behaviour, switching at t_1, and not, as might at first be expected, at a later time t_2. In this case straight competition from thirst would not appear to be able to explain the switch to drinking

occurs? Perhaps one's first guess would be that there is competition for dominance between hunger and thirst. As the animal ingests food so hunger decreases and thirst increases until the strength of thirst can oust hunger from dominance. Indeed, animals do sometimes switch activities by precisely such a process, but it would be wrong to assume that this is always the case.

An example of where we would have to admit other explanations is shown in Figure 3.2 (McFarland, 1974). Again hunger starts out at a higher level than thirst. The animal eats until time t_1 and then switches to drinking. We then repeat the experiment with the same level of hunger but a lower level of thirst. It surely follows that if drinking gains control by competition then the animal should eat for longer, say, to time t_2, before the switch occurs. If we were to find that the switch occurs, as before, at t_1 then competition cannot be the solution. We would argue that hunger disinhibits thirst at time t_1.

Under several conditions it is possible to demonstrate cases where changing

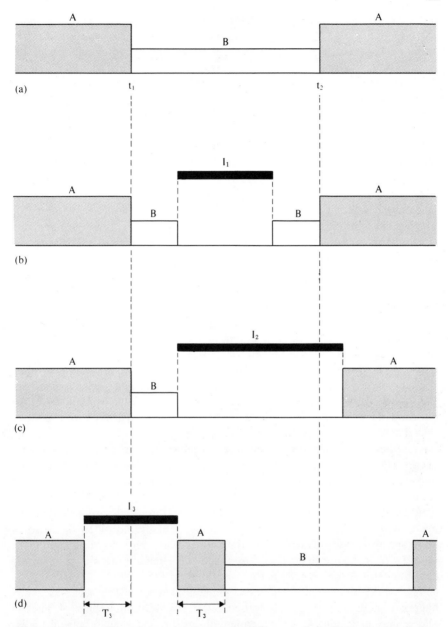

Figure 3.3 The effect of restraining an animal for a period of time *I* when it has activities A and B to which it can attend. For explanation see text (modified from McFarland, 1974. Reproduced by permission of Academic Press)

the strength of the causal factors for the second in priority motivation does not influence the time at which it manifests itself in behaviour. If the dominant behaviour controls the time at which the second in priority occurs and the length of time devoted to the second in priority then this is a process known as *time-sharing* (McFarland, 1974). In the case described, if hunger is increased then the animal may persist with feeding for much longer before switching to drinking.

Suppose no water is available in the experiments described above or that the animal is not thirsty. It may still interrupt feeding at the same time t_1 and in this case fill its time off from feeding with, say, washing or exploring the cage (McFarland and L'Angellier, 1966). The time is used for any lower-order candidates for the animal's attention.

In summary, to take the case when hunger is dominant, what appears to happen is that after a certain amount of feeding activity has been performed the feeding controller *allows* another activity to occur, i.e. it removes its inhibition. After a few seconds it then halts this other activity. By the word 'allow' I mean that timing of activities remains under the control of the dominant motivation. The dominant motivation lifts the inhibition exerted on the sub-dominant for a few seconds, and then re-exerts inhibition. The sub-dominant activity appears then by disinhibition. Following Falk (1969), McFarland uses the expression *adjunctive* behaviour to describe the sub-dominant behaviour.

A fascinating method of demonstrating the disinhibition of a sub-dominant motivation was reported by McFarland (1974). Figure 3.3 shows the theory behind the experiment. The animal first performs behaviour A. At time t_1 it switches to behaviour B and then, at time t_2, it reverts to behaviour A. Accept for the moment that in this case it is believed that A is dominant and B is disinhibited from t_1 to t_2. Suppose we devise a means of preventing the animal from engaging in either activity for a while. Will it perform A or B when its captivity ends? See Figure 3.3(b). If we impose the interruption I_1 the animal will still be in the duration of the time-off allowed by A, and so, presumably, it will resume B. Note that it returns to activity A at the same point in time as in the case when no interruption occurs. Having wasted time I_1, this is not compensated for when this occurs during the time allowed to a sub-dominant behaviour. However, if we impose the longer interval I_2 (see part (c)) the time when we allow the animal to resume activity now corresponds to when it would be back with A again. We would expect behaviour A to follow the interruption. That very little time has been spent on B cannot affect the timing. Finally, consider the case of interruption I_3 (see part (d)). Here the animal is not yet ready to disinhibit B; time T_3 must elapse first. Therefore when the interruption is over it should continue with A for time T_3 before disinhibiting B.

The way McFarland tested these predictions was as follows. A sexually aggressive male stickleback was housed in the tank shown in Figure 3.4. The fish had a nest at point N. An experimental session consisted of placing a jar containing a female fish at the other end of the tank. Thereby the male had two basic activities between which to divide its time: (1) courting the female through the glass of the jar; and (2) taking care of the nest by fanning and manipulating nest material. After the jar containing the female was placed in the tank the male performed courtship, but interrupted this at intervals to journey back to the nest (10–30 sec duration).

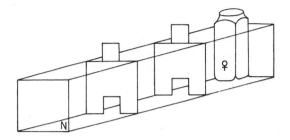

Figure 3.4 The testing tank employed by McFarland. ♀ indicates the glass jar containing a female stickleback, N indicates the nest. Note the two doorways that can be closed by remote control in order to trap the stickleback in the middle compartment. (Source: McFarland, 1974. Reproduced by permission of Academic Press)

In order to move between nest and female the fish swam through a middle compartment having doors at each end. It was possible to trap the fish in this compartment by quickly dropping these doors, and thereby impose the interruptions of Figure 3.3. The fish was observed following interruptions made during a transition from the nest to courtship and vice versa. Interval I was varied between 1 and 30 sec, thus covering the time range which the fish was observed to spend in the nest compartment when free to move through the tank.

In the case of an interruption during the transition from courtship to the nest the fish was likely to proceed to the nest if I was short, but return to courtship if I was long. When the interruption occurred during a transition from the nest to courtship the fish was likely to resume courtship irrespective of the length of interruption. Such evidence accords with courtship being the dominant activity and it disinhibiting nest inspection. Note that simple competition between demands cannot explain this result. It would suggest that the animal should continue in the same direction after an interruption irrespective of its length (within the time limits we are considering).

If an interruption occurs during performance of a sub-dominant behaviour then the sub-dominant behaviour is masked; in other words time is lost and not made up. By contrast, if the dominant behaviour is interrupted then it is simply postponed, since at the end of the interruption the dominant behaviour will always be resumed. This being the case, if an animal is interrupted at random times during the course of satiating, say, feeding (dominant) and drinking (sub-dominant), then drinking will suffer more than feeding. This is indeed the case (McFarland, 1974).

For what purpose could time-sharing have evolved? It may provide some stability to the animal's behaviour for the dominant activity to allow other behaviours to appear. If water is close then during feeding it makes sense to take time off from feeding for the occasional short drink. This way the physiological stress of large food loads is minimized. A rather different, but by no means

32

mutually exclusive, reason for taking time off from one activity was given by McFarland (1970):

'Functionally it seems important that long sessions of feeding or drinking behaviour should be broken up by pauses, during which the animal has the opportunity to scan the environment.'

3.5.3 Stability and persistence

On a visit to Sussex University in 1971 David McFarland illustrated the problem of persistence in the following way. On the previous evening he had been in doubt whether to visit the cinema or a Chinese restaurant; presumably the motivational tendencies were roughly equal. He decided upon the restaurant. On a competition model one might have expected that after a few mouthfuls hunger would have been lowered, at which point he would have left the restaurant and headed for the cinema. This, not entirely serious, example introduces an interesting problem, that of the mechanisms of persistence.

Positive feedback is one possible device for ensuring persistence. Suppose feeding and drinking tendencies are roughly equal. By chance, feeding takes control. Now if the first few mouthfuls of food actually *increase* rather than decrease feeding tendency the animal will persist at feeding for longer. After a while feeding tendency falls below drinking tendency and the animal switches activities. Wiepkema (1968) demonstrated short-term positive feedback in the feeding of mice. He first devised a measure of hunger drive, finding that the length of the first unbroken bout of feeding increased with the length of prior

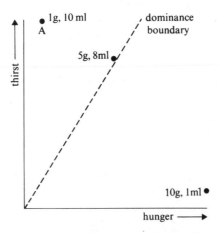

Figure 3.5 State–space representation of joint states of hunger and thirst. For points below the dotted line hunger is dominant; for points above it, thirst is dominant. Point 5g, 8ml is one where hunger and thirst are roughly equal

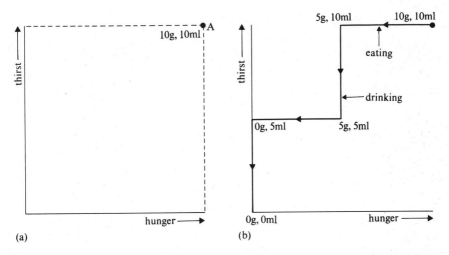

Figure 3.6 State–space representation. (a) a point of 10 g hunger and 10 ml thirst, (b) possible behaviour trajectory followed during the satiation of hunger and thirst

deprivation. Positive feedback was presumed to be present in that in the early stages of a meal bout length increased, indicative of increased drive. Similarly, in deprived rats, Le Magnen (1971) showed an initial increase in feeding rate followed by a sharp decrease.

Cue strength could provide stability in the natural habitat, and would form another example of positive feedback. Imagine an animal in the wild finding itself in a potentially ambivalent position between turning left for food or right for water. It turns left. This brings food nearer, which increases its cue strength, while water becomes more distant, which decreases its cue strength. Behaviourally the animal will gain 'momentum' by moving towards food. Given the geography of the situation a substantial amount of food will need to be eaten before drinking tendency (low cue strength) will exceed feeding tendency. This factor may even provide stability in the Skinner-box where two competing demands are present. Simply engaging in an activity such as pecking seems to involve a focus of attention which gives persistence (for details see Andrew, 1978).

To discuss persistence, it is necessary to digress and describe what to the reader may be a novel form of representation. Consider two behaviours, feeding and drinking in response to appropriate deprivation. We define deficit retrospectively in that an animal is said to have an x g deficit because it eats x g to satiety. Similarly a y ml water deficit would be defined retrospectively by observing the animal to drink y ml to satiety. (This is in some ways an unsatisfactory technique since eating creates thirst, but let's not worry about that now.) A criterion such as 5 min with no responding is used to determine when satiety has occurred, and the experiment terminated. If (defined retrospectively) the animal has an energy deficit of 2 g and a fluid deficit of 12 ml then thirst might be expected to show its

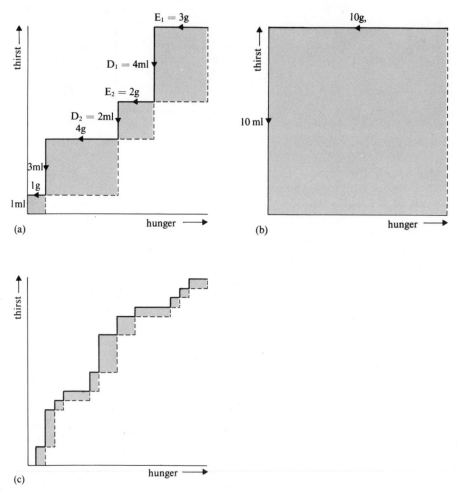

Figure 3.7 Behaviour trajectories followed in satiating hunger and thirst. (a) We multiply each bout of feeding (g) by the following bout of drinking (ml) and add them together in order to calculate a lock-on index. In this case $(3 \times 4) + (2 \times 2) + \ldots$. We arrive at 29 $(g \times ml)$. (b) This represents the maximum possible persistence on the part of the animal; it satiates one motivation and then switches. In this case the corresponding calculation gives a value of $10 \times 10 = 100$ $(g \times ml)$. Thus the lock-on index of part (a) is actual/maximum $= 29/100 = 0.29$ and part (b) is 1. (c) in this case the animal alternates frequently between eating and drinking. The grey area is correspondingly small

dominance by initially taking command of behaviour. Let us construct a diagram which shows both thirst and hunger states (see Figure 3.5).

Thirst is placed on the y-axis and hunger on the x-axis. Any state of simultaneous hunger and thirst can be represented on this diagram, and some points are shown. For example, the animal located at point A has an energy deficit of 1g and a fluid deficit of 10 ml. We imagine that a boundary exists somewhere on the diagram such that if the point falls on one side hunger is dominant (for example, 10 g, 1ml) and on the other thirst is dominant (for example 1g, 10 ml). The 5 g, 8 ml animal may be in an ambivalent position, but would of course opt for one or other alternative. This example introduces what McFarland (1971) calls *motivational space*; we plot motivational states on one diagram.

Let us suppose we take an animal that has an energy deficit of 10 g and a water deficit of 10 ml. As shown by point A in Figure 3.6(a), we can locate it in motivational space. When access to food and water is allowed (Figure 3.6(b)) it first (1) eats 5 g, then (2) drinks 5 ml, (3) eats 5 g and finally (4) drinks 5 ml. Time does not appear explicitly on these diagrams, but if it were relevant then landmarks in time could be placed by the side of the trajectory. Thus we do not know from the diagram how long it took to eat the first 5g, but we do know it ate 5 g before switching to drinking. A more realistic result is shown in Figure 3.7(a), indicating slightly more alternation between activities.

A measure of persistence of an animal is provided by what McFarland (1971) calls the *lock-on index*. If, in Figure 3.7(a), we add the shaded areas together we arrive at a value of 29 (g × ml). Now consider Figure 3.7(b) which is the ultimate in persistence, all food taken first and then all the water taken. Here the shaded area is 100. The lock-on index for the result of Figure 3.7(a) is $29/100$ or 0.29 (in other words the ratio of persistence to the maximum possible persistence). Figure 3.7(c) shows very much dithering and would yield a very low lock-on index, i.e. the shaded area is small.

Incidentally, persistence is not necessarily most reliably given by this particular measure. As the student will see if (s)he tries rearranging the order $E_1, E_2 \ldots$ and $D_1, D_2 \ldots$ in Figure 3.7(a) a different lock-on index will result. An index that avoids this problem is given by the number of transitions divided by the product of grams times millilitres taken.

Lock-on index provides a very convenient point to introduce the subject of optimal design, where we will need to take up the themes of this section.

3.5.4 Optimal design and cost

3.5.4.1 Cost of changing activities

Between the detection of internal physiological states and the execution of behaviour appropriate to those states there exist processes of decision-making. An experiment which shows clearly that factors additional to physiological state influence the choice and timing of behaviour was reported by McFarland (1971, p. 247) and developed by Larkin and McFarland (1978).

A hungry and thirsty bird was put into a Skinner-Box having two keys, one giving food and the other water. A barrier was located between the keys, and the animal had to negotiate its way around this in changing activities (see Figure 3.8). Lock-on index was calculated (Figure 3.9), and found to increase

36

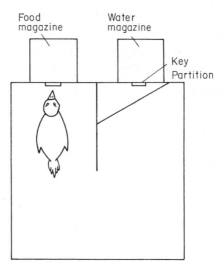

Figure 3.8 Skinner-box as used in the experiment of Larkin and McFarland. Note the partition around which the animal must move in order to change from feeding to drinking. (Source: Larkin and McFarland, 1978. Reproduced by permission of Ballière Tindall)

with increases in barrier length. This effect would not arise if the bird were basing decisions purely upon current physiological states, since it would switch after consumption of a certain number of grams or millilitres, irrespective of barrier length.

Larkin and McFarland (1978) speak of the *cost* involved in changing activities. Consider the bird feeding and drinking in a Skinner-box. When it switches activities a journey of only a few centimetres is involved. The use of energy and time is minimal in such a transition, and yet even here the bird takes this into account and persists longer as barrier length increases. Now consider a bird in the wild with several hundred yards between food and water. A switch in activity means investment of energy and risk of predation, etc. In other words its survival chances are lowered by the journey. However, if it persists for very long with feeding this will yield diminishing returns. Its weight will increase. Water balance will be disturbed by large amounts of food. After a while the cost of feeding with its decreasing benefits outweighs the cost of moving, and so the animal switches activity.

Let us return to the laboratory bird again. For the moment we ignore time-sharing and assume that feeding and drinking simply compete for dominance. Suppose in the Skinner-box it eats 3 g and then dominance switches to thirst, so it abandons feeding for a while. Now we consider a bird in the wild that finds itself in the presence of food and is in the identical physiological state to the laboratory bird. It eats more, say 4 g, before switching since at this point the costs of staying

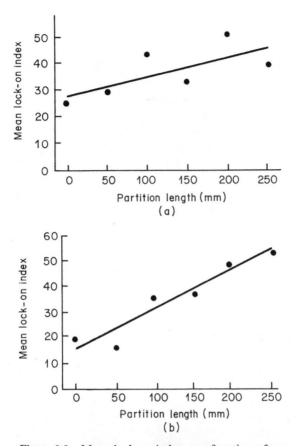

Figure 3.9 Mean lock-on index as a function of partition length. (a) transparent partition, (b) opaque
partition. (Source: Larkin and McFarland, 1978. Reproduced by permission of Ballière Tindall)

outweigh the costs of moving. The advantage of such behaviour is that the
animal makes relatively few trips between water and food when they are far
apart. It suffers more serious physiological imbalances as the price for
economizing on travel.

To summarize, suppose the bird were to eat all of its daily food in one large
meal and then drink all of its water in one drink. From the point of view of time
spent travelling this would be an economical design; but in terms of physiological
stress it would be bad design. It would endanger body-fluids to take an enormous
meal and suffer the resultant dehydration. The bird might get blown off-course
on the way to water, or the water source may have dried up. I must add that no
individual bird is necessarily a meteorologist. The bird's behaviour has been
shaped by evolution such that in making transitions it takes both physiological
and environmental cues (or memory of the environment) into account. Those
strategies that have proven useful have survived.

3.5.4.2 *Optimality in design*

For animals in the wild, life is a compromise between satisfying what are often conflicting and incompatible needs. Take for example an animal that is without water. It needs to search for water and this requires energy. Therefore as far as one requirement is concerned it needs to eat normally; indeed, the environment may be such that food shortage will be experienced next. However, if it eats certain foods these will pull water into the gut and dehydrate the blood still more. So what is the animal to do? Should it satisfy energy needs and risk severe dehydration, or should it, in the interests of its hydrational state, stop eating and risk death by starvation? In practice it makes a compromise; it eats less than normal. But how much less is optimal? An omniscient animal may know the answer. We assume that through evolution animals have tested various values and have arrived at a near optimum.

The reader interested in behaviour may say that the example of thirst inhibiting hunger is purely physiological—how about 'real' behaviour decisions? So let us take another example, an aspect of the behaviour of black-headed gulls (McFarland, 1976). Nests with broken egg-shells remaining after the chicks hatch are more likely to be found and the contents eaten by predators than are nests from which the parent removes the shell. Cost is attached to not removing the shell, and one might imagine that the parent would remove the shell immediately after the chicks hatch. In fact a delay of some hours occurs between hatching and shell removal. One therefore looks for a cost attached to early removal that outweighs the benefits. Newly hatched chicks are wet, and as such are likely to be the victims of cannibalism from neighbouring adult gulls. The risk of cannibalism is less after the chicks have had time to dry (see Figure 3.10). Beyond a certain time it is advantageous to remove the shell. Evolution is assumed to have shaped the bird's behaviour so as to maximize the chances of its genes being perpetuated.

Let us take an example from human society to illustrate what we mean by a decision-making criterion (McFarland, 1976). A university is to appoint an engineering lecturer. Two important behaviour characteristics have been given to the appointment committee: (1) ability to teach, and (2) ability to perform research. Each candidate is to be given a score out of 100 for his/her ability in each sphere. They consider a Dr Speak who is a brilliant teacher and scores 90% here, but has done no research for 10 years and scores only 10% here. Conversely, Dr Think scores 90% for research but is a hopeless lecturer (score 10%). The third candidate, a Dr Allrounder, is a fairly good teacher (score 50%) and researcher (50%). In deciding who gets the job, suppose the committee simply add up the scores. All would get 100. This represents an *additive* criterion. To break the deadlock, a committee member argues that teaching should be given extra weighting, so they adopt the criterion 2 × teaching score + research score. Now Speak wins with a score of 190. But suppose instead that the committee decided that a very low ability in either sphere was particularly harmful and could not be balanced by a good score in the other sphere. They

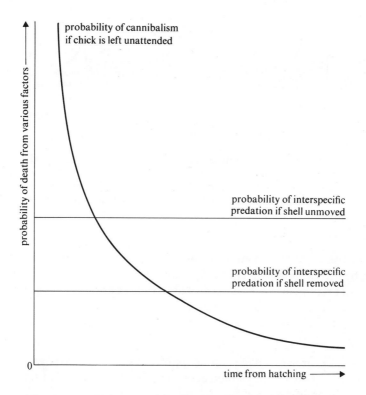

probability of cannibalism
if chick is left unattended

probability of interspecific
predation if shell unmoved

probability of interspecific
predation if shell removed

probability of death from various factors

time from hatching

0

Figure 3.10 Risk of predation for newly hatched gull chicks. Risk of cannibalism declines after hatching. Probability of predation from other species remains constant at a relatively high level if shell remains, and low level if it is removed

would probably adopt a *multiplicative* criterion. By multiplying each candidate's two scores, Speak and Think would each get 900, while Allrounder would be the undisputed winner with 2500.

I will now take the example further than McFarland does. Let us suppose that governments were to fund universities only on the strength of the results they produced in terms of capable graduates, and that universities were allowed to suffer extinction if results were not satisfactory. It could be that it is a disaster to appoint lecturers having low ability in either sphere, in which case those universities having a multiplicative appointment criterion would survive.

Now where do we apply the analogy to animals? Suppose two bits of information generate a feeding tendency in an animal and the resulting tendency must then compete with other potential activities. Certain possible combinations may have been tried out in the evolutionary sense. I don't believe that all combinations would have been, since the nervous system would impose constraints; but one possibility is as follows. Either the close proximity of food (high cue strength) or a high physiological hunger signal may very strongly goad

40

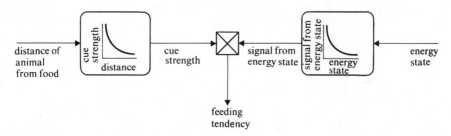

Figure 3.11 Model of how feeding tendency might arise. In this case, feeding tendency depends upon both food cue strength and internal energy state. As the animal comes close to food, cue strength rises sharply. Similarly, as energy state falls then the signal arising from energy state at first rises slowly and then more steeply. In other words, very heavy emphasis is placed upon (1) the close proximity of food, and (2) very low energy states

food ingestion. A combination of moderate hunger and fairly close distance would be less demanding. The reader may like to confirm that the diagram in Figure 3.11 represents such a system. This is only one possible combination rule, and numerous others could be suggested.

McFarland (1976) argues that two main factors contribute to feeding tendency:

'... (1) degree of hunger, in terms of physiological requirement for food; and (2) strength of food cues, in terms of the animal's estimate of the availability of food.'

Evolutionary selection will place more or less weight upon each of these factors in different species according to ecological and physiological conditions. McFarland argues that:

'... where food availability is erratic, more weight should be given to cue strength, while the weight attached to hunger should be related to the animal's physiological tolerance.' (Reproduced by permission of Cambridge University Press.)

We will investigate this further in the chapter on feeding.

3.6 SUMMARY, CONCLUSIONS, AND DISCUSSION

The nervous system of an organism receives information on internal and external events. It processes this information and the outcome is behaviour. We seek to find the decision rules.

I have tried here to show the kind of information that is supplied to the motivational decision-making processes of an organism. The picture is, of course, by no means complete, but we have some idea of how it is that chaos does not reign even though the animal has a number of potentially competing demands for its time.

At any point in time there may be an array of motivational tendencies within an organism, each one of which is appropriate for a particular behaviour. The strength of these tendencies will depend upon factors arising from both internal physiology and the environment. Sometimes 'internal event' means nothing more than event within the nervous system, while in cases such as hunger we can try to trace causality to its roots in the biochemistry of energy. By 'environment' I imply that complex organisms perceive the environment and form internal models of it. Unless this is appreciated the internal–external dichotomy can be confusing.

The strength of, for example, the feeding tendency depends upon the physiological energy state and the perceived distance to food. I use the expression cue strength to refer to the set of causal factors arising in the environment that influence a particular motivational tendency. Thus the sight or smell of food has a certain cue strength, which together with an energy state signal determines feeding tendency. Cue strength increases as food gets nearer and, correspondingly, so does the motivational tendency to feed or move towards food. Our model of behaviour is based axiomatically on the assumption that at any point in time one behavioural tendency will be dominant.

Let us suppose hunger is dominant. This means of course that the animal either feeds or shows appetitive behaviour such as moving towards food or bar-pressing for food. However, even while hunger is dominant other behaviours unconnected with feeding may occur. I think that it is helpful to consider that they occur by the permission of hunger. This means that hunger has not lost its dominance (though see below), but that it disinhibits another activity for a while. Shortly after disinhibition hunger inhibits again, and behaviour appropriate to hunger resumes. This is called *time-sharing*.

Thus a change from, say feeding to drinking need not be due to a switch of dominance through competition, though of course it may be. Sometimes the causal factors for hunger will decrease, those for thirst increase, and by straight competition thirst will oust hunger from dominance.

Mechanisms underlying behaviour have evolved so that animals generally spend their time economically and do not dither between their various competing demands (though a few notable exceptions are to be found in any standard ethology text). It would be hopelessly uneconomical and inimical to survival if animals were often to be in behavioural ambivalence, for instance taking a mere bite of food and then being pulled to water. We imagine that through evolutionary selection weighting has been given to various aspects of the causal factors. Thus a trade-off may be necessary between the two possibly undesirable features of wide fluctuations in energy state and frequent costly excursions to feed. The need to avoid frequent trips would carry particularly heavy weight if predation were high. Wide energy fluctuations may then be tolerated between meals. Potential physiological stress would give heavy weighting to the causal factor of low energy state.

Imagine an animal awaking from sleep (causal factors for sleep are lowered). It so happens that hunger emerges as its dominant motivation, so it either eats or seeks food. It is important that, having embarked upon a course of action, it should yield a substantial return. We therefore expect the animal to be equipped with mechanisms providing persistence. At least two such mechanisms have been postulated. If an animal ingests food then the first mouthful or so may increase its hunger, or at least, to be strictly objective, the animal behaves as though it were even more deprived than is the case. The other postulated mechanism arises from consideration of cue strength. If animals base their decisions upon representations of the environment, then as food is approached so the animal's tendency to persist with food-seeking increases. Both of these are positive feedback effects and would ensure stability. After eating for a while, internal physiological energy state will change such that even with the high cue strength of food the rival attraction of water will, by competition, pull the animal towards water. In the laboratory we can capture an aspect of this behaviour by noting that as it is made more difficult to change behaviours (by increasing barrier length) persistence increases.

In this area, as in so many others, the task of the biologist has profound similarities to that of the engineer. Of course, in the strict sense we are not designing anything, but we are constructing a theoretical model based upon the real organism. Like the engineer we must ask whether or not our 'design' would work. Have we made a sufficient number of viable postulates with parallels in the organism that the model would actually function realistically? By associating both cue strength and physiological state as determinants of behaviour we guarantee stability but now we must ask how time-sharing relates to this.

Toates and Bowles (1979) claim that the situation may have fundamental complexities beyond those described by McFarland (1974). Imagine a bird feeding in a field that is a considerable distance from water. We presume that food lowers the strength of feeding tendency and simultaneously increases drinking tendency. The high cue strength of food and low cue strength of water means that the bird will take a large meal before thirst can capture dominance. Let us envisage what happens before the point in time when thirst is able to offer serious competition. According to the behaviour of birds in the Skinner-box, one might expect the wild bird to time-share with sub-dominant motivations before dominance changes. Unlike in the Skinner-box though, it is not clear what will happen when it does this. Suppose that thirst is the sub-dominant motivation. It is disinhibited by hunger. So does the bird fly off, and then a few seconds later after a fruitless excursion, fly back to food in response to hunger re-exerting inhibition? Or alternatively, and metaphorically speaking, does hunger inform thirst of the very limited time at its disposal, so that the bird wisely does something else not involving departure from the food location, such as preening. In other words does it scan its lower priority candidates and select the most feasible? The answer is that we simply don't know.

There is another possibility that could arise, as follows. Hunger disinhibits thirst and the bird flies off towards water. As food recedes and water gets nearer food cue strength decreases and water cue strength increases. Because of the change in cue strength in favour of thirst it is possible that before hunger can re-exert inhibition there will be a change of dominance to thirst. If this is what occurs then the bird might, by a reciprocal process, alternate between food and water rather more often than would be the case if time-sharing did not occur. Again, in this respect we don't know what additional processes we need to postulate in order to give a convincing account of behaviour.

CHAPTER 4

Hunger

4.1 INTRODUCTION

The cells of the body need fuel to provide energy by metabolism. Figure 4.1 represents a cell of the body, for example a cell in the liver, stomach or skin. Cells are surrounded by a fluid matrix, through which fuel is transported to the cell. As a simplification Figure 4.1 shows a cell as a box with a hole through which fuel passes. Its inside is a miniature chemical process plant, and the processes require a constant source of energy. One fuel used by the cells of the body is glucose. The process of metabolism yields energy which performs work within the cell. Water and carbon dioxide are released as products of metabolism, and are returned to the fluid matrix which surrounds the cell. Heat is also generated. Thus Figure 4.2 shows glucose entering the cell, and water and carbon dioxide leaving it. This diagram illustrates the organism's need for energy regulation involving feeding—the cells of the body need fuel which can yield energy.

The problem might seem fairly straightforward then, and consists simply of ensuring a supply of glucose. Given that carbohydrates in food are converted to glucose in the process of digestion, a supply of glucose would seem to be all that is needed. Energy regulation is not so simple though, for several reasons. First, not all fuels eaten are in the form of carbohydrate, and so cannot easily be converted to glucose. Our diet consists normally of carbohydrate, fat, and protein. Secondly, most species have a storage problem to solve. Sometimes food is available in abundance, but scarce at other times. Therefore it is to their advantage that animals store a certain amount of excess energy during a period of plenty. Also animals often eat either by day or by night as the case may be, and

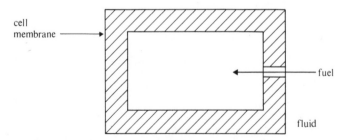

Figure 4.1 A living cell. The diagram indicates the fluid surrounding it, the membrane, and the passage of fuel across the membrane

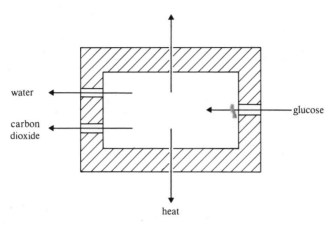

Figure 4.2 Glucose is taken up by the cell. Water and carbon dioxide are two products of metabolism, and heat is generated

depend upon the energy taken during the active phase for their supply during the inactive phase. One means of achieving such storage would be to hold a reserve of glucose in the body. In principle this is what occurs, but in fact glucose is first converted to another fuel and some of this then stored if there is an excess of glucose. If fat is taken in excess then this is stored as fat. At times the cells of the organism burn glucose which has recently arrived from a meal, but at other times they depend upon the breakdown of a stored fuel. Much of the discussion which follows concerns what determines the type of fuel to be used and the conversions within the body.

4.2 FUELS: THEIR CONVERSION AND CONSUMPTION

4.2.1 Digestion

Food enters the mouth and passes down the oesophagus to the stomach. On leaving the stomach it enters the first stage of the small intestine, known as the duodenum. During its passage through the stomach and along the intestine, chemicals secreted by the stomach and intestine break down the food into a form more easily handled by the body. This process is known as *digestion*. The piece of anatomy from mouth to anus is known as the alimentary tract.

4.2.2 Absorption

The construction of this section has been aided by the lucid account presented by Vander, Sherman, and Luciano (1975) to which the reader should refer for more details. Just enough of the biochemistry needed for understanding the essentials of feeding is given here.

Animals such as the rat and dog, as well as human beings, normally take

46

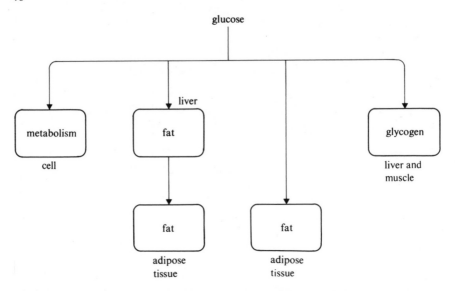

Figure 4.3 The fate of glucose after it is absorbed. Some is used as fuel by the cells, while some is converted to fat in the liver. A fraction is converted to fat at the level of the adipose tissue, and some is converted to glycogen and held as such in the liver and muscles. After conversion at the liver, fat is stored as adipose tissue. (Based upon Vander, Sherman, and Luciano, 1975)

distinct meals with long intervals between them. Shortly after eating, absorption of the meal from the alimentary tract starts, and it lasts for some time after the meal. This is known as the *absorptive state*. During this period part of the fuel contained within the meal may be put into storage; the remainder will be supplied directly to the cells of the body for metabolism. If the animal is eating in excess of its metabolic needs over a period of time then the excess will be stored. This is true of the rat during the dark phase of the light/dark cycle. When absorption of the meal from the alimentary tract has ended we speak of the organism being in the *post-absorptive state*.

In Figure 4.3 we consider the fate of glucose after its absorption. Glucose enters the blood and a fraction of it will be sent to the cells of the body for metabolism. Another fraction is converted into fat in the liver, and then carried by the blood to the sites in the body where fat is deposited. Collections of cells containing largely fat deposits are known as *adipose tissue* (the non-fat component of body tissue is known as the *lean tissue mass*). Another fraction of the ingested glucose is carried straight to the fat deposits, and there converted into fat. Conversion of a non-fat (or to be more technical, non-lipid), energy substrate into fat is a process known as *lipogenesis*. A fourth fraction of incoming glucose is converted into glycogen in the liver and muscle, and stored there as such. Storage of energy in the form of fat is very economical for the animal, since for its energy content fat is very light.

4.2.3 Post-absorption

We have considered the state when energy is actually arriving from the alimentary tract. In order to understand the pathways of energy supply to the cells of the body during the post-absorptive period, it is necessary to look more closely at the way in which cells use fuel. Here we must draw a distinction between the cells of the nervous system and the other cells of the body. The nervous system is very selective in that it can (except under severe conditions) consume only one fuel, glucose. By contrast other cells can burn fatty acids. Thus in the fasting state even though no glucose is being supplied to the organism it is essential for the nervous system that blood glucose concentration should not fall far, since this is the source of supply for the nervous system. During fasting, glucose is derived from the breakdown of glycogen in liver and muscle. Fat (triglycerides), located in the adipose tissue, is broken down into *glycerol* and *fatty acids* (a process known as *lipolysis*). Glycerol is converted to glucose in the liver. Figure 4.4 shows a summary of the main energy flows in the post-absorptive or fasting state.

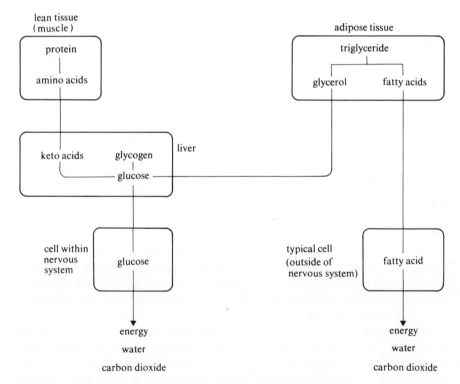

Figure 4.4 Chemical conversions when there is no source of energy coming from the gut, i.e. the post-absorptive state. Cells within the nervous system can, usually, only use glucose as a fuel. This is derived from sources within the liver. The typical cell outside of the nervous system can use fatty acid as fuel. The original source of fuels is shown to be lean tissue and adipose tissue. (Based upon Vander, Sherman, and Luciano, 1975)

Another source of glucose during fasting is from the breakdown of body proteins from the lean body mass (see Vander, Sherman, and Luciano, 1975), the final conversion to glucose being done in the liver. Protein enters the liver as amino acids.

In the post-absorptive state any glucose that is available should preferably be for the exclusive use of the nervous system, even though glucose is, of course, a potential fuel for any cell. The cells outside of the nervous system should help conserve glucose by burning the relatively abundant fatty acids, as is indicated in Figure 4.4. How is it that glucose is spared for the nervous system at such times and not seized by any cell? The answer is provided by the action of *insulin*.

Insulin is a hormone produced by the pancreas. It acts upon the cells of the body with the exception of those of the brain. The transport of glucose across the cell wall is dependent upon insulin. The nervous system can take up glucose in the absence of insulin. If insulin is injected into an animal in the fasting state, then: (1) glucose will move from the blood into the cells of the body; and (2) glucose will be converted into fat and glycogen. In other words it is possible to mimic the absorptive conditions of glucose abundance by insulin injection. Normally in the absorptive state insulin would be present in the blood, facilitating uptake of glucose by the cells and conversion of glucose to stored fuel. In the fasting state insulin is either absent or present at a low concentration. This allows fat to be converted to fatty acids and also glycogen to be converted to glucose. Further, since cells outside the nervous system need insulin to take up glucose then they are forced to burn fatty acids in the fasting state. The glucose which is available is for the almost exclusive consumption of the nervous system. Glucose concentration in the blood and insulin secretion form a negative feedback system. High glucose concentration causes insulin secretion which then lowers blood glucose concentration.

4.2.4 Metabolic rate

Most of the potential energy contained within the food that we eat is ultimately converted to heat in the body. Take for example the energy needed for pumping blood through the vessels of the body by the heart. Energy is supplied to the heart muscle which then gives momentum and kinetic energy to the blood. However, blood, of course, encounters friction in its journey through the vessels of the body and so the kinetic energy is converted to heat.

The *metabolic rate* of an organism is the rate of heat production and is the rate at which fuel is being burned. If the organism is at a constant temperature then this rate will be the same as the rate at which heat is emitted from the organism to its surroundings. This heat output can also be measured indirectly as oxygen consumption. Metabolic rate rises considerably when the organism is active or placed in a cold environment. For a standard reference we speak of *basal metabolic rate*, meaning the rate at which energy is converted in the resting organism under standard measurement conditions.

4.3   THE BIOCHEMICAL BASIS OF FEEDING

4.3.1   Introduction

The energy used by the body equals the energy entering, unless body energy is either being lost or gained, revealed by loss or gain of the non-fluid part of the body. In addition, there may be a small loss of energy by excretion. Thus energy eaten = energy used + energy stored + energy excreted over a period of time when lean or fat tissue mass is increasing, and energy used = energy in food + energy in lost tissues − energy excreted over a period when lean or fat tissue is being lost.

What gives rise to a food intake signal? Energy is constantly being consumed by the body, and yet many species of animal feed only at distinct intervals even in the midst of plenty. We must also ask: what terminates a meal?

The minimum essentials of a food ingestion mechanism may be illustrated by reference to a simple invertebrate nervous system. This was described for the snail *Helisoma* and marine sea slug *Aplysia* by Senseman (1977). *Aplysia* feed upon marine algae; amino acids which are found in algae stimulate the mouth-opening reflex. However, even in an abundance of food a satiety mechanism is evident if the animal has eaten a substantial amount. Meal size control in *Aplysia* is exerted by distension of the anterior gut; loading here with non-nutritive bulk causes satiety. So far then we have direct chemical stimulation from food exciting ingestion and ingested bulk inhibiting it. In terms of neural control an excitatory and inhibitory input to a synapse is implicated. The output of the post-synaptic unit would feed into a neural system controlling ingestion.

When we compare species the nature of the sensory input and the information on physiological state which jointly determine ingestion will vary, but we must always look for two such sets of controls. Not all species have an easy task in discriminating food. For some, such as frogs and hawks, it is a question of visually locating a moving target (e.g. bugs and birds) rather than automatically coming into contact with chemicals signalling food. For a species such as the rat, chemical discrimination by olfaction and taste is important but this will often need to be associated with complex predictive cues such as those involved in spatial maps of environment and food sites.

The discussion now turns to the possible physiological and biochemical bases of how a feeding signal arises, dealing mainly with mammals. It is this side of the story that receives most attention in textbooks. These days the question is usually of the kind: what aspect of internal energy has fallen at the time of a meal? Although we will need to look at this aspect we must never forget that two kinds of information are needed for an animal to feed: (1) information arising from the animal's internal physiological energy state; and (2) information on the presence of food in the environment.

4.3.2   Peripheral theories

One of the earliest attempts at answering these questions suggested that animals eat because the stomach is empty. They then take a meal of sufficient size to fill the stomach or at least to eliminate the signal arising from emptiness. It may well be true that some species generally eat when their stomachs are empty and stop eating when they are relatively full, but this of course does not necessarily mean that the causal factor is a signal from the empty stomach as such. There is some suggestion in the literature that, when their stomachs are empty, people experience contractions of the stomach which they report as hunger pangs. This does not, though, necessarily give the pangs a primary causal role in feeding, though the possibility of a conditioned association with hunger should not be ruled out.

Concerning satiety, moderate distension of the stomach is hard to detect, so distension on its own is unlikely to be a switch-off signal, unless the meal is very large. However distension may, as one of a complex set of cues, serve an anticipatory function (explained later) and help to terminate meals.

The empty and full stomach theory is an example of a double negative feedback system which would serve to give the body a supply of energy fluctuating within specified limits. However, although it is closed loop in so far as stomach content is concerned, it is open loop in so far as the body energy is concerned. If the need for energy in the body or the energy content of the food were to change up or down, the stomach might be blissfully unaware of the fact, and the organism could literally either shrink or explode around a well-regulated stomach. The only way it could possibly work would be for the stomach normally to supply the body with excess energy and for the body's metabolism to take care that the excess is burned (this may sometimes occur, but only under extreme conditions). We now no longer believe that the empty stomach theory has much to offer as an ultimate basis for feeding control. It is discussed for historical and heuristic reasons as an example of the early twentieth-century view that the seat of biological drives was peripheral. It was soon found that animals with stomach surgically removed (the oesophagus was connected directly to the duodenum) controlled energy intake almost normally. Some people have their stomachs removed surgically in cases of cancer and ulcers. They are then forced to take smaller-than-normal meals because of the limited capacity of the upper alimentary tract, but they control energy intake quite normally and experience hunger. Although few now believe that feeding is switched on by afferent signals from an empty stomach it remains possible that it can be switched off by a full one. Indeed in the extreme it would be most surprising if it were otherwise, since it would be remiss of nature not to provide a mechanism for preventing stomachs from bursting. Whether or not a mechanism involving fullness of the stomach comes into action at the end of normal meals is a subject for discussion later.

When the empty stomach theory was abandoned, attention shifted from the bulk property of food in the stomach to the question of how the quantity of energy held at a particular location within the body could be regulated (the alimentary tract is normally considered to be outside of the body proper).

4.3.3 Homeostasis—central theories involving negative feedback

Research and theorizing in the area of feeding from about 1916 was dominated by the explicit or tacit involvement of what we would now call a *set-point* value of a biochemical quantity. Somewhere in its body the animal was thought to be performing a comparison between (1) some actual state of an energy quantity, and (2) a set-point value. Hunger was believed to be aroused when (1) fell significantly below (2). Although the actual biochemical energy quantity which was assumed to be regulated differed from one researcher to another, none the less the underlying theoretical bases were similar. Some examples are given now. It is the view of the present writer that they serve very well to illustrate how a theory fits some of the evidence and contains a germ of truth, but then the contradictions give rise to a radically changed and better theory, and so on.

4.3.3.1   The glucostatic theory

This formed the first of the homeostatic theories, and was proposed by Carlson in 1916. Just to remind the reader, glucose is a form of readily available energy which is held in the blood. Its uptake by the cells of the body is dependent upon insulin. If insulin is injected then the above-normal level causes a rapid lowering of blood glucose level as glucose is taken up by the cells. It is known that animals can be induced to eat large quantities of food by insulin injections. This led researchers to the view that the stimulus for eating was provided by blood glucose level falling below a threshold value. The food taken enables blood glucose to return to the set-point value. However, this neat model fails to fit a very important observation concerning *diabetes mellitus*. This is a condition which afflicts some people, and is caused by a failure of the pancreas to secrete insulin. In the absence of insulin, glucose cannot enter the cells of the body and so the concentration in the blood rises to very high levels; so much so that it is excreted by the kidney. Thus fuel is wasted because it is building up and cannot be burned. However, people suffering from diabetes mellitus experience hunger sensations despite elevated blood glucose. Rats made diabetic by removal of the pancreas do not stop eating; on the contrary they eat large amounts. How is one to account for this? It could in some cases be continuation of an established habit, or it could be that the set-point is drastically increased, but such *ad hoc* speculation sounds quite implausible. Further, even in normal animals glucose infusions into the general circulation often fail to suppress a subsequent meal. Researchers slowly abandoned the theory, or rather modified it so as to accommodate more of the evidence. In fact, on close examination, it may not make good sense for a behavioural mechanism to derive a signal from blood glucose level. The fall is very slight even after a rather long period of food deprivation. This is because other sources of fuel, glycogen and fat, are converted into glucose in an effort to hold concentration near normal.

Derived from the glucostatic theory, Jean Mayer in 1955 proposed that the rate at which the cells of the body take up glucose is what controls food intake (Mayer, 1955). The distinction is important. With reference to Figure 4.2 the glucostatic theory proposes that a fall in the glucose concentration in the fluid medium surrounding the cell is what stimulates feeding, irrespective of whether or not the glucose is actually being utilized by the cells. Mayer proposed that feeding is in some way sensitive to the rate of uptake of glucose by the cell indicating that it is being metabolized. Normally glucose levels will be higher in arterial blood, which is the source of supply of glucose to the cells, than in venous blood which carries the products of metabolism from the cells. In the case of diabetes mellitus the difference in glucose level at the two locations in the blood will be minimal, indicating little or no glucose utilization. According to Mayer, this arterial–venous comparison, the A–V difference, generates a signal which determines feeding/satiety. A high A–V difference indicates a rich source of glucose and a high rate of utilization. Under such conditions no feeding is necessary. In normal animals as the circulating glucose is used up by the cells, then the amount in the arteries would fall and the A–V difference become relatively small. This small difference is hypothesized to be the feeding signal. In diabetes mellitus there is also a small A–V difference and feeding is excited. How is A–V difference measured? It must involve measurements at two different locations. As a first guess, ignoring some rather obvious problems, one might say a good site would be to look at the A–V difference across, say, a leg, since here maximal sensitivity might be expected due to the substantial number of energy-consuming cells involved. However, Mayer proposed that insulin-sensitive cells in the brain (exceptional cells for the brain because of their dependence upon insulin) examine their own arterial supply and venous return. As such they are a microcosm of the remaining insulin-sensitive cells of the body.

This theory has its problems (Novin and VanderWeele, 1977); intravenous glucose infusion does not always suppress feeding. However, rather than the theory having been rejected, interest has largely moved to other areas of investigation.

4.3.3.2 The lipostatic theory

This theory is often seen as an alternative to those which are based upon glucose levels, but some might prefer to see a lipostatic mechanism as an addition which accounts for more long-term regulation. The theory is that the size of the fat deposits of the organism are monitored and the information so obtained determines the tendency to feeding/satiety. When fat deposits fall the organism increases its food intake. Conversely, excess fat inhibits feeding. Some authors have argued that fat deposit size appears to take the form of a set-point controlled quantity. Loss of weight through chronic starving occurs largely through loss of fat deposits. Conversely, when an animal is force-fed the excess weight which results is largely fat. When force feeding is terminated animals refuse to eat for a time and weight is lost until the natural weight is restored. It is clear then why 'fat deposit set-point' and 'body weight set-point' are terms often used interchangeably. In the case of humans the difference between a skinny and fat individual of the same height is largely a difference in size of fat deposits. Johnsen (1973) reported that both fat and thin Danes sent to labour camps during World War II became thin, but at the end of the war they returned to their characteristic weights. There is a set of data then which lend some credence to a model of feeding in which fat deposit size is homeostatically regulated.

How would such a system work in practice? A suggestion once made was that the organism monitors the pressure on the soles of its feet (Johnsen, 1973), but that seems rather improbable to say the least. Let us suppose that the concentration of a chemical carried around in the blood is a function of the size of the fat deposits. The concentration could be compared with set-point value and feeding potentiated or inhibited. An alternative would be for the brain to contain a microcosm of the body-fat deposit and for the size of this microcosm to determine feeding (see VanItallie, Smith and Quartermain, 1977 for a discussion). By analogy, we know that the body cannot measure total cellular water content as a single entity, but rather appears to examine a hypothalamic cell, and take its size as being representative of total body-water (see the next chapter).

Another possibility which explains long-term stability of fat deposits without appeal to a set-point has been discussed (Toates, 1975; Booth, 1976). It is more parsimonious in that it is based upon known biochemical properties, rather than hypothetical set-points.

Figure 4.5, derived from Toates (1975, p. 250) indicates how it would work,

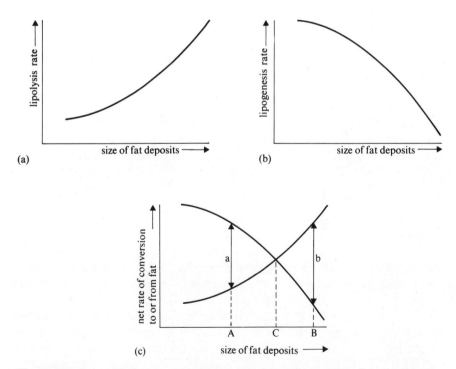

Figure 4.5 An equilibrium process which could account for stability of the fat deposits. (a) lipolysis, the rate at which the fat deposits are broken down, increases as a function of fat deposit size; (b) lipogenesis, the rate at which fat deposits are built up, decreases as a function of fat deposit size; (c) both functions, as they might average over a period of time. When fat deposits are at size A there is a *net* lipogenesis of magnitude a. At B there is a *net* lipolysis of magnitude b. The system will arrive at equilibrium point C

and see also Russek (1971) for a very similar argument. We suppose that lipolysis (the rate at which the fat deposits are broken down) is a function of the size of the fat stores, as shown in Figure 4.5(a) (such a mechanism was discussed at least 24 years ago (Mayer, 1955)). This shows the average values; at any given fat deposit size, both lipolysis and lipogenesis rates would fluctuate throughout the day. The relationship of Figure 4.5(a) is an example of intrinsic negative feedback: the rate of fat breakdown increases with the size of the fat deposits. It could be due to the fact that large adipose cells have a larger component of readily mobilized triglyceride. Figure 4.5(b) shows a possible relationship between lipogenesis (the rate at which non-lipid based energy is converted into fat) and the size of the fat stores. Again a negative feedback effect is apparent in that lipogenesis is high for low fat deposits. A possible explanation for the shape of this function is that when the fat deposits are large a given insulin level is relatively ineffective at inducing glucose uptake in adipose cells. At a given level of fat deposits then both net lipogenesis and net lipolysis can be expected to occur at various times of day. Figure 4.5(c) shows on one graph both functions as they might average over a day. When fat deposits are of size A there is a net rate of gain of fat having magnitude a. Imagine a starved animal having food restored. A large proportion of incoming energy will be converted to fat. Fat deposit size will increase until size C is reached at which point loss equals gain on average and weight is stable. Conversely, if fat deposits find themselves at B through, for example, forced overfeeding, when overfeeding is terminated fat deposits will be broken down at rate b over their level of formation. The animal will tend to refrain from eating. Fat deposit size will fall to C and remain there. In other words any tendency to move the size of the fat deposits from C will cause action to be taken such as to return them to C. No set-point mechanism is involved and yet fat deposit size and hence body weight will be stable and return to normal following a disturbance.

4.3.4 The liver as a detector site

The ability of the organism to detect energy state at the liver has been suggested by several researchers, the most notable being Russek (1971). It has been suggested that the liver is capable of measuring the rate of glucose supply from the gut (Russek, 1976; Friedman and Stricker, 1976), possibly working in parallel with glucose receptors in the duodenum (Novin, 1976; VanderWeele and Sanderson, 1976; Novin and VanderWeele, 1977).

Figure 4.6 shows the fate of nutrients after leaving the alimentary tract. They are absorbed into the capillaries in the wall of the intestine and travel via the hepatic portal vein to the liver. The role of the liver in feeding was reviewed by Carlson (1977, p. 349). Clearly the liver is ideally placed to detect what is entering the body from the gut. Russek's (1971) seminal observation was that injections of glucose into the portal vein induced satiety in dogs. By contrast, intrajugular injections were relatively ineffective. In other words, Russek placed glucose either (1) just upstream from the proposed detector, or (2) in the general circulation. The portal injection caused inhibition of food intake, whereas the

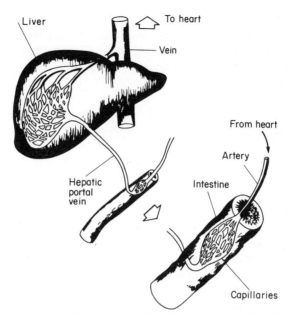

Figure 4.6 Nutrients are absorbed into the capillaries in the wall of the intestine, and travel, via the hepatic portal vein, to the liver. (Source: Carlson, 1977. Reproduced by permission of Allyn and Bacon, Inc.)

jugular injection did not. In dogs starved for 22h, intraportal glucose inhibited the feeding performed in a 45–75-min session. Intraportal isotonic saline had no effect (Russek, 1971).

We have argued (see later) that the feeding system is sensitive to the rate of supply of energy. It seems to be possible during a meal to trick the organism into 'believing' that a load is very much larger than it really is. A relatively small caloric glucose load in the duodenum is, as far as energy flow is concerned, equivalent to a much larger load in the stomach since stomach emptying is a limitation in absorption (Booth, 1978). An intestinal load results in a large absorption from the intestine. Such loads can be very effective in suppressing feeding relative to their size (VanderWeele and Sanderson, 1976), though numerous qualifications of any conclusions are needed (Novin and VanderWeele, 1977). As might be expected they must be made at the time of feeding in order to be really effective. At other times they will not be able to mimic large loads. A tentative analogy may be proposed. If one wants to prevent a room-heating system from working, even a small amount of heat applied to the thermostat at the correct time will be effective.

Russek's (1971) postulated receptors could work in basically two ways, either (1) they are sensitive to rates of glucose and amino acids (Russek explicitly mentions this addition) arriving from absorption (if either is present in sufficient amount, the glucoreceptor discharge (hunger signal) is low or absent); or (2) the actual liver reserves of glycogen and protein determine the hunger signal (high reserves, low hunger signal). Incidentally, conditioning is also possible. An auditory stimulus applied during intraportal infusion of glucose can, it is claimed, become conditioned to induce satiety (Russek, 1971).

Novin and VanderWeele (1977) concluded:

'Visceral processing of nutrients and the information derived from the innervation of the viscera play an important role in feeding. This statement is clearly our major proposition and is a nontrivial one in view of the recent history of hunger research, much of which has overlooked the natural order involved in the handling of nutrients. Food is encountered (i.e. smelled and tasted), ingested, digested, absorbed, reassembled in storage, or metabolized. Temporary stores are depleted, and the motivation to eat is again activated. At every stage in this sequence there are opportunities for the periphery to inform the brain of the changes, and it seems unnatural that this information would not be utilized in controlling feeding. The presumption seems unfounded that the relatively invariant conditions maintained by the CNS would take precedence over the continually varying conditions of energy metabolism in the viscera in the control of so dynamic a condition as feeding.' (Reproduced by permission of Academic Press.)

Even though we may wish to place the energy detector at the liver it is difficult to associate the final decision to feed with anything other than the brain. In some way the brain must be informed of the energy state as measured by the liver, and use this information along with that derived from other sources to make a decision as to whether to feed. How does the liver communicate with the brain? Russek (1971) found that the satiating effect of portal glucose injections was eliminated if neural transmission along the vagus nerve was blocked. Presumably this nerve would normally carry a satiety signal. As an alternative, or perhaps more likely an additional, means of communication, a hormonal messenger might be employed between liver and brain. Novin (1976) discusses how the brain could integrate information on nutritional state arriving from the liver, and also from the duodenum (Novin and VanderWeele, 1977).

4.3.5 An energy flow model

In 1972 David Booth proposed that feeding is under the control of the rate of supply of readily metabolizable energy to the organism (Booth, 1972b, 1976, 1978). For example, at the end of a meal energy is being supplied to the body from a relatively full gut at a high rate and feeding is therefore inhibited. Booth argues that an energy detector examines this rate of supply. Later refinements included energy supply from some internal sources. Thus if an animal is force-fed until it is grossly overweight and then force-feeding is terminated it will refrain from eating for a while. According to Booth's argument, energy is being made available by the unusually great rate of breakdown of inflated fat deposits. The system detects this abnormally large rate of energy supply from fat, and feeding is inhibited. Booth (1972a) observed that if glucose solution is either drunk by rats or given by stomach tube then a reduction in the animal's subsequent intake of maintenance food is made. The magnitude of the reduction precisely matched the calories loaded or drunk. Booth (1972b) investigated whether the effect was specific to carbohydrate stomach loads, or whether the animal would make a compensatory reduction in subsequent food intake in response to other normal metabolizable dietary components. It was found that subsequent feeding was inhibited whatever the substrate of energy. Under normal circumstances, whether the fuel is glucose from the gut or fat from an internal source it is not important. If flow is sufficient, feeding is inhibited. If the animal has been starved no energy is entering through the gut and the amount coming from intrinsic fat stores is relatively low, so when food is made available there is a high probability that the animal will eat.

Booth's theory is somewhat revolutionary in at least two respects. First, energy rate is the crucial factor for the initiation and termination of meals. Most previous theories have been based upon examination of an energy quantity in the body, such as glucose concentration or fat deposit size. Mayer's version of the glucostatic theory involved glucose uptake but in practice as an A–V difference only. Glucose uptake rate is in fact A–V difference × blood flow. As glucose is a major energy source in lab chow, the energy flow theory may therefore be a modification of a strictly stated glucose uptake theory. Secondly, the system, as detailed by Booth, takes a broad and comprehensive view of its energy state. Supply of glucose, fats, or amino acids can influence the decision of whether to feed or not.

In 1973, the present author began collaboration with Booth to construct a model of feeding in the rat (Toates and Booth, 1974; Booth and Toates, 1974; Booth, Toates, and Platt, 1976). We hoped to develop not only a fully explicit picture of what we believed was important in the control of feeding in the rat, but also a quantitative computer model which would mimic the rat's behaviour. For the sake of simplicity, I will first divide the model into several parts and later assemble them. Figure 4.7 represents a single piece of information: the net rate at which energy enters the gut is given by the rate at which energy leaves the gut

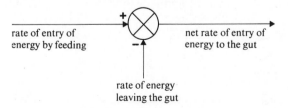

Figure 4.7 Net rate of entry of energy into the gut is given by rate of entry by feeding minus rate of energy leaving the gut

58

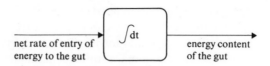

Figure 4.8 Energy content of the gut is given by the integral of the net rate of entry of energy to the gut

subtracted from the energy rate at which the animal feeds. Figure 4.8 represents the fact that if we take the integral of the net rate at which energy enters the gut we will obtain the actual quantity of energy in the gut.

The rate at which energy leaves the gut to enter the blood, i.e. absorption rate, depends upon the amount of energy actually present in the gut. It is known from experimental observation that when the stomach is full it will pump energy at a relatively high rate to the intestine. As stomach contents become small then so this rate is reduced.

After the stomach is empty then of course absorption rate is zero as soon as the small amount in the intestine has been digested and absorbed. The rate at which the stomach empties is the limiting factor which determines the rate at which energy can be absorbed across the intestine wall (see Booth, 1978). Thus we are justified in considering the rate at which energy leaves the stomach to be reflected in the rate at which energy leaves the gut after a short delay for digestion. For our present purposes it will suffice merely to show an operator which relates flow out of the gut to its energy content, in the knowledge that energy flow decreases with gut energy (Figure 4.9).

We worked on the assumption that the organism examines the net rate at which energy is being made available from the gut and from intrinsic fat mobilization (lipolysis). Under normal circumstances rats eat relatively little in the light phase, but lipolysis is supplying energy and so, for much of the day, hunger is answered this way. In the dark phase a large proportion of any energy coming from the gut is shunted off into the fat deposits. Therefore in the dark phase the animal must eat relatively large amounts in order to maintain energy supply to non-lipid tissues. Thus most of the energy for metabolism has either arrived shortly before as food, or is derived from mobilization of fat. Suppose we deprive an animal of food. We might conclude that the breakdown of intrinsic fat deposits would be such that the animal would not experience hunger until they

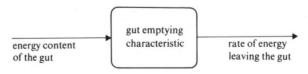

Figure 4.9 The rate at which energy leaves the gut is dependent upon the energy content of the gut and the gut-emptying characteristics

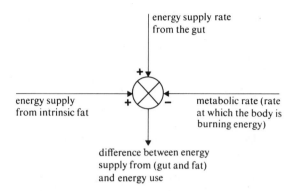

Figure 4.10 The energy balance sheet of the organism. One source of energy is from the food eaten, i.e. via the gut. Another source is from intrinsic fat deposits, i.e. lipolysis. If these two sources do not match the demand for energy, a source such as lean tissue will supply the difference

were depleted. This is not the case, and would be very maladaptive if it were so. Apparently the maximum rate of lipolysis from normal-size fat deposits with protein and glycogen mobilization from lean mass is not a sufficiently high energy flow in an active rat on an empty gut to stop it from getting hungry. Thus even though reserves are substantial, the lack of absorption provides a goad to seek food. What arouses hunger? How does the animal detect the slowing of absorption and insufficiency of fat mobilization? The answer is that we do not know, but we are in a position to make some intelligent guesses. It seems that researchers are gradually homing in on the answer. Look at Figure 4.10 for a moment. The rate at which the non-fat cells of the body use energy must obviously be equal to the rate of supply of energy. Now suppose it is the light phase and the animal is eating very little. Energy derived from fat is the primary energy source. Now imagine the animal after a period of food deprivation. The gut supplies nothing. Fat mobilization supplies some of the energy but it does not supply all. Glycogen in the liver supplies only a little energy at any moment. There is essentially only one other source for the missing calories and that is that the body will have to burn lean tissue, muscle being a major source. To be precise, body proteins are converted to amino acids released from the cells and converted to glucose in the liver. Formation of glucose in the liver from a substrate such as amino acids is known as *gluconeogenesis*. The fact that the liver is receiving amino acids from an internal source means that fat is not supplying all of the needed calories. The animal's lean body mass is literally shrinking and so a hunger signal should be generated. Whether or not the appearance of the gluconeogenic amino acids at the liver is what generates the feeding signal remains to be shown (Newman and Booth, 1979). As nobody knows how it works, we have to ignore the details of the detection, and proceed on the basis that the organism is able to detect the difference between, on the one hand,

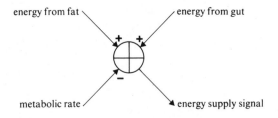

Figure 4.11   Block diagram representation of the suggested derivation of an internal energy state signal. If supply of energy from the gut and from the intrinsic fat source exceeds demand, this is a satiety cue. Deficiency of this supply is a cue for promoting food ingestion

metabolic rate, and, on the other, the rate of availability of energy from the gut and fat stores. Figure 4.11 shows this. A positive signal indicates an abundance of energy and is inhibitory upon feeding. Some energy will be converted to fat. A negative signal indicates that supply from gut and fat cannot meet demand and this provides the hunger signal. Averaged over the light phase, the signal marked 'from fat' takes a positive value indicating energy supply (lipolysis). Averaged over the dark phase, this signal takes a negative value, indicating a removal of energy from the pool into the fat deposits (lipogenesis). An absolutely fundamental question is: what determines whether net lipogenesis or lipolysis will occur? What gives this energy exchange a circadian rhythm? That it occurs has been very clearly demonstrated (LeMagnen and Devos, 1970; LeMagnen *et al.*, 1973). A plausible explanation might be that a hormone determines this cycle and is secreted in the form of a circadian rhythm. Such an intrinsic biochemical rhythm might seem the most obvious guess. Booth (1978) considers this possibility but places little faith in it. He notes that the gut contents are absorbed more quickly in the dark phase than in the light (Booth and Toates, 1974; Booth and Jarman, 1976), and this is due to more rapid stomach emptying.

Booth argues that a net lipogenesis occurs in the dark phase because, due to rapid gut emptying, energy is arriving at the liver and the pancreas at a faster rate then. Net lipolysis occurs in the light phase because of slower stomach emptying then. That gut emptying rate is crucial is suggested by the data of Le Magnen and Devos (1970). Even in the light phase, lipogenesis occurs after a meal. In the dark, lipolysis occurs before a meal.

To recapitulate then, energy supply rate is taken to be rate from gut plus rate from fat. If we subtract the rate at which energy is being used from this rate of availability then we obtain a picture of the current energy supply state of the animal, as distinct from the energy store state in gut, fat, liver, and muscle. It was assumed in the construction of the model that this rate is examined and the decision of whether to feed or not is based upon it. Figure 4.12 shows such a decision mechanism. The vertical axis is eating rate, and in this simplified model is taken to have just two values: 1000 cal/min, which is the rate at which rats

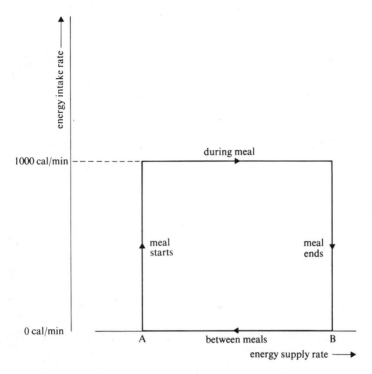

Figure 4.12 The internal energy state component of the feeding decision mechanism. This diagram relates to an animal surrounded by food in the laboratory. When energy supply rate falls to value *A* a meal is initiated, and energy intake rate rises from 0 to 1000 cal/min. During the meal, energy supply rate increases, and at value *B* the meal is terminated. Note that in this case food cue strength is constant. If food were not always readily available, the feeding decision would reflect fluctuations in availability of food

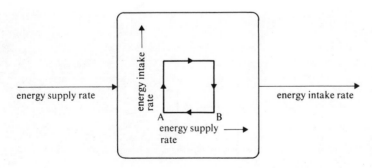

Figure 4.13 Block diagram representation of the feeding decision mechanism. Energy intake rate, whether to feed or not, depends upon energy supply rate

normally eat laboratory chow and corresponds to about 0.3 g/min, and zero, which represents not eating. When net energy supply rate falls to value A, eating is initiated. This puts energy into the stomach, which means it is being supplied from the gut a little later. The animal eats until this energy supply rate reaches value B, and then eating is terminated. The arrows indicate the flow of information. At point A, eating state switches suddenly from zero to 1000 cal/min and at point B, it drops suddenly back to zero. In Figure 4.13 we put this function into a box with energy supply as the input and feeding as the output. We have now completed the components of the diagram and in Figure 4.14 we assemble them to give our model of feeding.

We found from our computer simulation that the model 'ate' at roughly realistic intervals and took meals of the correct size. The model explicitly incorporates the view that feeding can be terminated by absorption of energy. It was not found necessary to appeal to short-term feedback from stomach stretch receptors, except when stomach contents get very large. This runs counter to the assumption, commonly expressed in textbooks, that absorption is so slow that it could not possibly account for the termination of meals. Such an assumption is reasonable when we think in terms of conventional homeostatic theories. For example, if the animal eats a 4 g meal it would, in these terms, be argued that it is correcting an energy deficit corresponding to 4 g. If at the end of the meal most of the 4 g is still in the gut it is necessary to postulate a short-term inhibitory effect which restrains further feeding until absorption has been completed.

In terms of our alternative model it might be said that in looking at rate the system is anticipating the energy quantity which is about to follow, if rate of

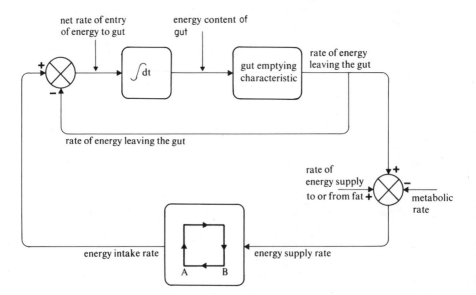

Figure 4.14 Complete block diagram of the feeding controller

absorption is dependent upon the amount of energy in the gut yet to be absorbed. The organism takes a sample and feeding is suppressed on the basis of this sample. There can be little doubt that absorption of energy from the gut is sufficiently fast that information reflecting gut contents is available at the end of a meal. Thus, within 5 min from the start of a meal, some radioactivity from the starch content has had time to be digested, absorbed, and carried to the brain of the rat (Pilcher, Jarman, and Booth, 1974).

4.3.6 Integration of internal and external stimuli

Except for unusual experimental conditions which we discuss later, ingestion of food is under the joint control of detectors of internal energy state and food in the environment. To be precise we have argued that a low rate of energy supply goads ingestion, and now we must add that this, accompanied by the stimuli from food, triggers ingestion. That both factors are always jointly involved is often overlooked, and one occasionally finds statements of the kind that animals at times eat for the caloric value of food and at other times for the taste of food. Similar unwelcome dichotomies occur in other areas, for instance genetic and environmental determinants of behaviour. Although few would argue that a person's behaviour is sometimes the result of genetics and sometimes that of the environment, there is nonetheless a tendency to ask questions of the kind: how much is it one or the other? All behaviour is *perfectly* dependent upon genes. All food ingestion is *perfectly* dependent upon taste; the ability of sensory factors to arouse ingestion is a function of energy state within the organism. It does not seem helpful to say that an aspect of behaviour, such as intelligence, is $x\%$ the result of genes and $(100 - x)\%$ the results of environment. However, it may be meaningful to say that if two individuals have an identical environment and yet behave differently then a genetic difference is implicated. Analogous statements concerning physiological state and sensory input may be made about food ingestion. That is to say both factors are always involved, and both in practice and theory it is impossible to tease apart their roles. Instead the parameters of their interaction need to be determined.

Consider an animal ingesting a normal *ad lib* meal of chow. One might say that it needs the energy, that its physiological state is determining its ingestion. But this would only be part of the picture. Why eat chow? Why not any substance that happens to be in the cage, such as sawdust? Chow must be capable of selectively arousing ingestion because of the effects of its chemical composition upon the animal's sensory receptors. In addition, though, past associations between ingestion of this substance and its post-absorptive consequences contribute to the sensory effect, which therefore is not simply added to an independent effect of physiological state. Sensory information from the food and from the animal's body jointly determine ingestion. To rephrase this in the language of Chapter 3, food has a certain cue strength by virtue of its taste, availability, etc.

Some researchers have tended to pay little or no attention to the taste, texture, or olfactory qualities of food. As long as food is known to be either present or absent then that is the only concern and the explanatory weight is placed upon internal events. It is salutary therefore to read Milner's (1977) comments made as part of an argument having something in common with that advanced here:

> 'It does not seem to have occurred to anybody at the time that a changed bodily state could reinforce learning only by influencing the nervous system in some way, and that it is no more reasonable to assume that an increase in blood sugar should influence the nervous system than that sugars in the mouth should do so.' (Reproduced by permission of Plenum Press.)

Other researchers consider taste to be a distinct causal factor separable from internal state. The reason that taste is sometimes discussed as a distinctive causal factor over and above regulation is that animals having *ad lib* food and water can be persuaded to ingest large quantities of sucrose, glucose, and saccharin solution. These are said to provide a taste incentive which can override the homeostatic regulatory processes and, in the case of sugar solutions, cause obesity. However, such an argument falls into the trap of assuming that there is something absolute about chow diets and the body weight which results from their ingestion. Depending upon the diet, there are other possible bodily composition equilibria that can be established, though chow may be closer to the kind of diet available in the natural habitat than is a sugar solution.

When given sugar solutions animals do not abandon regulation. They ultimately slow or cease ingestion, presumably because of reaching a new weight/energy state commensurate with the elevated steady-state caloric ingestion. Rats made obese by infusion of calorically dense liquid diet reject saccharin solutions. They lick the tube occasionally and then withdraw from it (Mook and Kenny, 1977).

In other words some average physiological energy state x_1 is associated with caloric intake y_1 of chow whereas average energy state x_2 (higher average body weight and detected energy state) is associated with ingestion y_2 of a sweeter diet. In each case an interplay between the sensory stimulus and energy state determines ingestion (see Mook and Kenny (1977) for a similar argument). Ingestion of highly palatable glucose solution is reduced by stomach loads of glucose, indicating the dependence upon energy state of the power of sensory factors to elicit ingestion. Glucose solution intake can be suppressed by concurrent intragastric, intraduodenal, or intraportal glucose solution infusion but not by intrajugular infusion, which implicates an hepatic glucose receptor in the inhibition of glucose solution ingestion (Mook and Kenney, 1977).

It is helpful to view the organism in terms of its evolution, the environment in which it has normally found itself and the food availability. The subjects of our studies are the product of long periods of evolution in a natural environment where neither chow nor sugar solutions were available. A certain weighting (cue strength value) would have needed to be applied to the presence of food in such an environment. If we then provide an abundance of readily available calorically

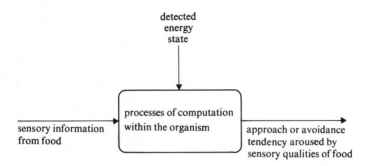

Figure 4.15 The two primary causal factors for feeding—energy state and sensory information from food. Sensory information from food can arouse approach or avoidance by the animal, depending upon the detected energy state

rich food to an animal in a small metabolism cage, then it is hardly surprising that it should increase its weight. It is wrong to invent processes in addition to the normal ingestive ones in order to explain behaviour that arises in an abnormal environment.

At first one might be tempted to consider a simple algebraic summation of sensory and energy state signals as determinants of intake. The assurance given by parents (at least when I was a child) that 'you would eat anything if you were hungry enough' might suggest this. Thus an animal depleted in energy state might show little discrimination (accept low-intensity food stimulus) whereas only more palatable food would be eaten when not so deprived. The evidence is compatible with such an addition but I believe only at a superficial level, and that a more complex model is needed. This would be approximately as follows.

An animal may either approach or avoid (or, intermediately, show indifference towards) caloric substances depending upon its internal energy state. In other words, the same incoming sensory information has different effects on the organism according to its energy (and hydrational) state. In the case of humans the subjective hedonistic rating of sweet solutions depends upon the person's hunger or satiety at the time of ingestion (Cabanac, 1971). What, to a hungry person, elicits approach (i.e. is said to be pleasant) is less pleasant or even unpleasant to one who is satiated. An extrapolation from such subjective reports leads to a model of the kind shown in Figure 4.15, though our understanding of these processes is by no means complete (Mook and Kenney, 1977) and Figure 4.15 should be regarded only as a provisional sketch.

Mook and Kenney (1977) doubt whether one can ever distinguish between eating for calories and eating for taste:

'A hungry rat which "needs" food, may drink a sweet solution and refuse water. In such a case the response to need *is* to respond to taste — that is, to the taste that identifies the needed commodity. States of deprivation and repletion can then be seen as affecting the parameters of that response (its threshold, its persistence, its susceptibility to inhibition by other factors, etc.' (Reproduced by permission of Plenum Press.)

Mook (1974) argued that substances such as saccharin solutions are ingested only in so far as they are predictive of subsequent energy yield. The expression 'releasing property' is very aptly employed. When the rat discovers that saccharin does not live up to its initial promise then the ability of the substance to stimulate ingestion falls. Mook raises the question of why, if rats ingest saccharin solutions, do they continue to eat chow, and suggests that whereas saccharin may mimic the taste of sugar, chow would be eaten for its protein component. Possibly another way of viewing this is to regard the rat having chow *ad lib* as being in a state of deprivation in so far as an easily ingested nutrient (e.g. a sugar solution) is concerned (Mook, 1974). Solutions can be drunk with little effort on the part of the rat, i.e. they have a high rate of energy (or perceived energy) yield (Mook and Kenney, 1977). This could explain part of the attraction of saccharin solutions.

Oral factors determine which behavioural controller is to be recruited by the animal. As Mook and Kenney note, a hungry but not thirsty rat which would decline water will avidly ingest saccharin solution. It would do so by the behavioural processes and mechanisms normally associated with drinking, even though it is the hunger motivational controller that underlies its behaviour. Taste is able to recruit the drinking controller to serve the ends of hunger. Rats will drink saccharin solutions until seriously over-hydrated, this being particularly apparent if they are injected with ADH so as to prevent loss of water by urination (Rolls, Wood, and Stevens, 1978). The inhibition of ingestion arising from the drinking control system (i.e. from over-hydration) appears weak when placed in opposition to the excitatory effect on ingestion from the feeding controller (i.e. ingestion aroused by saccharin).

The response of the animal to sweet solutions is useful as a demonstration of two other aspects of its ingestive behaviour: learning (see next section) and sensory adaption. Booth, Lovett, and McSherry (1972) showed that rats fed chow *ad lib* and also allowed a choice of 10% and 35% glucose initially showed a preference for the more concentrated solution. However, after a period of exposure to both solutions the preference reverses. It was argued that this was the result of an association between the 35% flavour and some aversive post-ingestional consequence, possibly hyperglycaemia. No such 'phobia' would be present for the 10% solution, and so its intake would increase relative to the 35% solution.

Why should rats reduced in body-weight by starvation drink only a certain amount of saccharin and then stop (Mook and Kenny, 1977)? We imagine that no post-ingestive change following saccharin consumption could have been responsible. Saccharin brings no caloric restoration. Stomach filling as a satiety signal was ruled out since concurrent (one-for-one) infusion of the same saccharin solution into the stomach did not change the amount drunk. Some form of sensory adaptation is implicated.

The discussion now turns to the role of learning in the termination of normal meals.

4.4 LEARNING

Most researchers have accepted that a biochemical factor such as a fall in blood glucose helps to initiate a meal. Most have also argued that it cannot be a reversal of this biochemical event which terminates the meal. It would appear to take too long. Therefore stretch receptors in the stomach detecting bulk have been postulated. These serve to lower the motivation to feed until the meal has been digested, by which time of course the original stimulus has been corrected. Normally the implication is that neural signals travel in fairly fixed pathways to inhibitory regions of the brain. We could imagine say x amount of stomach stretch exciting a signal equal and opposite to the biochemical correlate of food deprivation. When the stretch is excessive, feeding is totally inhibited irrespective of hunger level.

Total inhibition represents a safety mechanism—clearly animals do not commonly burst their stomachs. Under more normal conditions stomach bulk may inhibit feeding but by a rather different mechanism than that of fixed inhibitory pathways from gut to brain. The mechanism appears to be one of conditioning.

Le Magnen (1955, 1971) appears to have been the first to suggest that conditioning plays a role in the termination of a meal. Although his initial experiment is open to alternative interpretations, the message of this paper has recently found confirmation from several groups of researchers. He argues that specific oro-sensory signals characteristic of the diet concerned, and gastric signals, are associated with subsequent post-absorptive satiety. By a process of conditioning the oro-sensory and gastric signals therefore acquire a satiety potential of their own. In terms of classical conditioning the biochemical detection is the unconditioned satiety stimulus (US). By repeated association with the US, stimuli which would otherwise be neutral, or at least imprecise in their satiety potential, i.e. the oro-sensory and gastric signals, induce satiety. These are known as conditioned stimuli (CS), their effectiveness is *conditional* upon the association with the primary satiety stimulus. The signal concerning filling of the alimentary tract (i.e. the CS) may be neural in origin. In addition, a hormonal connection may be involved. The hormone cholecystokinin (CCK) has come under investigation recently, and may play a role in conditioned satiety (see Hawkins, 1977 for a recent review). It is as though the animal says 'A meal of this size has been satiating in the past so I will stop eating now'. Let us translate this into terms appropriate for the model being advanced here.

Normally the maximum flow of energy from a meal will occur slightly after the end of the meal and it appears that the rat becomes conditioned to stop eating in anticipation of this maximum rate (Booth, 1977, 1978; Booth, Toates and Platt, 1976). In other words a certain rate of absorption x at the end of a meal (on a particular diet) is followed, a little after the meal, by rate $x + y$. By past association between the meal and the post-meal rate $x + y$ the rat has learned to terminate the meal at rate x. According to the argument of Booth and associates, post-absorptive biochemical as well as oro-sensory factors are integrated to terminate a meal. It seems wrong then to speak of a dichotomy between short-term feedback involving the mouth and long-term feedback restoration of an energy state, which is often the way in which these processes are described.

Suppose a particular flavour of a caloric solution is associated with a relatively low yield of nutrients passing beyond the gut. This could be caused by pairing a certain distinctive odour with a low caloric value solution. When this distinctive odour is paired with another caloric solution in the future the rat takes a large meal, even if the new solution is of relatively high caloric yield (Booth, 1977). In time, it would 'recalibrate', take smaller meals and develop a new association, i.e. the taste and a high energy yield. Booth and Davis (1973) argue that a conditioned reaction to a distinctive oral stimulus is established. If the ideas expressed here are correct then during the course of feeding a complex change is initiated in the way sensory information is used. Booth (1972a, 1977) suggests that this is achieved by the learning of sensory preferences dependent on bodily state—a preference in hunger can become an aversion in satiety. Many experiments show signs of learned control of feeding although until Le Magnen's (1955) and Booth's work, the experiments were not designed to distinguish learning from physiological adaptations. There is a broad agreement that external chemical stimulation by taste, smell, and texture, as well as internal events related to energy state, both influence feeding. Booth (1977) argues that much of the control exerted by these two sets of stimuli is not innately reflexive but rather is the result of learning. The reaction of the organism to food depends conjointly upon external and internal stimuli. In other words, it is impossible to view either set of stimuli in isolation, the reaction of the organism depends upon a past history of association between diets of a particular flavour taken at a particular energy state. The association established depends upon the post-absorptive consequences of the particular diet. Booth continues:

> 'It will not be possible to elucidate the role of taste and smell in normal food intake by purely neurophysiological work on these senses, or by psychophysical studies. Any of the common experimental disruptions of the normal relations between smelling or tasting and feeding is liable to destroy the phenomena needing investigation. Similarly, biochemical or physiological studies of interoceptive hunger or satiety signals are unlikely to be illuminating as to normal feeding unless they are designed to allow for control of current feeding as much by *past* contingencies between foodstuff and body state as by *current* state of gut, blood, tissues, or neuronal activity.' (Reproduced by permission of Japan Scientific Societies Press.)

Booth (1977) reported an experiment which showed that preference was a function of internal energy state. Rats were given high- and low-concentration starch diets over a period of time. One distinctive flavour (F_1) had been associated with a high-concentration starch diet (S_H), while another flavour (F_2)

had been associated with a low-concentration starch diet (S_L). So as to control for flavour preference, an equal number of rats obtained experience with (F_1, S_L) and (F_2, S_H) associated, as with (F_1, S_H) and (F_2, S_L) associated. On the test day, animals were given a choice of two diets, having flavours F_1 and F_2. It was found that at the beginning of the meal, rats took more food having flavour which in the past had been associated with high-concentration starch. Later in the meal, the preference shifted and more of the sample of the flavour previously associated with low starch concentration was taken.

What seems to be happening is as follows (Booth, 1977). The hungry animal derives greater immediate satiety from high-concentration starch than from low-concentration starch. Energy flow is high soon after eating. Therefore the animal in a state of hunger develops a conditioned preference for the flavour associated with high-concentration starch. The brain classifies the flavour as one appropriate to the body state. However, because of the lag between eating and maximal energy flow, if the animal were to ingest such a diet when it was near to satiety, it would tend to overshoot the optimal level of energy flow. Therefore, restraint is indicated when confronted with such a diet in a state near to satiety.

A diet of low energy yield would in the experience of the animal have produced a less strong energy flow signal at the start of meals and so in a straight choice by the hungry animal would not be so favoured as the high-yield diet. However, the animal nearing satiety could afford to take relatively large quantities of such a diet with no risk of overshooting the optimal maximum energy flow.

Janowitz and Hollander (1955) investigated the effect upon subsequent feeding of making stomach loads of various percentages of dogs' normal daily food intake. Animals were allowed 45 min in which to feed each 24 h. Six hours before feeding either 50, 100 or 175% of the control caloric intake was placed in the stomach. In the case of the 50% load then food intake was reduced so that compensation was almost complete, but for the other two loads the amount eaten took several weeks to come down to an asymptotic value. One wonders if the period of decreased intake represents a learning curve: the animal learns not to eat because of the loaded calories. It is interesting from the same point of view that when loading is ceased the animals at first seem to eat less than normal and take a week or so to increase intake to normal levels.

If rats are deprived of food for 23 h and then allowed 1 h eating, and this is repeated for several days then the latency to eat after food presentation decreases and the amount eaten increases over trials. The same thing happens if they are water-deprived for 23 h and given water for 1 h (Ghent, 1957). There may be two distinct processes at work here. First, the latency reduction could represent the fact that the rats are learning to overcome an initial reluctance to consume what takes some of the properties of a novel substance even though it is the return of a familiar diet (see later). Secondly, the fact that more of each commodity is taken as trials proceed could represent learned desatiation. If the consequences of ingestion fail to answer the physiological need at, say, 1 h post-ingestion, then possibly the satiety threshold is elevated for subsequent meals. Undeprived rats exposed to a daily 3 h session in which milk may be taken show an increase in intake over days (Williams, 1968). Davis and Campbell (1973) appear to see an analogous learned desatiation.

It could be argued that in the experiment of Ghent the animals were simply getting hungrier as the trials proceeded. Indeed, on such a schedule, rats' daily acitivity increases over a period of days and weeks (Reid and Finger, 1955), and activity is known to increase with severity of food deprivation (Misanin, Smith and Campbell, 1964). Also, a further question is: could the animals be eating more because of their greater activity? I believe the evidence supports a learning explanation rather than one in terms of animals getting hungrier. Following various lengths of food deprivation, when presumably hunger was at various levels, 2 h food intake shows little or no dependence on the length of deprivation (Dufort and Wright, 1962). Le Magnen and Tallon (1968) showed that the size of the first meal increases with repeated deprivation unconfounded with increasing hunger.

We return now to the theme of conditioned satiety. Starting from different premises to Booth and associates, and considering mainly the case of humans, Stunkard (1975) arrives at similar conclusions. Stunkard believes that the metabolic signal which is the unconditional stimulus for feeding/satiety is central. Yet, as he observes, feeding is terminated long before a substantial proportion of the ingested energy is absorbed. Stunkard reminds us of the ability of rats to adjust intake in response to caloric dilution of the diet by addition of non-nutritive bulk. Such compensation is achieved by gradually taking bigger meals. He argues that stomach volume receptors amongst other things terminate a meal, and the consequences of the meal are detected perhaps 2 h later. If a mismatch occurs between expected consequences and actual consequences, by, for example, the meal unexpectedly being calorically dilute, then adjustment can be made when the next meal is taken so that more is eaten. Stunkard notes that the interval between the conditional stimulus (gastric filling) and the unconditional stimulus (central determination of total ingested energy) is up to 120 min, and Pavlovian conditioning has normally been thought to work with CS–US intervals of up to a second or so. Therefore Stunkard appeals to more recent studies (discussed later) on taste and food intake which show that in the particular case of food ingestion, enormous CS–US intervals are possible.

Hawkins (1977) accuses Stunkard of somewhat over-emphasizing gastric filling as a CS and understating the importance of oro-sensory factors. He notes, in the context of human feeding, that satiety may be specific to certain oro-sensory properties. An example of this is the 'dessert effect' in humans. Rearousal of appetite by presentation of gateaux after meat may represent a temporary release from conditioned satiety due to a change in oro-sensory properties.

Whatever theory of feeding we propose, it is clear that a large degree of inertia in the feeding pattern must be accounted for. Once established on a regular diet, rats continue to eat on the basis of constant intake of grams rather than constant intake of calories for a while after caloric density is changed (Janowitz and Grossman, 1949).

4.5 BRAIN LESIONS

A large proportion of the literature on feeding is devoted to brain lesions. In particular, the effect of lesions at two brain sites, the lateral and ventromedial hypothalamus, have been investigated (for recent account of lesion studies see Friedman and Stricker, 1976; Blundell and Latham, 1979).

When the ventromedial nucleus of the hypothalamus is lesioned, rats put on weight because they eat considerably more than unoperated controls. Rats lesioned in this way are sometimes called VMH or VMN rats. There are dynamic and static phases of this syndrome. In the dynamic phase, the 4–12 weeks following the operation, eating is excessive and weight increases. Weight then reaches a plateau and feeding returns to near normal levels, the static phase. In the static phase if the animal is starved it will, when food is returned, eat excessively until it regains its previous elevated plateau weight. Conversely, if weight is pushed above the new plateau by forced feeding, when *ad lib* conditions are restored it will reduce intake until the plateau is regained. Such results have encouraged the assumption that two mechanisms are at work, and some books take these to be established facts. First, it is argued that the body weight set-point is increased by the lesion. The evidence that the animal defends the elevated weight by behavioural means is used to support this. Secondly, it is argued that the VMH is a satiety centre, since its destruction leads to overeating (we will say more about this in a moment). There is cause to question the assumption that excessive eating is due to a sudden disparity between a set-point value (now elevated by the lesion) and an actual value of body fat. What appears to occur is that the biochemical balance of the organism is disturbed. Following the lesion a relatively large percentage of any energy available is seized by the adipose tissue and builds up the fat reserves. Now it could be that two factors are at work:

1. a behavioural effect, that the animal has a tendency to eat more irrespective of the fate of the energy; and
2. a physiological effect which sees to it that a larger than normal percentage of energy is converted into fat.

However, factor 2 will account for the evidence on its own. As Friedman and Stricker (1976) point out, in the dynamic phase the VMH rat acts as though it is permanently in the absorptive state as far as conversion of energy to fat is concerned. This conversion to fat means that less glucose is available for the cells. In addition, they are denied the energy which normally comes from fat in the light phase, since energy is now being converted to fat all through the cycle. Therefore the VMH rat overeats because of the need for fuel, not because of a primary motivational change. The increased conversion of energy to fat and the hyperphagia of the VMH lesioned rat is mainly evident in the light phase. In the dark phase, a fraction of incoming energy is normally converted to fat in any case, and is released in the light phase. It is this release which does not occur in the VMH lesioned rat. Carlson (1977) expresses it as follows. Rather than fat being a 'reversible sponge', it is a sponge which simply soaks up fat during the dynamic phase (see also Booth, 1978).

It is not merely the case that it is more parsimonious to account for the VMH syndrome in terms of a primary metabolic change. The VMH rat simply does not behave as though its hunger drive is suddenly and dramatically increased by the lesion, which would surely be implied by a set-point explanation. They sometimes refuse a diet less palatable than normal which intact rats accept. They sometimes work less hard for food reward than controls (see Friedman and Stricker, 1976, for a review).

The results are in keeping with the interpretation that they overeat because the normal consequence do not follow from feeding. Friedman and Stricker (1976) argue that in addition to biochemical changes caused by the lesion the VMH rat experiences a change in sensory reactivity and emotionality. This is manifested by finickiness towards food, and heightened reactivity towards any aversive stimulus.

Lesions made in the lateral hypothalamus (LH) cause the animal to lose weight. After a while it reaches a subnormal plateau, and an attempt to move it from here brings corrective behavioural action. Inevitably, among some writers it became the view that whereas VMH lesions move the set-point up, LH lesions pull it down. Again, we would wish to question this assumption. As Friedman and Stricker (1976) point out, a wide variety of motivated behaviours is disrupted by LH lesions, such as feeding, drinking, maternal, and thermoregulatory behaviour. They suggest that a non-specific activational component of motivation is affected. The equilibrium established as a result of sensory and bodily effects will be one of a lower body weight.

4.6 MEASUREMENT OF HUNGER AND RECOVERY FROM DEPRIVATION

If we use the amount of food eaten in a test as a measure of hunger we can expect anomalous results. After about 24h of deprivation the amount eaten does not increase with deprivation but may even decrease; but if we use the amount of quinine which must be added to the food to deter the animal from feeding as a measure, it appears that hunger level increases up to at least 54h. The same result is obtained if bar-pressing is used as an index of drive (Miller, 1955).

Students seem surprised as to how little rats eat following a period of food-deprivation. It might seem logical that if they normally eat 15g in 24h, then following a day of food deprivation they should eat 30g in 24h. In fact they may eat only about 20 g. It may be wrong to think that they have lost 15 g equivalent of energy which must be corrected. In fact during food-deprivation and immediately after there is an adaptive reduction in their metabolism (Westerterp, 1977), which should be compared with the renal response of decreased urine flow during water deprivation (see next chapter).

Another reason for the animal's failure to compensate precisely is that during starvation fat deposits will have been mobilized. Recovery of fat deposit size to normal may take several days after return to *ad lib* feeding.

In order to hold hunger constant in motivation studies it is a tradition to maintain rats at a fixed percentage of their *ad lib* feeding body weight (see Bolles, 1975). Daily ration size is adjusted accordingly. The assumption is often made that loss of weight *per se* generates the feeding signal. We would prefer to say that as the animal is food-restricted then fat deposit size is reduced, and this means a decrease in the rate of energy supply from fat. Loss of body weight means then reduction in lipogenesis and dependence upon amino acid as a fuel. We would associate the animal's behaviour on a food-deprivation schedule with shifts in energy supply rather than loss of weight *per se*.

4.7 ECOLOGICAL ASPECTS OF FEEDING

So far we have mainly considered two factors which influence the decision to feed: energy state and palatability of food. In this section we examine a third, the ease with which food is obtainable, which when applied to the natural habitat is known as the ecological factor.

Collier, Hirsch, and Hamlin (1972) and Hirsch and Collier (1974) argued that traditional research on feeding carried out in laboratories is dominated by a simple model of behaviour, that of reflex homeostasis. That is to say, departure from optimum of a physiological event inevitably leads to ingestive behaviour. Collier *et al.* see such thinking as following firmly in the restricted traditions of Descartes' hydraulic model of animate motion. They make the claim that a homeostatic feeding model inevitably ignores ecological variables which are instrumental in shaping a species' responses, and it pays no regard to the evolutionary history of the species.

Collier *et al.* argue that it is unlikely that some species ever experience short-term depletion, an example being ruminants. Here the gut acts as an enormous reservoir, and input of energy across the intestinal wall is nearly constant. Collier *et al.* claim that the energy content of the stomach of freely feeding laboratory housed rats and guinea pigs is never zero.

Collier *et al.* and Hirsch and Collier discuss the nature of the feeding response. Diet and the animal's adaptation are reflected in its speed and manner of eating. Herbivores are usually surrounded by food of low caloric value. They eat often and take meals of long duration. Carnivores eat relatively infrequently since considerable effort is involved in catching prey. Lions and tigers, who are themselves not preyed upon, eat at a low rate of intake and take large, long meals. Obviously having exerted effort they should take as much food as is available. Hyenas, who are not in such a favourable position and must compete for food, eat large meals at a high rate of intake. Availability of suitable prey combined with some measure of energy state would be the cue for killing prey in the case of lions and tigers. In all probability, a meal is not ended by a change in energy state in such species. The animal may simply eat until no more food is available irrespective of the consequences. Carlson (1977, p. 338) suggests that the wolf

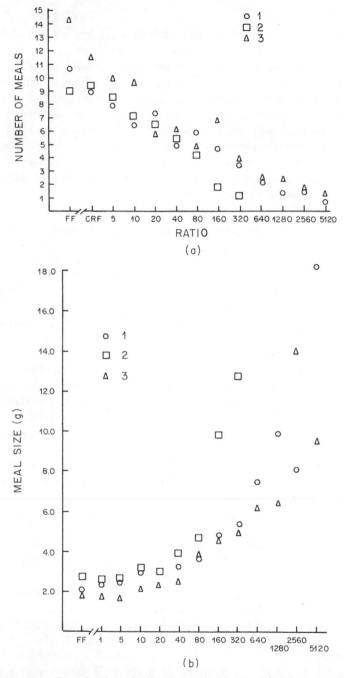

Figure 4.16 The effect of making feeding 'expensive' by requiring the rat to press a bar a number of times to gain access to food. 'Ration' means the number of presses needed to gain a period of access. (a) daily number of meals taken by three individual rats as a function of the ratio size employed (FF—free feeding, CRF—continuous reinforcement). (b) amount of food consumed per meal by three individual rats, as a function of fixed ratio size. (Reprinted with permission from *Physiology and Behaviour*, **9**, G. Collies, E. Hirch, and P. H. Hamlin, 'The ecological determinants of reinforcement in the rat, Copyright 1972, Pergamon Press, Ltd.)

behaves as follows. We assume that it gorges itself following a kill. The pack eat everything available. If it was a large prey, the length of time until the next kill will be relatively long. It is not the case that they hunt every few hours and take a meal having a size proportional to the length of time since the last meal. Meal size depends upon the prey, and initiation of meal-searching upon an energy consideration.

In studies on the laboratory rat Collier *et al.* attempted to simulate certain aspects of a natural habitat where food was not readily available and where effort was needed to obtain it. They reasoned that if an animal must expend a significant amount of energy in getting to food then it might eat less often and take larger meals. In one experiment rats had to press a bar in order to gain access to a food cup. Access, once obtained, was for an unlimited amount of time provided that the rat actually remained in the tunnel leading to food. If it interrupted feeding to, for instance, drink, then if 10 min elapsed when it was out of the tunnel the food cup was withdrawn, and the animal had to work again for its return. The number of presses needed to obtain access to food was increased from 1 to 5120 over several days, i.e. a fixed ratio (FR) schedule was in operation on any given day. Figure 4.16 shows the number of meals taken against the FR employed, and the mean meal size. Perhaps we should focus our attention on the low-ratio end of the scale since at very high ratios the animals had some difficulty in maintaining energy intake. When, for instance, 20 to 40 responses are needed to obtain food then the number of meals taken is about half of normal, a clear demonstration of the involvement of effort in feeding pattern. By reducing the number of meals and increasing their size the animal was able to conserve energy. Experiments reviewed by Bolles (1975, p. 198) suggest that the rat can be deterred from taking frequent meals simply by requiring that it moves from its living quarters to the food source.

An alternative way of imposing a constraint on the animal was to use an FR schedule such that every, say, five bar-presses gave one pellet of food. Unlike the experiment just described, the rat was unable to conserve energy by altering the frequency and length of meals. However, it was found that here also rats took fewer and larger meals on high FR schedules. Effort appears to decrease meal frequency, even if to do so may not help the animal's energy balance. Collier *et al.* concluded:

'. . . the nature of the ecological niche, that is, the availability of food, its caloric density, the work and danger required to obtain it, the animal's relation to other species occupying the same habitat, and its position in the food chain, determines the feeding patterns of animals.' (Reproduced by permission of Pergamon Press, Ltd.)

Peck (1978) found that body weight decreased when animals were placed on ratio schedules for obtaining food. Steady-state body weight was a function of the schedule with the highest ratio associated with the lowest weight.

4.8 DEVELOPMENTAL ASPECTS OF FEEDING

So far we have looked at the adult organism which is under the control of internal physiological signals indicating energy state, availability of food and sensory information concerning food. It is appropriate now to investigate how far the animal is innately programmed to respond to certain tastes and smells, and how much they acquire their power to evoke feeding by association with their metabolic consequences. In other words, to what extent is early feeding a trial and error process?

Le Magnen (1971) raises the question of whether innate feeding reactions to certain olfactory stimuli exist. He notes that they could be specific to very particular foods, and gives maternal milk as one obvious example. Other cases of innate preference appear to reflect ecological factors. Wyrwicka (1976) sees feeding and drinking as early instrumental reactions which are strengthened by their consequences. She notes, for example, that foods which are initially rejected can become acceptable if accompanied by rewarding brain stimulation. Under more natural conditions, it appears that chicks learn to drink water gradually, a start being made by pecking at it, and only later is the accidental discovery made that the chick must lift its head in order to swallow. Finally, she reports that naive kittens who have beem removed from their mother suckle both nutritive and non-nutritive nipples, and slowly increase the percentage of time spent with the former at the expense of the latter.

Hogan (1977) considers the issue of how ingestion and hunger motivation develop an association. Hogan found that recently hatched chicks peck at a variety of objects. If the object of the peck is food then pecking frequency shows an increase. If it is, say, sand, then pecking frequency decreases. The difference in pecking rates appear at about 1h after each substance was first ingested. This suggests that detection of the chemical nature of the food substance is registered and this potentiates pecking, as a kind of positive feedback system. The animal develops a preference for the food substance because of its metabolic consequences, but appears not to develop an aversion to sand.

There are at least two processes by which wild post-weaning rats can come to select the same diet as their parents (Galef and Henderson, 1972; Galef and Clark, 1972). At a distance from the nest infant rats approach adult feeding conspecifics and feed in their vicinity. Also rats recognize certain characteristics of food which had been associated with mother's milk. This was demonstrated in a controlled laboratory study. Suckling mothers were fed either Purina Lab Chow or Turtox 'fat sufficient diet'. Subsequently after weaning their pups showed a preference for the diet given to the mother, the only channel of communication to facilitate this preference being shown to be via the mother's milk.

As in most aspects of behaviour, taste and diet selection is a complex function of genetic blueprint and its subsequent modification by experience. Clearly some innate taste- and odour-recognition process is present in the organism, as tastes such as saccharin seem to be highly favoured. However, flexibility is also built in so that experience can alter preferences. Booth, Stoloff, and Nicholls (1974) argue that in rats very early experience of diets is associated with either high or low energy flow as a consequence of ingestion. They show that by association, the animal rapidly comes to acquire preferences for nutritious food, and aversion for non-nutritive bulk. Again, we return to the theme that the animal's behaviour is a complex function of innate sensory factors acting conjointly with the energy consequences of each diet.

4.9 HUNGER FOR SPECIFIC CHEMICALS

Earlier in this chapter we spoke about energy regulation by feeding as though it were a matter of some indifference in what form the energy actually entered the body. In some respects this is true; the body has a remarkable ability to synthesize fuel from different substrates. Glucose which is essential for the nervous system can be derived from protein substrates. Also of course the diet we normally consider, laboratory chow, is well balanced between protein and carbohydrate.

However, we do not eat only to obtain energy to burn. Important vitamins, amino acids, and minerals which are essential for the functioning of the body are obtained in the diet. An *essential* amino acid is one which cannot be synthesized in the body. A diet lacking one of the essential amino acids is an imbalanced diet. If a diet is inadequate in an essential component, rats may lose their appetite for that diet even though it is perfectly capable of meeting their energy needs (Harper and Boyle, 1976). Booth and Simson (1971) brought rats into a state of protein deficiency. Rats were then stomach-loaded with either a balanced mixture of amino acids or an imbalanced amino acid mixture. After each load, a protein-free diet having a distinctive odour was allowed. It was later found that rats preferred food having the odour associated with the balanced load of amino acids to the imbalanced load. There was aversion to odour-labelled food which had been paired with the imbalanced amino acid mixture.

Biologically such behaviour makes very good sense. It would obviously yield diminishing returns for an animal to continue eating normal amounts of a diet lacking an essential amino acid. To do so seems to cause discomfort. If such a diet is avoided, it induces a state of energy need coupled with a specific aversion to the deficient diet. If the animal then encounters a diet having a different flavour it will have the energy state appropriate to ingest it. The diet may prove to contain the essential amino acid. If it does not, a new aversion will develop.

Similar results apply to vitamin deficiency. First a rat is placed on a thiamine-(vitamin B_1)-deficient diet until the effect of the diet becomes apparent in the animal's behaviour. It is then presented with a choice: the old or a novel diet. A clear preference for the novel diet is shown. This is true even if the old diet has had thiamine added, and the novel diet is thiamine-deficient at the time of the choice.

If it turns out that the novel diet, which is eaten for a few days to the total exclusion of the old diet, contains thiamine then a learned preference appears. If it lacks thiamine learned aversion occurs. In other words rats do not have an innate thiamine recognition system (Rodgers and Rozin, 1966). If the diet contains thiamine then some physiological change occurs following ingestion of the diet. This is associated with the rat appearing to 'feel better'. Rozin and Rodgers (1967) emphasize that learned aversion to a familiar but deficient diet produces an energy deficiency which allows ingestion of a novel diet. It may appear to be an active *neophilia* (reaction due to novelty *per se*), but it is better explained by a combination of hunger signal and a diet for which no aversion exists. Rats in fact seem to be somewhat hesitant to eat novel diets, except under the abnormal conditions just described, and this phenomenon is known as *neophobia*. It serves an obvious adaptive value for the animal to sample a small amount of a novel diet and see what happens.

The experiments of Rozin and associates are part of a very exciting development in animal behaviour seen in the last ten years or so. One aspect of this is as follows. First an animal is allowed to eat a distinct and novel substance; in the case of rats its flavour characterizes it. Then the animal is made 'ill' by a drug such as lithium chloride. In the future, it will either totally avoid the food of this flavour or eat less than control animals (see the collection of readings edited by Milgram, Krames, and Alloway, 1977). A number of particularly interesting and unusual features surround this phenomenon, and perhaps explain why its discovery was at first greeted with suspicion (Revusky, 1977b). A single trial is all that is needed to produce a long-lasting aversion. The interval between tasting the food and the injection can be several hours and aversion still results. In both of these respects the demonstration is spectacular among learning experiments. Finally, the phenomenon is powerful ammunition to those having an ethological or natural history orientation, in that species differences are large and reflect the adaptive value of the mode of learning (though it is possible to overstate species differences (Revusky, 1977a)). Rats easily develop an association between taste of food and subsequent ill-effects, whereas birds form an association with the visual characteristics of the stimulus. Learning is then specific to the combination of stimulus and its consequence, and the species concerned.

Rozin (1976) reviews specific hungers in a natural history context. Monophagic animals (those eating only one specific food) simply require a sensory system wired to recognize this particular food and a motivational mechanism which generates ingestive behaviour. However, the monophagic is an inflexible and vulnerable animal, liable to extinction at a time when the food source is temporarily unavailable. Thus most species are prepared to sample a variety of foods. The larger the selection, though, the more difficult it is for the nervous system to recognize it by innately wired mechanisms. Further, the chance that the diet will be deficient in an important component increases with the animal's flexibility. This is particularly evident in the omnivore and the herbivore.

Rozin notes that for the animal two solutions are available to this problem: (1) multiple specific systems, or (2) an open-ended, or trial and error, mechanism which allows learning from experience. Those species so far most studied seem to employ a mixture of these two. In the next chapter we will consider sodium appetite which is at least in part a pre-wired discrimination. A pre-wired preference for sweet taste is present in rat and man, as is an avoidance of certain poisons. These innate biases give a general direction to food-seeking behaviour, but experience then provides more refined selection programmes. As Rozin expresses it, in the case of rats:

'It selectively connects internal metabolic consequences with chemo-sensory input; put colloquially, it associates stomach aches with tastes and not lights and sounds.'

4.10 SUMMARY, DISCUSSION, AND GENERAL CONCLUSIONS

Animals need to feed in order to obtain (1) energy which they can subsequently use, and (2) substances such as amino acids, vitamins, and minerals which are necessary for the functioning of the body. With regard purely to energy the actual substrate need not matter too much. Glucose, the essential fuel for the brain, is obtainable by post-digestive chemical conversions.

For rats in the dark phase when food is available the animal will derive most of its energy from recently ingested food. In the case of laboratory chow this will be largely glucose derived from starch. In the light phase the animal eats less than its metabolic rate and so in addition to energy from recent meals it depends upon breakdown of fat deposits. We have proposed that an energy rate detector is aware of the flow of energy from fat, and consequently the animal eats fewer and smaller meals in the light phase. During total starvation, energy is available only from glycogen stores (a very limited reserve), fat, and lean mass. Exactly how a feeding signal arises from the animal's energy state is as yet unclear, though there is a convergence of theories upon energy supply, with the liver as the focus for the detection site. Possibly the liver examines the rate at which glucose arrives from the gut and initiates feeding when this rate falls below a certain value.

In addition, though, we would need to explain why the animal refrains from feeding under some circumtances even though there is no energy coming from the gut. For instance, after gross fattening by force-feeding, little or no feeding occurs for a while. It is necessary to postulate that the flow from intrinsic lipid stores is also taken into account. Possibly the algebraic sum of energy flow from gut and lipid exchange is compared to energy demand, and feeding or satiety is the result. If there is an abundance of energy from the gut, the liver's glycogen stores will be rich. Similarly if lipolysis is high then liver glycogen may remain at a high level. However, when supply from gut and adipose tissue is relatively low, liver glycogen stores will be called into service and hence depleted. The size of these stores may contribute to the feeding signal. Alternatively, or perhaps in addition, the animal could examine the rate at which lean tissue is being broken down. After a period of fasting the animal will be heavily dependent upon amino acids as a substrate for glucose; that is amino acids from the breakdown of lean tissue. Dependence upon such amino acids is clearly a condition which in some way or another should be associated with a drive to feed. As yet, the precise signal or signals is unclear.

However, we feel confident that in some form an energy supply theory is nearer the truth than any of the older homeostatic theories, such as the glucostatic and lipostatic. The glucostatic is clearly inadequate, since blood glucose level is a very poor predictor of feeding/satiety. At one stage a thermostatic theory was also discussed: that animals eat when body temperature falls and stop eating when it rises. Although under some conditions animals may possibly be induced to eat by cold temperature *per se*, it is doubtful whether temperature is normally the stimulus to initiate a meal. Animals, such as rats, eat more in a cold environment than a warm. This can be explained by a higher metabolic rate in the cold, and consequently the need for a higher energy supply rate.

The lipostatic theory correctly points to the relative stability of the animals' fat deposits. However, a number of mechanisms could yield such stability, and it is unnecessary to postulate that feeding is anchored to fat deposit size *per se*. Artificially low fat deposits cause abnormally high conversion of any incoming energy to fat and hence cause hyperphagia, whereas unnaturally high fat deposits inhibit feeding by lipolysis. Energy supply predicts stability of fat stores. Energy supply can, in addition, better explain feeding on a meal-to-meal basis than can the lipostatic theory.

The constancy from day to day of body weight (largely dependent upon body fat) commonly leads to the assumption that a *set-point* of body fat is present. This may in fact exist. We don't know, but there is no necessity to postulate its existence. Remember the village pond that showed constancy in the absence of a set-point. Wirtshafter and Davis (1977) use the term *settling-point* in preference to set-point. This expression is neutral with regard to the mechanism by which weight/fat size is maintained relatively constant, but it does convey the meaning that the system settles at some value. Consider the fact that body-weight settling-point increases as the diet is made more palatable. It is difficult to imagine that a control system set-point is a function of the diet. However, it is easy to imagine that a new equilibrium arises between the excitatory effect of food and some inhibitory (i.e. energy supply from lipolysis) influence from fat size. If the old diet is returned then the original body weight may be restored, but there is not necessarily anything privileged about this weight, it is merely the outcome of the equilibrium. It is sometimes argued (e.g. Mrosovsky and Powley, 1977) that the increase in weight shown by birds prior to migration is a demonstration of an increase in set-point. It is not difficult to appreciate the adaptive significance of a bird putting on weight at this time, but it does not necessarily argue in favour of a set-point and its adjustment. As Davis and Wirtshafter (1978) point out, the changing photoperiod, which is the necessary stimulus to bring them into the migratory phase, could cause a change in lipid metabolism. In other words lipogenesis would increase relative to lipolysis until a new equilibrium of higher fat stores was attained. The fact that a larger percentage of ingested calories was being converted to fat would mean increased food intake during the weight gain period.

Peck (1978) wrote:

'. . . it has been common to contrast organic or internal with sensory or peripheral determinants of ingestion and to assume the former are "physiological" and homeostatic and the latter are "psychological" and nonhomeostatic. To distinguish in these ways between organic and environmental determinants of ingestion is in effect to assume that there is one normal value for the caloric reserve of fat and to deny that different values of this reserve may be appropriate in different environmental circumstances, as when foodstuffs of different incentive qualities are available.'

Peck (1978) postulated that 'rats defend different body weights depending on palatability and accessibility of their food'. He found that rats increased body weight if the diet was made palatable and decreased it if it was adulterated with quinine. In the steady state a range of body weights was observed according to diet. The interpretation which the present author would place on this result is that body weight is a consequence of caloric intake. It is perhaps misleading to say that they 'defend' a particular body weight, a reservation which Peck would acknowledge. Suppose the diet is changed from quinine-adulterated to normal, and animals increase their weight to one characteristically associated with the normal diet. It is not that the animal in some way recognizes that it is at a weight incommensurate with the new diet and eats to compensate. Rather, it eats more because of the sensory qualities of the new food and inevitably reaches a higher steady-state weight.

We have argued that feeding tendency arises as a function of the interaction of several factors, one of which is a measure of the organism's energy state. The sensory properties of the available food also influence when a meal is taken and how much is taken. In the future we will hopefully be able to construct programmes to show how the animal's oral and olfactory detection interacts with energy state in determining feeding tendency, taking into account its past history of association between diets and their caloric consequences. Availability of food is another source of information to the animal which influences when and how much it eats. If, in the wild, a long journey to food is a factor in determining the timing of meals, then it seems that the animal must employ some kind of neural model of the environment not only to find food but also to influence the decision to seek food (this would also be true of water-seeking).

If we consider the laboratory rat then the cue strength of food would remain fairly constant—food is rarely more than about 20 cm from the animal's nose. Probably energy state, if we were able to measure it, would provide a very good prediction of when intake would occur. In the natural habitat cue strength of food would vary widely. We imagine that various energy states may accompany the initiation of a meal, this being due to fluctuations in food availability and sensory cues from potential food.

Some species, notably herbivores, may be surrounded by food; whereas others, for instance, carnivores, may have no fixed food location. In the latter case energy state would be causal in food-seeking, but less so in meal-termination.

We have only begun the study of the ecology of feeding and how it relates to the physiological psychology of energy intake. Animals will place different weights upon different aspects of the feeding situation. For some the cue strength of food will be high. This subject needs careful attention in the context of results which show that it is difficult to suppress feeding by infusing caloric substances. For animals which feed occasionally and invest a large amount of effort in obtaining food then heavy weighting should be placed upon the presence of food. The wolf is reported to be able to eat up to 17% of its body weight in meat in a single meal (Mugford, 1977). In addition social facilitation of feeding has been reported in a variety of species (Mugford, 1977). In carnivores hunger intensity should be particularly aroused by the flesh of the prey. This is suggested by some lines of evidence. In cats ingestion of stale chow can be strongly revived when 'satiation' appears, by blowing meat odour through the diet (Mugford, 1977).

Carnivores have adapted for conditions of alternate feast and famine, their evolutionary history being one of taking occasional meals. It is perhaps not surprising that domestic carnivores occasionally show obesity.

In their energy ingestion, animals show something analogous to what economists call elasticity (Lea, 1978). If the price of a commodity is increased and the amount sold decreases this is an elastic demand curve. If quantity sold is unaffected by price there is said to be zero elasticity. A ratio-schedule is analogous to a high price for food, whereas food freely available is analogous to a low price, and indeed animals reduce consumption when the price is high. Such flexibility of ingestion, and consequently body weight, may come as a surprise to the student. Adult organisms of different weights are different mainly in terms of the amount of fat contained in the body. Thus rats on a palatable diet and/or low cost schedule have a larger quantity of body fat than those on low palatability/high cost, but lean tissue mass is not very different except under extreme conditions (Peck, 1978). Fat is an energy reserve of variable size according to circumstances. It is obvious that large fat reserves can be an advantage since they are a buffer against starvation. In one sense the larger the better. However, there are trade-offs to be performed. If food is hard to obtain it may be to the animal's disadvantage to invest vast amounts of time and energy in obtaining it. By cutting down on food intake it saves on time and on weight. Its metabolism will also fall a certain amount, thus at first preventing loss of lean tissue.

Thus we must view the organism as under the control of a complex decision mechanism which has been subject of selective pressures. It is flexible in that different strategies can be employed according to environmental circumstances.

Although animals on an adequate diet may have the same lean mass but differ in fat weight, clearly if energy supply from food and breakdown of body fat is inadequate to meet metabolic needs the animal must burn lean tissue. In so doing it will, if pathological conditions are not imposed, reach a lower than normal steady-state value of lean tissue mass. One way of viewing the situation is that the animal shrinks in size until it reaches a lean tissue mass which can be adequately supported by the reduced energy supply from food.

The factors which serve to terminate a meal present a problem and cause for additional theorizing, for all researchers in the feeding area. The demand for short-term feedback is perhaps less apparent in an energy flow model than in a traditional homeostatic theory. However, we argued that a combination of a post-absorptive energy rate signal accompanying an oro-sensory signal could best explain meal termination. Whatever mechanism is involved then it must be capable of showing adaptive control. Caloric dilution of a diet leads to a gradual increase in the size of meals, implying that the animal is recalibrating its switch-off cues.

Some diets, such as that of the carnivore, may be guaranteed to yield all essential nutrients; whereas for others a certain trial-and-error process may be employed. However, it is likely to be the case that an innate preference for the most likely energy-yielding substrates in its environment exists even in a species having a very flexible trial-and-error facility.

CHAPTER 5

Thirst

5.1 THE BODY-FLUIDS

5.1.1 Introduction

This chapter concerns control of the organism's intake of water and sodium. In looking for the roots of this behaviour we must examine some aspects of the biochemistry and physiology of the body. Water is the medium through which the chemical processes of the body operate. About 69% of a rat's body weight is water. In humans it is about 60%. Cells in the body contain various percentages of water according to their type; lean tissue cells typically contain about 80% water. Thus the energy processes of the cell, described in the last chapter, and other intracellular processes occur within a fluid medium. The cells of the body require a supply of numerous substances. Consider, for example, potassium: an essential dietary component. This is carried to the cell dissolved in water; it crosses the cell wall, and again is within a fluid medium. Water also, of course, gives the blood its fluidity, enabling it to be pumped by the heart. The red blood cells, though very efficient carriers of oxygen, are only transportable because of the fluid phase.

With reference to Figure 5.1, it is convenient for some purposes to consider the water of the body to be located in distinct and fairly homogeneous compartments (sometimes called phases). We speak of the *cellular compartment*, known also as the *intracellular compartment*, meaning the total quantity of water housed within all of the cells of the body. In most respects this water, located in various tissues such as liver, heart, and skin, does behave homogeneously, like, say, the contents of a well-shaken test-tube. However, this convenience of description should not disguise the fact that diverse tissues are involved. When we speak of typical cells and their water content they may be typical only under certain conditions. All the fluid outside of the cells may be lumped together, for the purpose of description, as the *extracellular compartment*. This does indeed behave as one unit in many respects. Thus we speak of the two body fluid compartments: the cellular and extracellular.

For some purposes it is necessary to look more closely at the extracellular compartment and consider the two slightly different sub-compartments of which it is composed. The blood flows in the veins, arteries, arterioles, and capillaries of the body. It is through the minute capillaries that water enters and leaves the blood in its exchange with parts of the body which lie beyond the circulation. The

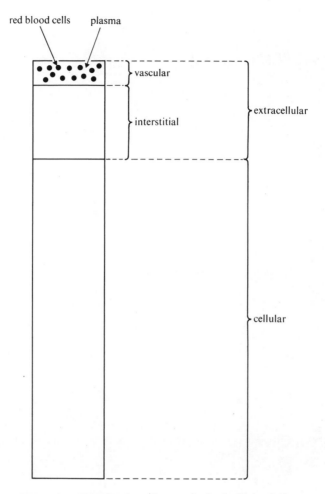

Figure 5.1 Distribution of water in the body (excluding the gut). Water is either in the cellular or extracellular compartments. The extracellular may be further divided into vascular and interstitial

whole system of vessels filled with blood is termed the vasculature. Its contents are known as the vascular compartment. This is composed of water and red blood cells. The fluid part of the blood is termed the *plasma*, and forms one of the two extracellular compartments. Sometimes the terms 'vascular' and 'plasma' are used almost interchangeably, but it must be remembered that water is only part of the vascular compartment.

Most of the extracellular water is outside the vascular compartment, in the *interstitial compartment* (see Figure 5.2). The capillaries, the finest branches of the vascular compartment, do not make direct contact with the cells. Cells are surrounded by interstitial fluid which forms the link between them and

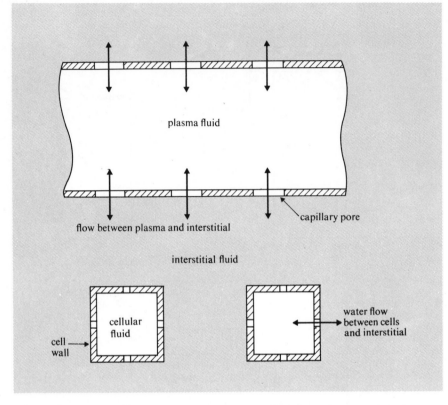

plasma fluid

flow between plasma and interstitial

capillary pore

interstitial fluid

cellular
fluid

cell
wall

water flow
between cells
and interstitial

Figure 5.2 Relationship between plasma fluid (the fluid component of
the vascular compartment), the interstitial fluid, and the cellular fluid

the blood in the capillaries. The capillary has pores through which fluid is
exchanged with the interstitial phase. Water is able to slip through the pore, but
the red blood cells cannot. A very subtle mechanism causes the exchange of water
between the plasma in the capillary and the interstitial. There is not space to
examine this here; the reader who is interested should consult Guyton (1976). As
far as water is concerned, for our purposes plasma and interstitial can be
considered simply to comprise one homogeneous extracellular compartment.
However, we need to focus attention upon mechanisms by which water is
exchanged between cellular and extracellular compartments.

5.1.2 Cellular–extracellular exchange

The cell wall or, as it is usually called, cell membrane, allows water to pass
through with ease. If there is a net flow in one direction then the cell must either
shrink or swell. Normally there are large flows of water in both directions across

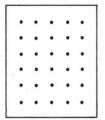 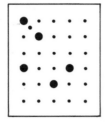

• water molecule
● solute (for example sodium)

Figure 5.3 Water concentration. The left is pure water, i.e. maximum water concentration. On the right a solute has been dissolved in water—the solute takes up space and so water concentration is less than on the left. (Note. The solute molecule is not necessarily larger than the water molecule. This is merely to aid the description)

the cell membrane, but these are equal and opposite, so cell volume remains constant. However, it is possible to cause an imbalance in flow.

It is necessary to consider the term *water concentration* (see Vander, Sherman and Luciano (1975) for a full account). Imagine first a litre of pure water. The water concentration is a maximum, meaning that all the available space in the fluid is occupied by water. If sodium chloride is dissolved in it this displaces some water, and 1 litre of the resulting solution contains less water than the original 1 litre. Its water concentration is less (see Figure 5.3).

Osmosis refers to the net movement of water across a membrane when there is a difference in water concentration between the two sides. *Osmolarity* (I will not draw the distinction here between osmolarity and osmolality. For our purposes we can treat them as the same) refers to the quantity of solutes in a solution. If we add sodium chloride (NaCl) to water, we increase the osmolarity and decrease the water concentration. Figure 5.4 shows two solutions of equal volume but different osmolarity immediately after being placed in a container and separated by a *semi-permeable membrane*. This is a membrane which allows water freely to pass through but resists (for several possible reasons) the passage of solutes. On the left is a high-osmolarity, low-water-concentration solution, and on the right a low-osmolarity, high-water-concentration solution. Figure 5.4(b) shows the situation some time after the two solutions are side by side. Water migrates from the low to the high-osmolarity solution. This net movement comes to a halt when the increased weight of water in the left arm is sufficient to prevent further osmotic shift. This illustrates the meaning of *osmotic pressure* rather clearly—the hydrostatic pressure difference is equal and opposite to the pressure exerted by the remaining osmotic difference. It is rather like putting a weight on a spring. It finds an equilibrium where the weight is equal to the tension of the stretched spring.

The reason water moves in response to an osmotic gradient is as follows. Water molecules are in a state of constant random movement. Therefore a molecule is sometimes travelling at sufficient speed and in the right direction to pass through the semi-permeable membrane. If the solutions on the two sides of the membrane

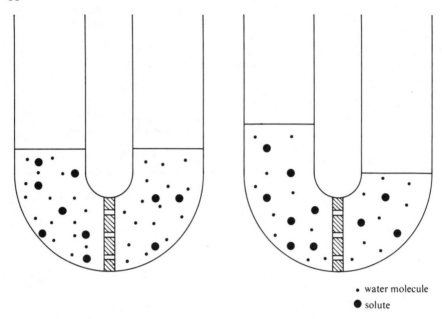

• water molecule
● solute

Figure 5.4 A semi-permeable membrane with solutions of different osmolarity placed on either side. The membrane is permeable to water but impermeable to the solute. (a) immediately after introduction of the two solutions; (b) some time later when equilibrium has been established

are of equal osmolarity then as many water molecules cross one way as the other, and no net flow occurs. However, if as in Figure 5.4(a), we introduce excess solutes which cannot cross the membrane on one side, then in effect they 'get in the way' of potential migrating water molecules on that side. Since water in the low-osmolarity side encounters less obstruction, a net flow into the high-osmolarity solution results.

In the case of the cells of the body, water can pass freely through the membrane but a substance such as sodium finds difficulty. If, say, concentrated sodium chloride solution is injected into the blood this disturbs the equilibrium across the cell membrane, and water moves from the cells into the extracellular space until a new equilibrium is attained. (The extracellular fluid necessarily contains a large amount of sodium, while the intracellular fluid contains a large amount of potassium.)

A solution is said to be *isotonic* if it has the same osmolarity as extracellular fluid. A solution of NaCl of approximately 0.9% by weight is isotonic with plasma. If an organism is injected with, say, 1 ml of 1.8% saline it needs an extra 1 ml of water to restore isotonicity. Solutions with higher osmolarity than plasma are called *hypertonic*; those with lower osmolarity, *hypotonic*. In considering the osmolarity of plasma we often discuss only the sodium and chloride ions. Sometimes only sodium is mentioned; by implication chloride is also present. In the case of intracellular ions, often only potassium is considered; it is tacitly assumed that negative ions are also present.

5.1.3 The fate of ingested water

After entering the mouth, water passes along the oesophagus and into the stomach. It leaves the stomach and enters the intestine. From the intestine it moves into the blood by osmosis. If pure water or hypotonic solutions are in the intestine then water moves from the intestine to the blood. However, if a solution more concentrated in salts than the blood is ingested, the osmotic pull is in the opposite direction, and the blood is dehydrated. This explains the intense thirst which follows the drinking of sea water. Unless the subject either vomits or suffers diarrhoea, the original sea water plus that pulled from the blood will ultimately be returned to the blood. Sodium chloride is actively transported across the wall of the intestine, and water passively follows it.

5.1.4 Loss and gain of water

Water is lost from the body by a variety of routes. The kidney filters the plasma of the blood, removes waste products such as urea and excretes them in urine. In so doing, it inevitably has to lose water as well. The lungs are a source of loss by evaporation. According to the species concerned water may also be lost as sweat, or in grooming and spreading saliva on the skin. In each case it is lost from the plasma compartment. Although a redistribution of fluid occurs within the body so that the plasma does not ultimately bear the full weight of the loss, in the long term, constancy can only be maintained by taking in fluid. The need to replace fluid lost from the body forms the basis of the motivational mechanism of drinking.

Sodium is also lost in the urine. The loss of water and sodium may be either increased or decreased according to the sodium and water state of the animal. If hypertonic saline is injected then the kidney accelerates loss of sodium. Conversely, if the animal is sodium-deficient the urine is of low sodium concentration due to conservation mechanisms which employ aldosterone.

Apart from the water intake by drinking, there is another source which must not be ignored. It will be recalled from the last chapter that, in the oxidation of carbohydrates and fats, water is released as a product of metabolism (see also next chapter). Over a period of time intake must equal output, and a balance sheet is shown in Figure 5.5.

GAINS	LOSSES
Water drunk	Urine
Actual water in food	Sweat
Water of metabolism	Saliva spreading
	Evaporation from lungs

Figure 5.5 A balance sheet for water exchange in the organism

5.2 INTRODUCTION TO INGESTION: WATER AND ENERGY INTAKE MECHANISMS COMPARED

In considering behavioural mechanisms of water intake it is worth comparing with energy intake. Sometimes in discussions of homeostatic regulation of water and energy, textbooks show that important parallels exist in terms of mechanism and function served. This is partly true, but it is worth making one distinction.

It is often said that man is what he eats. Scientifically this means that the chemical structure of the body depends upon ingested food. Specifically proteins are the body's building blocks. However, the problem of food intake control is usually discussed primarily as a problem of obtaining metabolizable sources of energy. If there is not enough to meet metabolic needs the body will be forced to burn itself. It is also true that food must supply components, such as vitamins and amino acids, for needs other than metabolism. Yet the instigating stimulus for a meal has traditionally been associated with some aspect of metabolizable energy, such as blood glucose or rate of energy supply. Those postulating a glucostatic mechanism are aware that a critically defended blood glucose level is only important in that below this level the energy cannot be supplied to the cells at a sufficient rate to meet metabolic need. A regulated glucose concentration *per se* is of no importance. Fluid regulation has something in common with this, in that water must be supplied in order to be lost as urine and sweat. However, even if all water loss from the body were to be halted the body would still need a well-regulated internal fluid milieu. In the absence of the fluid matrix of the cells and blood the organism could not survive.

Thus, whereas an energy supply model of feeding, involving a balance between rate of energy metabolism and supply, sounds feasible, a comparable system for drinking would sound not only impracticable but also illogical (though water supply from the gut may play a minor role as part of a bigger regulatory system).

Explanations of drinking are traditionally given in terms of the hydrational state of the animal. In order to understand why it drinks, when it drinks, and how much is taken, it might appear that we need only to probe the state of its body-fluids and ultimately we can illuminate all aspects of water ingestion. I have argued (Toates, 1979b) that insufficient attention has been paid to the water in the environment that the animal actually drinks, and that availability and location of this water is an important causal factor in determining when and how much is taken. Given the two causal factors, environmental and body water, how do they jointly act to generate ingestive behaviour? This is a fundamental question that we need to consider later. For the moment I will follow the traditional line of concentration upon the water within the organism. Not only should this give a feel for the historical development of the subject, but of course it is one aspect of ingestion that any theory must discuss. Later the other aspects will be introduced.

Traditionally it has been assumed that the animal's nervous system is supplied with information from a sample of body-fluids, and bases drinking decisions upon the state of this sample. It would seem impossible to examine total body-water content. How could it do it? Just as the thermostat samples temperature at one site in a room, so it is believed that the drinking mechanism samples the fluid environment at one or more sites. Room temperature control can go seriously wrong if it is arranged that the temperature at the site of detection is not typical of the room as a whole. By analogy, experiments can be arranged so that the drinking controller detects a signal not typical of body-fluids. In the next sections we examine detection of fluid state.

5.3 THE DRY MOUTH THEORY

Grossman (1967, p. 399) and Fitzsimons (1973) review the historical develop-ment of theories of thirst, and the student should consult these texts for details. Apparently the idea that thirst and drinking are aroused by sensations from the mouth and throat can be traced back to Hippocrates; it has frequently been reiterated since then. According to this theory, known as the dry mouth theory of thirst, the mucosa becomes dry and this stimulates water ingestion. Closely associated with the theory is the American physiologist Cannon. No one would deny that a dry mouth sometimes causes humans to drink water. The issue is whether such a mechanism unaided by any other detector can account for an organism's fluid intake. In a sense the dry mouth theory represents the process of taking a sample of the cells of the body. Indeed, severe dehydration is reported to be associated with a dry mouth; according to anecdotal evidence very dramati-cally so (Wolf, 1958). The amount of saliva secreted decreases sharply during dehydration in dogs (see Grossman, 1967), and correlates with body-fluid volume. It therefore represents a possible mode of feedback control.

There is little doubt that animals can be induced to ingest fluid on the basis of local dryness and lack of saliva flow. However, even if the mouth is flooded with water this will not satiate a water-depleted animal. Such an experiment has been performed by surgically cutting the animal's oesophagus and bringing it to the outside. Water drunk fails to change the state of body-fluids but of course wets the mouth. Such animals show only limited and short-term satiety. Other less drastic procedures which cause a profuse saliva flow fail to quench the effect of what is clearly an additional and central drinking command.

5.4 CENTRAL CONTROLS OF FLUID AND SODIUM INTAKE

5.4.1 Historical background

In the interests of space I omit much of the historical background to this subject which can be found in either Grossman (1967) or Fitzsimons (1973).

In advocating that mouth dryness stimulates thirst, researchers in the early 19th century argued that this reflects events in the blood. They suggested that dehydration increases viscosity of the blood. In 1900 A. Mayer reported that the osmotic pressure of the blood increased as a result of water-deprivation. He believed that the increased solute content of the blood is detected by receptors in the blood vessels, and this serves to excite thirst.

Another major contributor to the evolution of our current view of thirst was Wettendorf. He observed a substantial increase in osmotic pressure of the blood only after an animal had been deprived of water for longer than 24 h. Yet animals show compensatory drinking in response to water-deprivation periods of less than 24 h. If water is being lost from the body disproportionately to solutes, but osmotic pressure shows little or no increase, this can only mean that water is migrating from the cells to the extracellular space.

Could this cellular loss, rather than an extracellular osmotic change, be the stimulus for drinking? Wettendorf felt that it could, this view surviving to the present day with only a little dissent. However, Wettendorf believed that all the cells of the body reflected their water loss and contributed to the arousal of thirst. We now believe that only a minute sample of the cells is effective.

5.4.2 A cellular sample

One of the most cited experiments in the thirst literature is that of Gilman (1937); its result is of fundamental significance. He compared the effect of concentrated sodium chloride and urea injections made into the blood. The osmolarity of the two solutions was equal. The sodium chloride injection caused enough drinking to return extracellular osmolarity to normal. The urea injection produced considerably less drinking, insufficient to restore osmolarity to normal. The ions of Na encounter resistance at the cellular membrane, and are therefore largely confined to the extracellular compartment. The urea molecule meets little resistance and is able to penetrate the cells. It therefore establishes equilibrium throughout the cellular and extracellular compartments. In contrast to NaCl, it causes no long-term concentration difference across the cell membrane which would dehydrate the cells. That animals do drink a certain amount in response to urea injections is probably due to a small transient osmotic shift of water which occurs before urea has time to be fully distributed. Brain cells may be transiently dehydrated to a slight extent.

Gilman's study is widely cited as one of the strongest pieces of evidence in favour of the view that cellular dehydration, rather then an increase in extracellular osmolarity *per se*, is what initiates drinking. Figure 5.6 contrasts the effect of urea and NaCl after equilibrium has been established.

If NaCl is injected into the blood, extracellular osmolarity rises, water is lost from the cells, and drinking ensues. This then returns extracellular osmolarity to normal, and water returns to the cells. The process of course involves swelling of total body-fluids which is tolerated in the interests of restoring cellular fluids to normal. Excess NaCl will later be excreted, together with an obligatory quantity of water. If we inject concentrated NaCl it is easy to calculate how much water the animal needs to drink in order to neutralize it. However, drinking is not the only corrective action initiated by the disturbance. The kidney will start to excrete a hypertonic urine, which means that the animal needs to drink less than we would estimate. Fitzsimons (1971) examined the drinking response to hypertonic saline, without allowing the kidney to affect the result. To do this he

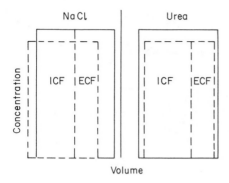

Concentration

NaCl   Urea

ICF   ECF   ICF   ECF

Volume

Figure 5.6 A comparison of the effect of (left) hypertonic sodium chloride (NaCl) injection and (right) injection of urea of the same osmolarity. Injections are made into the extracellular compartment. ICF = intracellular fluid; ECF = extracellular fluid. Broken lines indicate pre-injection identical isotonic conditions of body fluids in each case. Solid lines show the result some time after injection. NaCl increases ECF volume but decreases ICF volume. Unlike sodium, urea freely diffuses into the ICF, and in consequence the injection (urea + water) increases the volume of both ICF and ECF. Note that concentration is increased by the same amount in both compartments in both cases. That is to say, the injections were of identical osmolarity and in the steady state there is no concentration difference across the cell boundary. (Source: Blass, 1973. Reproduced by permission of Hemisphere Publishing Corporation, Washington, D.C.)

used nephrectomized rats (rats with the kidneys surgically removed). The animals drank precisely enough water to restore isotonicity, resulting in excess isotonic fluid in the extracellular compartment.

Other substances which are excluded from the cells also provoke drinking when injected in hypertonic doses; the effect is not peculiar to sodium (see Blass (1974) for a review). However, there is a suggestion that in addition to the osmoreceptor, a specific extracellular sodium concentration detector may also be involved in drinking and anti-diuretic hormone (ADH) control (for discussion see Peck, 1973; Andersson, 1973; McKinley, Denton and Weisinger, 1978). At present the existence of such an additional control is still a subject of debate. Hopefully, concrete evidence will soon emerge, but I will discuss only the osmoreceptor.

We do not believe that it is possible for the organism to measure accurately a reduction in size of the entire cellular compartment, and initiate a drink of corresponding size. Rather, drinking is determined by the state of a specific cell (or cells), which serves as a sample or microcosm of cellular water. This cell is thought to be located in the hypothalamus, and is probably a mechanoreceptor which is excited by shrinkage. It may exhibit a tonic level of activity under normal conditions of fluid balance; increased firing might be caused by shrinkage and reduced firing by swelling. Amongst other factors directing attention to the hypothalamus is the result of Verney (1947), that osmoreceptors in this area control secretion of ADH. Working in Sweden, and using goats as subject, Andersson (1953) found that hypertonic saline injections of minute size made into the hypothalamus elicited copious drinking. Such results are open to more than one possible interpretation since to apply hypertonic saline to the brain is

94

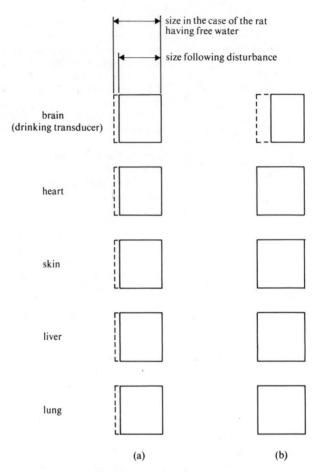

Figure 5.7 Comparison of water deprivation and intracranial hypertonic sodium chloride injection on various cells of the body: (a) water deprivation; (b) small intracranial injection of NaCl

obviously an unphysiological stimulus, but a consensus of opinion favours the view that osmoreceptors are located there (see Blass, 1973).

Figure 5.7(a) shows a fairly representative sample of cells from the body under *ad lib* conditions (dotted) and following water deprivation (solid). It may be seen that the proposed detector cell accurately reflects the state of the other cells. Figure 5.7(b) shows the effect of a minute local injection of hypertonic saline at the site of the detector cell. Local shrinkage, quite untypical of the rest of the cells of the body, is produced; and copious drinking would be expected.

Figure 5.8 is a model of what is believed to be happening. Even if the details are wrong the essentials are surely correct. Figure 5.8(b) shows in systems terms the relationship of Figure 5.8(a). Commonly we would represent it as shown in

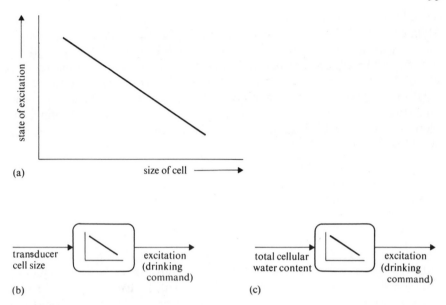

Figure 5.8 Suggested model of the relationship between ICF and the excitation of an intracellular fluid detector: (a) relationship in graphical form; (b) as part of a block diagram; (c) the model is generalized to cover the whole ICF compartment

Figure 5.8 (c) since we take the transducer cell to be typical of the body cells. We can estimate the state of total cellular water, but, as we have just shown and will return to later, such a model is seriously misleading when brain fluid balance is not an accurate reflection of the whole cellular fluid state.

So much for the constancy of the cellular environment, and indirectly extracellular osmolarity. We seem to be able to account for the response to an acute rise in extracellular osmolarity by hypertonic injection. Water deprivation leads to loss of water, a slight rise in extracellular osmolarity, and compensatory movement of water from the cells. Thus the mechanism which we have described can explain drinking following water deprivation. In correcting deprivation-induced cellular loss we would automatically correct any extracellular loss; water can only repair cellular loss by first being introduced into the extracellular phase. Another situation in which drinking is elicited by cellular depletion is potassium deficiency. If potassium, the major intracellular electrolyte, is lost from the body, water will migrate into the extracellular phase. However, there are circumstances where water is lost from the organism but of which a cellular detector would be unaware.

5.4.3 An extracellular stimulus

For sometime it has appeared that, in addition to the osmoreceptor, detectors of extracellular volume can stimulate drinking. Indeed, as Fitzsimons (1971) notes it would be remiss of evolution not to provide an extracellular drinking stimulus. Maintenance of circulatory volume is crucial for survival, and yet being a relatively small compartment and one through which exchange of water with the outside world occurs, it is particularly vulnerable. Loss of blood through injury is obviously a problem which adaptive mechanisms have been forced to encounter.

Historically there have been anecdotal reports suggesting that loss of isotonic fluid from the extracellular compartment elicits thirst. Diarrhoea, vomiting, and blood loss have all been associated with thirst. Cannon, from experiences in World War I, wrote:

> 'After a battle the universal cry of men who have been badly wounded and who are suffering from haemorrhage or shock, is the cry for water (Cannon, 1947, p. 58).'

Cannon believed that this represented a homeostatic response to blood loss, and although evidence collected under these conditions cannot be unambiguously interpreted, controlled studies are congruent with it. Excess of NaCl in the extracellular space causes movement of fluid from the cells, and conversely deficiency of extracelluar NaCl causes an extracellular deficit. In order to lower extracellular osmolarity the animal (or human) is fed a sodium-deficient diet for a while. McCance (1936) performed such an experiment on human subjects who reported thirst.

An alternative procedure for lowering extracellular volume is as follows. To remove extracellular fluid an intraperitoneal injection of a solution having a high protein concentration is made. This pulls isotonic fluid into the peritoneal cavity. Water and sodium are removed, but proteins and red blood cells are not affected (Fitzsimons, 1971; Stricker, 1973). As the concentration of injected solution is increased then so does the plasma deficit, and the amount of compensatory drinking also increases (Stricker, 1973).

Haemorrhage causes drinking in rats (Oatley, 1964; Fitzsimons, 1971), as would be expected in the context of other results discussed here. However, the effect is not always reliably obtained; it may be traumatic and is not so easy to interpret as other procedures (Almli, 1970).

As we saw earlier, the extracellular space consists of two components: vascular and interstitial. It seems that detectors look at the size of the vascular compartment while interstitial volume passively follows the regulation of vascular volume. It is possible that detectors of interstitial fluid volume exist. Indeed the present writer has felt tempted to postulate them (Toates, 1979a), but the evidence is perhaps more strongly against their existence (Fitzsimons, 1971). As Fitzsimons notes, the circulation is well-equipped with devices to measure its degree of filling which could provide a cue for drinking. Probably stretch receptors in the walls of the blood vessels are implicated. The veins hold a very large amount of the total volume, and are very distensible. They would be an excellent site for detection, since a relatively large signal would be generated by any change in blood volume acting through the degree of stretch of a vessel. By comparison arteries would be insensitive, except under extreme conditions (Fitzsimons, 1971). Behavioural evidence fits this interpretation. If the amount of blood returning to the heart is decreased by blocking one of the veins (the vena cava) drinking occurs. We conclude that stretch receptors detect the state of filling on the venous side of the circulation and initiate drinking (Fitzsimons, 1971).

Drinking occurs following loss of extracellular fluid irrespective of whether the animal has kidneys present or not. It is important to establish this, because one mechanism of water intake is known to operate via the kidney. This is the *renin–angiotensin* mechanism which is now described. The subject is extremely complicated, and some comprehensive reviews are available (Fitzsimons, 1976; Epstein, 1978), so just the essentials follow. Fitzsimons (1971) restricted the amount of blood entering the kidney by partially constricting the renal arteries. This caused no change in venous blood pressure but elicited drinking. Thus in addition to detectors in the venous side of the circulation, a detector system at the kidney appears to be implicated. This seems to be good biological design. If blood volume is decreased then renal arterial pressure would drop, and drinking is an appropriate response. How could the system detect a fall in renal arterial pressure and inform the brain?

A reduction in renal blood flow causes the hormone *renin* to be secreted by the juxtaglomerular cells of the kidney. There is a substance in the blood called angiotensinogen which is converted to angiotensin by renin. Angiotensin is a powerful stimulus to drinking. If only minute amounts are injected into the brain copious drinking results (Booth, 1968; Fitzsimons, 1972, 1976). Angiotensin has other powerful effects within the body which serve to maintain the circulatory state near normal.

Figure 5.9 shows a representation of the most popular view of extracellular

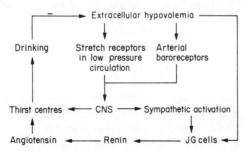

Figure 5.9 Mechanism for extracellular thirst, as proposed by Fitzsimons. Loss of extracellular volume (hypovolaemia) causes secretion of renin by the juxta-glomerular cells (JG cells). Renin is converted to angiotensin which promotes thirst/drinking so as to counter the hypovolaemia. Additional routes, not involving the renin–angiotensin system, whereby hypovolaemia causes drinking are also shown. (Source: Fitzsimons, 1971. Reproduced by permission of Academic Press)

thirst. Reduction in renal blood causes drinking because of direct activation of the renin–angiotensin mechanism. In addition the volume detectors on the venous side of the circulation cause renin release by a more indirect route. That is to say, sympathetic neural activation of the kidney initiated by blood volume reduction causes renin release. However, since nephrectomized rats, that is rats which have no renin source, drink in response to reduced venous filling there is presumably also a direct neural connection from the volume detector on the venous side to the brain.

According to a recent report the importance of the renin–angiotensin system may have been overestimated in so far as excitation of drinking is concerned. Stricker (1977) argues that the major responsibility for drinking following ligation of the inferior vena cava does not reside with the renin–angiotensin system. The increase in plasma renin concentration following this treatment was found to be insufficiently large. This places weight upon the direct connection from volume detector to brain. How does one explain then that the drinking response to such ligation is strongly attenuated by removal of the kidneys? Stricker showed that nephrectomized rats subjected to ligation go into hypotensive shock, and this undermines the ability of the animal to exhibit any behaviour. Stricker's argument shows that what may appear to be a neat physiological/behavioural result may in reality work through a pathological route.

5.4.4 Additivity of thirst stimuli

We can feel fairly confident that our experimental stimuli produce either purely cellular (e.g., salt injection) or extracellular (e.g., polyethelene glycol) loss of fluid. In severe dehydration the animal might be expected to suffer a loss of fluid in both compartments. How does the organism handle this? From experiments which stimulate both extracellular and cellular thirst it appears that the stimuli are additive (Fitzsimons and Oatley, 1968). Assume a rat drinks x ml in response to an NaCl injection, y ml in response to haemorrhage and z ml in response to a period of water deprivation. It will drink

$x + z$ in response to NaCl injection combined with deprivation;
$y + z$ in response to haemorrhage combined with water deprivation;
$x + y$ in response to NaCl injection combined with haemorrhage.

5.4.5 Cancellation of a thirst stimulus

If both cellular and extracellular phases have been depleted then clearly drinking is the appropriate response, and there is compatibility between cellular and extracellular interests. However, if extracellular space is depleted of isotonic fluid a conflict of interests between the two compartments can arise. As far as the extracellular compartment is concerned the animal should drink, preferably a saline solution to repair sodium as well as water, but failing this then water would be desirable to repair volume. The cellular compartment is not depleted and so drinking is inappropriate to its state. However, if the animal drinks water in response to an extracellular cue, most of this water will be pulled into the cellular compartment. In order to restore extracellular volume it would need to drink an excessive amount, which would partly be excreted and partly swell the cells. The risk of water intoxication is present if the cellular phase is seriously overhydrated. Stricker (1973) found that the rat stops drinking before the extracellular deficit is repaired; excess water in the cells inhibits further drinking. In this way the animal, so to speak, balances the excitatory effect of the extracellular stimulus against the inhibitory effect of the cellular. If isotonic saline is available for drinking the situation is different. No conflict occurs since the ingested fluid is confined to the extracellular space. With such a fluid available the extracellular space is restored to normal.

It is interesting that an extracellular excess has no effect on the drinking provoked by a cellular stimulus (Corbit, 1965). This was suggested by the response to hypertonic saline injection where the animal drinks, despite the fact that any cellular loss stimulus is accompanied by an equal and opposite extracellular excess. Even large exogenous loads of isotonic saline exert no inhibition on the drinking caused by hypertonic saline injection. The animal has little to lose by an extracellular excess of isotonic fluid, in contrast to the dangers of water intoxication associated with cellular overhydration.

Following a disturbance to the body-fluids, drinking of water may be only one of the necessary responses for correction to be attained. In the natural habitat with the help of thirst and the kidney, the animal would need to concoct isotonic saline from water and sodium chloride following isotonic fluid loss. Therefore at this point a digression is necessary and appropriate to consider the behavioural mechanism of sodium appetite and ingestion.

5.4.6 Sodium appetite

Sodium and water appetites are appropriate to an animal which has lost extracellular fluid. Animals are equipped with mechanisms for identification of sodium which, environment permitting, ensure a sufficient supply of it. Compensatory sodium ingestion occurs when they are sodium-deficient and then find sodium (see Stricker, 1973). Rats show a memory for where sodium is located even though at the time of encountering it they were not sodium-depleted. In a cleverly designed experiment, Krieckhaus (1970) tested water-, but not sodium-deprived rats in a T maze. Either distilled water or hypertonic saline was given as reward. Rats avoided the hypertonic saline side and developed a preference for the water-reward side. When sodium-depleted but not thirsty they exhibited a preference for the hypertonic saline side. Controlled studies ruled out the possibility that the sodium-deficient rats were simply avoiding water, and the conclusion emerges that rats remember where sodium is located and use this memory when the need arises.

One situation of sodium deficiency is loss of extracellular fluid by polyethelene glycol treatment. Rats will ingest highly concentrated NaCl solutions in response to this; solutions which they would normally reject. By taking both water and concentrated NaCl in appropriate amounts rats are able to restore extracellular volume, and leave osmolarity unchanged.

What are the cues for sodium ingestion? At present it is not entirely clear, but it could be one of several such as hypovolaemia, or a low extracellular sodium concentration or possibly elevated aldosterone concentration (the hormone which causes sodium to be reabsorbed in the kidney). The immediate response to hypovolaemia is for the animal to drink water, and sodium appetite develops a little later (Stricker, 1973). It could be that under these conditions cellular over-hydration accompanied by an extracellular deficit is the cue for sodium ingestion. Possibly movement of water from the cellular to the extracellular causes a lowering of sodium concentration some time after removal of isotonic fluid. Drinking in response to hypovolaemia would produce hyponatraemia. Hypovolaemia and hyponatraemia in concert may excite sodium intake. Further suggestions were made by Stricker and Wolf (1969) and Wolf, McGovern, and DiCara (1974), such as the possibility of a sodium reservoir (for instance in bone tissue) which ultimately reflects body sodium content and which feeds into behavioural mechanisms. Denton (1972) believes that sodium depletion is detected by a fall in intracellular sodium ion concentration in a population of neurons (or their surrounding neuroglia) which have responsibility for arousing sodium ingestion mechanisms.

It is wrong to consider sodium appetite being aroused only when there is some observable need state. Rats in normal body-fluid equilibrium and obtaining adequate sodium in the diet still drink large amounts of weak sodium chloride solutions (see Wolf, McGovern, and DiCara, 1974). The mechanisms that normally guarantee adequate sodium in the natural environment appear to be 'overwhelmed' by the presence of the joint reward of sodium and water in one mouthful. This joint activation of two control systems can explain why rats normally seem to prefer isotonic saline to water as their drinking fluid.

However, the rat will reverse its preference if exposed only to isotonic sodium chloride solution for a while and then given a two-bottle choice (Devenport, 1973). Rats normally obtain hypotonic saline in so far as body-fluids are concerned by sampling both water and isotonic saline in the two-bottle choice. The ratio of amounts taken changes in favour of water after exposure to isotonic saline solution alone.

Let us refer to Figure 5.10 and give a summary of the picture we have at present for ingestion of water and sodium in so far as fluid compartments are concerned. Intracellular dehydration (ICF ↓) causes excitation of thirst and water intake to eliminate dehydration. The arrows show that it inhibits sodium appetite. It is appropriate to ingest only water in response to such a stimulus. We should not ignore, though, the possibility of a learning component in the thirsty rat's rejection of hypertonic saline. A duodenal osmoreceptor may be present, or even a systemic change may be sufficiently rapid that the rat can sample such solutions, find that their effect is harmful, and reject them. This would reinforce

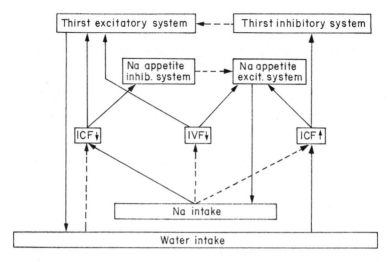

Figure 5.10 Summary diagram of the factors implicated in water and sodium appetite. Solid arrows indicate stimulation, broken arrows indicate inhibition. IVF = intravascular fluid, ICF = intracellular fluid. (Source: Stricker, 1973. Reproduced by permission of Hemisphere Publishing Corporation, Washington, D.C.)

any initial prejudice against them arising from taste alone (see also Mook and Kenney, 1977). Loss of intravascular fluid (IVF ↓) excites water and sodium appetite, both appropriate for vascular restoration. Finally the effect of intracellular over-hydration (ICF ↑) is to inhibit water intake and excite sodium appetite. Clearly cellular over-hydration is made worse by water ingestion and ameliorated by sodium ingestion.

Although Figure 5.10 represents the essentials of the appetite mechanism, rats do not always behave in the way which seems most appropriate to their body-fluid state. Thus, in response to loss of cellular water by hypertonic saline injection, given a choice of water or isotonic saline, one would expect rats to drink water, while in response to hypovolaemia they would be expected to drink isotonic saline. In fact they take a mixture of both solutions in response to either treatment (Stricker and Wolf, 1967), ingesting rather more isotonic saline than water. Such a hypotonic mixture may possibly be optimal for rapid absorption. More likely, though, for an explanation we should view their behaviour as being under the complex control of both immediate body-fluid state and inbuilt taste factors appropriate for sodium and water. Rats simply have an innate preference for the sensory properties of mild saline solutions (Epstein, 1978). It appears that while a rat which is slightly over-hydrated in the cellular space and under-hydrated in the vascular will reject water, it will accept isotonic saline (Blass and Hall, 1976). Under these conditions hypotonic and isotonic saline may have ambivalent effects: the thirst-controller (or at least its systemic component) may inhibit ingestion but the taste component of fluid ingestion and the sodium-controller would stimulate intake.

Rozin and Kalat (1971) consider sodium appetite in the context of their general theory of preference/aversion, discussed in the last chapter. In contrast to learned preferences, the sodium-deficient rat is genetically programmed to respond positively to sodium (and similar tasting salts) on first taste. Animals which have been deprived of sodium for the first time show an avid appetite for sodium on first exposure (Denton, 1972). However, while admitting a very strong initial approach tendency towards sodium, we should not ignore the strengthening of this appetite which would be expected to follow systemic gain of sodium by the depleted animal (see Smith, 1972). Accuracy of replacement following depletion also appears to depend upon learning (Denton, 1966). Rozin and Kalat remind us of some differences between sodium and other substances in so far as living organisms are concerned:

1. A large and fairly constant quantity of sodium in the body-fluids is vital, and yet sodium is often scarce in the environment. This could have placed particular pressure on evolutionary processes responsible for ingestion.
2. The stimulus of sodium is well defined in so far as recognition by taste is concerned. As Denton (1972) points out, salt fibres form part of the primary taste modality in all higher vertebrates.

It is important to add additional factors. Sodium excess or deficiency can make itself felt through changes not only in the composition of the body-fluids but also in their volume and distribution. It is easy for us (and presumably evolution also!) to see the kind of cue which should be associated with sodium appetite, i.e., hypovolaemia, cellular over-hydration and/or hyponatraemia. No such obvious physiological cues are necessarily available to goad and steer, say, the thiamine-deficient rat. Perhaps though, as Denton (1972) points out, the most important factor is that sodium:

' . . . does occur in relatively pure form in nature, and thus the capacity to locate, recognize and ingest it could carry significant survival advantage in sodium deficient ecological conditions. To state the seemingly obvious, Vitamin B_1 or any of the other B group does not occur in relatively pure form in nature, and hence there is unlikely to be any selection pressure favouring development of a specific recognition and an innate appetite for it.' (Reproduced by permission of the Royal Australasian College of Physicians.)

The problem of sodium deficiency is very much dependent upon geographical factors and upon diet. Carnivores have little problem, even when living in regions of low salt, since their prey will contain sodium (Denton, 1972).

5.4.7 Termination of ingestion

The internal stimuli which excite water and sodium ingestion, and the inhibitory stimulus on water ingestion of cellular over-hydration, have been described. In this section we consider drinking termination and the problem of terminating sodium ingestion. In the long-term, satiety involves elimination of fluid displacements, but is drinking normally terminated by correction of the fluid deficit which initiated it? The assumption is commonly made that restoration of the internal milieu is too slow to account for the termination of drinking, and that short-term feedback signals from mouth and stomach inhibit water ingestion until correction of fluid imbalance occurs.

The problem may be restated as follows. If an animal has a fluid deficit of x ml in the cellular/extracellular space then it needs to drink x ml and no more. Depending upon the species, animals drink at various speeds. Some correct a deficit in one continuous burst, whereas others interrupt their drinking with several pauses. Either way, it may be argued that drinking is relatively fast compared to time which water takes to leave the stomach, be absorbed across the intestinal wall, and (depending upon which compartment is in deficit) pass across the cell membrane. Water is an easy material to ingest and it may be of adaptive significance to drink fast and leave the vicinity of water. But it does pose the problem that by the time x ml have been ingested only some of this will have reached the compartment that initiated the drinking response. Hence the idea arises that messages from the alimentary tract indicating passage or presence of water serve to inhibit further drinking. Traditionally two such sources of information have been implicated: mouth and stomach receptors.

In the case of rats, Hatton and Bennett (1970) questioned the assumption that absorption of water from the gut at the time of drinking termination is of insufficient magnitude to be a causal factor in satiety (Corbit (1969a) advanced a similar argument). They adopted the criterion of satiety as an intake of 0.2 ml/min or less for 3 consecutive minutes, and found that this was reached in about 10 min from the start of drinking following $23\frac{1}{2}$ h water-deprivation. By this time the plasma osmolarity was the same as that of control rats having food and water $ad\ lib$. They suggested that cellular rehydration could explain drinking termination. The argument is powerful but raises a number of questions. If, after some 10–20 min, cellular dehydration is no longer present, where does the 8–14 ml still present in the alimentary tract go following its absorption? The answer could be that it repairs the extracellular deficit, if one exists. But could it do this without most of it spilling into the cells and causing cellular over-hydration? This doubt rests on the assumption that the fluid in the gut at drinking termination is pure water. In fact by then sodium could possibly have had time to move from the plasma into the gut, contributing to a decrease in extracellular osmolarity. This would be another factor tending to move water into the cellular space. At present we have no precise answers to these questions.

The common assumption of inhibition from stomach receptors was questioned by Feider (1972). Feider showed that, following a stomach load of up to 15 ml of isotonic saline, subsequent drinking in rats was unaffected. By contrast, water loads depressed drinking and hypertonic saline loads increased it. Stomach bulk in isolation appears not to inhibit drinking. It appears that inhibition is either purely a function of post-absorption hydrational change, or that stomach bulk inhibits only in *association* with the motor act of drinking carried out just prior to the arrival of fluid in the stomach. Adolph (1967) found that dogs drank very substantial amounts even though their entire estimated deficit had been loaded by stomach tube. From Feider's results one might suggest that a stomach osmoreceptor(s) detects the presence of water and hypertonic saline loads, but is unaffected by saline loads. However, Feider found that removing the vagus nerve so as to eliminate such feedback (and presumably distension feedback also) made no difference to the result.

When animals were required to bar-press for water-reward following water, isotonic or hypertonic loads, a rather different effect was found. Increasing size of gastric pre-load decreased rate of bar-pressing irrespective of the salinity of the load when a VI 1min schedule (see Table 5.1) was employed. In other words stomach bulk, which had no effect on thirst as measured by the drinking response, gave the impression of lowering thirst when measured by rate of responding on a VI 1min schedule (see also Miller, 1967). Feider concluded that stomach distension had an aversive or distractive effect upon bar-pressing for water, but left drinking unaffected. Bar-pressing is an unnatural response for any animal and in this case it was being asked to respond to the equally unnatural stimulus of stomach-loading. Other examples are available of animals failing to respond 'appropriately' to suddenly imposed disturbances when the response is not the natural one of straight water ingestion. Interestingly, in contrast to

Table   5.1

INTERVAL SCHEDULES		RATIO SCHEDULES	
A certain amount of time must elapse between the delivery of one reward (e.g. a food pellet) and the delivery of the next. The first response (e.g. a bar-press) made after this time has elapsed earns a reward. Responses made before the time elapses are fruitless.		The animal must respond a certain number of times before reward is earned. Every response therefore brings the animal nearer to gaining reward.	
FIXED INTERVAL (FI)	VARIABLE INTERVAL (VI)	FIXED RATIO (FR)	VARIABLE RATIO (VR)
The interval is a constant length of time say, 1min (FI 1min)	The interval varies on each occasion of reward. It may be 20sec or 2min, but will be characterized by its mean value of, say, 1min (VI 1min).	The number of responses needed to earn a reward is constant. For example, an animal must press a bar 30 times to get one pellet (FR 30).	Number of responses needed for each reward is variable. Sometimes it might be 5 and sometimes 60. Characterized by its mean value, say, 30. (VR 30).

others, Ramsey, Rolls, and Wood (1977) found in rats a depression of deprivation-induced drinking by gastric loads of isotonic saline. The reason for the effect was not clear, but was not due to the excretion of hypertonic urine by the kidney.

It may be that: (1) metering of the passage of water through the mouth; and (2) the presence of water in the stomach, act synergistically to inhibit drinking, a view I will develop shortly. Whatever the details, the strength of short-term inhibition must be rather well calibrated, so that animals ingest a quantity of water appropriate to their need.

A similar problem to drinking termination appears in connection with sodium ingestion (Denton, 1966, 1972). Detection of sodium deficiency occurs somewhere within the organism. When sodium is made available the animal rapidly ingests a quantity appropriate to the size of the deficit, and yet at the time ingestion is complete it is unlikely that any substantial reversal of deficiency will have occurred at the detection size. We do not know what information is used in the short-term feedback.

5.4.8 Self-injected intravenous drinking

An experimental design is to deny the rat access to drinking water but allow it to infuse itself intravenously with water by pressing a bar for water infusion. They take less water than normal, and experience a permanent dehydration relative to drinking controls, though infuse enough to stay alive and function fairly normally (Nicolaidis and Rowland, 1975). An animal self-infusing in this way does not make any additional compensation to acute thirst stimuli such as salt-injection (Nicolaidis and Rowland, 1975). Rowland (1977) notes that only natural stimuli of slow onset can motivate non-oral water intake. Sudden disturbances such as salt injection are ineffective, whereas heat exposure, fluid deprivation, etc., motivate increased effort by such animals. Could it be that salt injections cause some pain or trauma over a period following the injection? Furthermore, is it too wild a generalization to look at the work on sensory aversions discussed in the last chapter for some possible lead in this area? It is of particular interest in this connection that whereas rats subjected to normal laboratory conditions, but with only dilute quinine available for oral intake, regulate body fluids satisfactorily, they fail to increase intake of such solutions as an immediate response to hypertonic salt-injections (Rowland, 1977). Again we have a juxtaposition of an unnatural test stimulus and an unnatural means of answering it.

Corbit (1965) was unable to get sustained bar-pressing for the reward of intravenous water in rats, whereas Nicolaidis and Rowland reliably obtained it. Rowland (1977) proposes the following explanation. Corbit first trained his animals to bar-press for oral water reward accompanied by intravenous reward, and then omitted the oral reward. Intravenous reward was satiating when an intravenous bonus was added to the already established oral reward. In other words total intake in response to a dipsogenic stimulus, oral intake plus intravenous intake, was the same as when only oral reward was allowed. The bar-pressing habit was not continued, though, when oral reward was omitted and only intravenous reward remained. Rowland argues that the expectation of oral reward in this experiment was such that when it was omitted the animal paid no attention to the intravenous reward and merely searched the reward cup frantically. Without such 'priming' oral reward, the purely systemically rewarded response can be learned.

Intravenous 'drinking' by bar-pressing occurs close in time to meals. As Rowland (1977) argues, the following two interpretations are possible: (1) they 'drink' in response to or in anticipation of meal-associated dehydration; or (2) they have learned a ritual such that the bar-press is part of a pre-meal or post-meal sequence involving exploration and grooming. Rowland favours (1) on the grounds that when the amount of water earned by each bar-press was halved responding immediately doubled in intensity. Rowland argued that the animal actively meters the amount of water delivered by the injection. However, Toates (1978) claimed that some meal-associated intake may be more of a ritual; even grossly over-hydrated animals in the Rowland and Nicolaidis experimenter-controlled infusion paradigm often drink before meals.

5.4.9 An early model of drinking: reflections upon a simplified approach

Figure 5.11 shows the model of drinking which was presented by Toates and Oatley (1970). It now looks naive, but represented our understanding then of the mechanisms underlying drinking in the laboratory rat. A neural signal proportional to cellular deficit adds to a neural signal proportional to extracellular deficit. The addition gives the total excitatory thirst stimulus (a negative signal may be generated by a cellular surplus). Inhibitory influences arise from water in, or having just passed, the mouth, and in the stomach. If the net effect of the excitatory signals is significantly greater than that of the inhibitory signals the animal will drink. If body-fluids are slowly displaced a minimum departure from

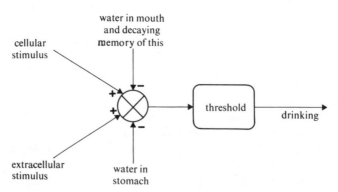

Figure 5.11   An early model of drinking, as proposed by Toates and Oatley (1970)

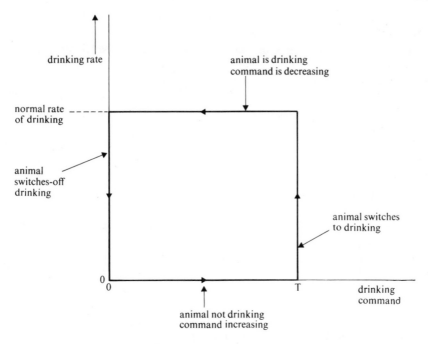

Figure 5.12   Hysteresis in the drinking controller

normal (i.e. the threshold) is needed to initiate drinking. It was felt that without a threshold the animal would constantly be making excursions to the spout for small quantities. This would of course not be a problem in the wild where thirst would be in competition with numerous other candidates for the animal's time.

From the behaviour of the laboratory rat the threshold appeared to have the characteristics of hysteresis (Figure 5.12). Although the drinking command must reach value T to initiate drinking the animal does not stop drinking when it drops below T, but continues until it is zero. Of course, on reflection, it could hardly be otherwise. If the animal started drinking when it was just below T then it would not be a threshold at all. The body-fluid volume corresponding to value T would then be the equilibrium point of the system. There is nothing absolute about the value we called 0. It is only that we have observed a particular value to be that which body-fluids normally assume, and from this some choose to employ the expression set-point.

When the signal is of magnitude T it takes command and moves the animal to seek water. Having made the effort to initiate a drink the animal then entirely corrects the deficit.

Rats drink at a rate of roughly 6 licks per second in response to a variety of eliciting stimuli (Corbit and Luschei, 1969). If it is very thirsty the rat will spend a relatively long time before switching to another activity, but in any given bout it will not show a higher frequency of tongue contacts in comparison with a less thirsty animal. Rats lick at this rate either until their behaviour is interrupted by another activity or satiety ensues.

It is necessary to include short-term feedback in order to solve the problem of stability. Without such feedback we estimated that the animal would over-drink. It would persist with drinking until the fluid compartment in deficit was replenished, by which time the gut would hold water in excess of need. This would later enter the body-fluids and have to be excreted; obviously an uneconomical process.

By cutting the animal's oesophagus we can investigate the inhibition exerted by oral factors alone. Any water drunk simply leaves the oesophagus, thus precluding gastric and post-absorptive feedback. Such drinking is known as sham drinking. Blass, Jobaris and Hall (1976) found that intact rats which drank x ml after a period of water deprivation, took roughly $4x$ ml when sham drinking. Only by grossly over-drinking does the rat appear to be able to generate inhibition equal and opposite to excitation.

Figure 5.11 now appears over-simple. Toates and Oatley suggested that drinking is the outcome of a simple algebraic sum of excitatory and inhibitory components. However, we now believe that taken in isolation the inhibition from mouth (i.e. the sham drinking results) and stomach may be rather weak (pure stomach distension may operate for pathologically large loads). Hall and Blass (1977) argue that it is deceptive to consider peripheral feedback in isolation from the hydrational change which follows water ingestion. It may be that a combination in the order {mouth signal}→{stomach signal}→{change in hydration} is much more powerful than the sum of the three components taken in isolation. The experiments of Hall and Blass tend to support this.

The question is raised as to how the animal's hydrational state immediately post-absorption combines with orogastric signals in inducing satiety. Suppose an animal drinks x ml and some time later $\frac{1}{2}x$ is in the intestine and $\frac{1}{2}x$ has been absorbed. Possibly the change in general hydrational state by amount $\frac{1}{2}x$ potentiates the inhibitory effect of the $\frac{1}{2}x$ still in the stomach and the oral trace of the x ml. Alternatively, water leaving the gut is detected at the portal level, and in combination with oral and gastric factors, inhibits drinking. Osmoreceptors in the portal circulation have been implicated in the control of water excretion (Haberich, 1968) where they interact with central osmoreceptors, and provide advance warning of the fluid exchange with the gut. Yet another alternative is that a duodenal osmoreceptor is involved.

5.5 ECOLOGICAL AND ETHOLOGICAL FACTORS

A cage with water-spout provides a very convenient and effortless way of obtaining water. By contrast, an experiment devised by Logan (1964) involved exertion of effort in obtaining water, and can be compared to the feeding studies of Collier discussed in the last chapter.

In Logan's experiment either (1) force required to depress the bar, (2) number of presses to get reward or (3) size of water reward were

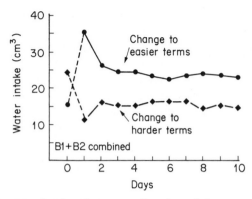

Figure 5.13 Daily water intake of rats as a function of the ease of obtaining water. For explanation, see tax. (Source: Logan, 1964. Reprinted from *Nebraska Symposium on Motivation* (ed. D. Levine) by permission of University of Nebraska Press. © 1964.)

varied. It was found that daily water intake could be reduced by changing to 'harder terms' either by (1) increasing force, (2) increasing the ratio unrewarded to rewarded presses, or (3) decreasing reward size. Figure 5.13 shows the effect of changing from easy to hard terms and vice versa. When the terms were changed there was at first an overshoot before reaching an equilibrium. Steady-state intake was roughly 23ml (easy terms) and 14ml (hard terms). Why should there be overshoot, particularly in going from hard to easy terms? Rats may acquire a habit on a given set of terms. After a schedule is established much of the rat's ingestive behaviour may be the result of well-tried sequences in which physiological signals concerning body-fluid compartments play only a modulatory role. Thus if we suddenly make the terms easy the animal continues with behaviour appropriate to hard terms and overshoots in its fluid intake. It then recalibrates over a period of days. Conversely, when the terms are made harder it overshoots in the opposite direction. We would expect misalignment to be more quickly corrected in the case of underestimating fluid intake; a deficit would goad the animal more than a surplus which would be easy to excrete.

Although body-fluid measurements were not made by Logan, the animal on hard terms probably has a lower fluid volume in the compartments which are detected than that on easy terms. For such a large difference in intake, daily urine loss (and possibly insensible loss) would be expected to be lower for the hard terms animal. This implies increased ADH secretion which in turn would be due to a lower body-fluid volume. If this is true then the 'hard terms' animal needs more intense stimulation to work for water, and adopts a strategy of low water level and the lower total effort which this allows. I am in no sense suggesting that the animal consciously adopts an appropriate strategy; it simply means that larger internal displacements are needed to goad it. In other words, we should get away from our fixed set-point model and instead think of a range of tolerable body-fluid volumes each appropriate to a given 'ecological' situation. An alternative possibility which deserves consideration was given by Peck (1979). He argued that rats may be able to maintain the same value of body-fluids over a range of different intakes. This would be permitted by the animal grooming less and hence cutting down on this expenditure when water is not so readily available.

Marwine and Collier (1979) reported an experiment in which rats were required to bar-press a number of times to gain access to water. This, perhaps more closely than the experiment of Logan, captures a feature of the natural environment, since the wild rat would, after a journey, presumably obtain unrestricted access to water. Increasing the number of bar-presses to obtain water may be analogous in some respects to making water more distant, though of course the cues associated with water reward in the Skinner-box are constantly present even if the schedule is hard. Marwine and Collier found that when water was made particularly expensive, i.e. the rat had to make 300 responses for access to the drinking tube, only one large drink was taken each day. The feeding pattern was largely unaffected by placing restrictions upon water availability, indicating that although the two activities are normally closely associated (see next chapter), feeding in response to energy cues can show considerable autonomy. These authors suggest that:

' . . . the external economics of water acquisition are as important as the internal economics of water balance in determining the patterns of drinking.'

Recently, in a review of drinking behaviour, Rowland (1977) delivered an appeal to experimental psychologists to interpret their data in the context of the evolutionary history of the species concerned. He argued:

' . . . the neural organization of behaviour confronting us today is the outcome of natural selection processes, and has thus been dictated by ecological contingencies and not by expectation of having to cope with an experimental psychologist. . . . '

Rowland presents several instances of where unnatural test stimuli have been employed in an unnatural environment, and therefore the animal's failure to respond according to expectation may have been classified quite wrongly as a regulatory failure. The brain-lesioned animal which fails to drink in response to a salt injection is seen as an aberration, and yet as Rowland notes there may be instances in the natural habitat of where animals fail to respond to deficits. A wild rat is probably:

'. . . unprepared (even counterprepared) to leave its burrow to drink during the daytime, even when intensely thirsty. . . ' (Reproduced by permission of ANKHO International Inc.)

This could not be classed as a regulatory failure, although it certainly does not conform to simple reflex-like homeostasis. Drinking by day in laboratory-housed rats may be permitted only by the close proximity of the drinking spout.

5.6 TOWARDS A NEW THEORY OF FLUID INGESTION

Toates (1979b) presented the rudiments of a new theory of drinking, and this paper was published along with commentaries, ranging from very favourable to very hostile, by leading researchers in the area. The student should consult the article and commentaries for full discussion of the issues. The arguments advanced are a mixture of empirical observation and speculation which I hope will prove fruitful.

The theory is based upon the premise that in order to understand fluid ingestion we must always consider at least two sets of causal factors, one set arising from the animal's internal fluid condition and the other from its perception (or memory) of water availability in the environment. It is meaningless to imagine an animal instigating a drink in the absence of perception of water availability. Now obviously I am not suggesting that any researcher has allowed himself/herself to ignore that water is associated with drinking. But there is a tendency to see the presence of water as all or none. If water is present when, on the basis of physiological state, the decision to drink is made, then the animal ingests; otherwise it doesn't drink. This applies to conditions in the laboratory where water is usually either readily available or absent. The link between the laboratory and the natural environment is sometimes unclear. I argued that water availability is not an all-or-none condition permitting drinking once a motivational decision has been made, but rather availability is instrumental in making the decision. How do we characterize the processes by which responsibility is shared between the sets of causal factors? There may be several parallel, but not mutually incompatible, levels of explanation.

Let us first consider the relevance of the geographical location of water and the ease with which it is obtainable. Imagine an animal normally living some distance from water and setting off on a journey to a particular water location. Surely the minimum causal factors needed to explain this are related both to body-fluid condition and spatial memory of the location of water (e.g. its direction and distance)? Convincing evidence is not available, but an animal located some way from water would probably suffer a substantial fall in body-fluids before seeking water. In other words, low cue strength would have to be associated with a high internal signal to generate thirst-motivated behaviour.

Contrast such a natural environment with the laboratory metabolism cage. The spout is very close; water easily obtainable and therefore cue strength high. It would be surprising if a relatively low body-fluid volume would occur in this situation. The adaptive significance of such a flexible 'settling value' is illustrated by Figure 5.14. As body-fluids fall then so ADH secretion rate rises. Urine loss is minimized (though the kidney is often very efficient even under laboratory conditions), and insensible loss may also be reduced. In the next chapter we will see that if body-fluid level falls the animal eats less, and in the steady state this means decreased renal need. Such mechanisms would be economical, the animal needing less water overall when water is not so easily

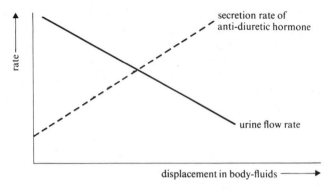

Figure 5.14 Graph showing how secretion rate of ADH increases and urine rate may fall, if body-fluids are displaced downwards from some arbitrary norm observed in the laboratory. Thus, if water is distant or available only on 'hard terms' the animal may drink less and consequently body-fluids would fall. This would economize on water loss. However, the interaction with feeding is important here (see next chapter) and theorizing must be cautious

available. However, it is important to reiterate that this is still a provisional theory awaiting more data.

Imagine that in the laboratory cage the rat's entire normal daily water intake were supplied subliminally by an intragastric or intravenous infusion controlled by the experimenter. A tube can be chronically implanted, through which the rat is infused. Would we expect the rat to drink nothing? Presumably on strict homeostatic terms we would expect this, though a variety of qualifications might need to be added (Toates, 1979b). In practice rats often show rather little reduction in daily intake even if enormous quantities are infused. To expect perfect compensation is to consider only the factor of body-fluid state and to ignore the high cue strength of the water tube. If we can stop thinking about a fixed set-point of body-fluids we may be better able to appreciate this result. Even if body-fluid level is increased slightly by an artificial method then the mere presence of water is still enough to motivate ingestion.

Another way of viewing the joint cue strength–fluid state determination of drinking is to consider a positive or, as some might say, hedonistic (Wyrwicka, 1979) value of water. This factor may be overlooked because water is often regarded as neutral or indifferent as far as taste is concerned, unlike food with its obvious taste and flavour (see Bartoshuk, 1977 for a discussion of water taste). However, it would obviously be maladaptive if animals were not equipped with mechanisms for recognizing and responding favourably to water. In other words we are ultimately seeking a model in which the reinforcing effect of water in the mouth is some function of body-fluid state. Water must be selected by its sensory qualities before it can serve any systemic need. As Mook and Kenney (1977) remind us, writers who have argued that the animal can regulate even when taste has been by-passed have restricted the discussion to experimental situations where the animal is not confronted by problems of identification and selection.

Another important aspect of the model that is being developed here is that animals may adjust their drinking behaviour on the basis of body-fluid state that results some time after drinking; i.e. learned associations are formed over long intervals. Certain aspects of this will be pursued in the next chapter. For the moment suffice it to remind the reader that if an event such as taking a meal is associated with subsequent disturbances to the body-fluids then the meal and dehydration may form an association so that the animal drinks and pre-empts the disturbance, i.e. feedforward. Given the skeleton of a model it is now appropriate to consider some other topics in detail and to attempt to relate this to the model.

5.7 LEARNING

Obviously animals come into the world equipped with a certain amount of neural machinery to be used for drinking. They must have some basis for the recognition of water; for example they seem to react to its coldness (Gold and LaForge, 1977; Mendelson, 1977). Mouth cooling may be an instant form of reinforcement for the thirsty newborn animal making its first contact with water.

In the last chapter we discussed one of the major theoretical and experimental questions posed in this area of psychology: how genetic and early experiential factors determine subsequent behaviour. In the context of thirst the role of learning has received minimal attention but was discussed by Booth (1979) and Milgram (1979), in the context of the thirst model advanced here. Given that the animal has some initial reaction to water, its visual, taste, tactile, and temperature qualities can then be associated with subsequent hydrational change. This association would give positive value to water. It is not clear where the animal's experiential and developmental history is apparent in its drinking response when adult. However, it is interesting that Rowland (1977) reported:

'. . . only ten out of twelve naive rats drank in the first three hours of water restoration (with food absent), yet in subsequent tests in the same situation all twelve showed short latency drinking.'

We do not know how typical these rats are, but early experiences are almost always lost in the adaptation period or in the response to the 'unrecorded habituation injection'.

If an animal were raised on a hydrating fluid other than water it might form particularly durable associations between this fluid and the restoration of body-fluids which the fluid causes. Indeed, rats do appear to form a particularly strong affinity for isotonic saline if they are raised with it as their sole drinking fluid (see Toates, 1979b for review). A similar affinity applies to the temperature of the familiar fluid upon which the animal is raised (Deaux and Engstrom, 1973). Thirsty rats allowed coffee-flavoured water, and then only plain water under *ad lib* conditions, showed a preference for the coffee flavour in a two-bottle choice long after their initial exposure to it (Revusky, 1974).

It would seem very likely that learning is involved in thirst satiety. The animal could associate, on the one hand: (1) passage of water through the mouth and oesophagus, (2) water in the stomach, and (3) a small rehydration with, on the other hand, full rehydration some 30 min or so later. The rat learns that relief of dehydration is on the way and stops drinking in anticipation of it. If this is what occurs then it could explain the apparent synergistic dependence between mouth, stomach, and post-ingestional satiety stimuli. Taken out of context a potential cue such as oral feedback would be ineffective (e.g. Blass and Hall, 1976).

Conditioning might help explain why there is such variation in the effectiveness of intragastric infusions in inhibiting drinking. Intragastric infusions can be either via a tube passing through the mouth and down the oesophagus or through a tube passing through the stomach wall (Rowland and Nicolaidis, 1976). The most effective way to inhibit drinks is to infuse the animal by nasopharyngeal tube at the time it takes meals. Here the suppression can reach almost 1 ml for each millilitre infused; oral temperature and mechanical cues may be given.

If we consider a learning component in water ingestion then a reduction in fluid intake following loading may be partly learned. For this to happen the rat must have some identifiable cues between which to form associations. If water is loaded at times near to when the rat might normally be expected to drink (i.e. at meals), and, after the appropriate delay, systemic fluid gain is detected then an association could be formed. Thermal and mechanical cues at the mouth appear to allow detection of the infusion. The experiments of Rowland and Nicolaidis (1976) permit such an interpretation (though other explanations can also be given (e.g. Deutsch, 1979)). Slow continuous intragastric and intravenous infusions (giving the same daily load) are relatively ineffective at inhibiting drinking. They may provide no distinctive cues either at the 'sensory' or systemic ends.

Other experiments also pose challenges. Wild rats brought into the laboratory drink enormous amounts of water at first but reduce intake substantially over a period of days (Boice, 1971). Toates (1979b) suggested that this may reflect a slow learned readjustment to the fluid opulence of the laboratory. The rat in the wild would need to attach relatively high importance to the presence of water, and if this strategy were to persist in the laboratory it could become over-hydrated. Interestingly, following a 23 h deprivation–1h drinking schedule the intake becomes excessive again when *ad lib* water is made available. Subsequently intake falls over several days of *ad lib* drinking. It seems reasonable to suggest that the animal increases the importance of water cue strength during the schedule.

5.8 SUMMARY, CONCLUSIONS, AND DISCUSSION

The fluid content of the organism, its blood, interstitial space, and cells, must be held within certain bounds, despite loss of water through a variety of routes.

Water is gained by drinking and to some extent by eating (see next chapter). Clearly the condition of an animal's body-fluids is influential in determining water intake. Loss of water either from the cells or blood motivates drinking of water in amounts quantitatively appropriate to the size of the disturbances. The unambiguous reaction of the animal to such stimuli as salt-injection (cellular loss of fluid) and polyethylene glycol dialysis (extracellular loss) forms the basis of our contemporary view of thirst. The natural tendency is for theorists to extrapolate from this to *ad lib* drinking behaviour. The extrapolation would lead to the conclusion that each drink occurs when body-fluid level falls below a threshold value. An early model of drinking (Toates and Oatley, 1970) was based upon just such assumptions.

In the modified theory which has been presented here in no sense is this negative feedback aspect denied. In any given environment a fall in body-fluids does tend to arouse drinking. However, the idea of a fixed set-point and threshold in all environments is challenged. The assumption that drinks *necessarily* occur only in response to shifts of body-fluids is also questioned, though a significant downward shift in body-fluids for a constant water cue strength is a *sufficient* condition to initiate a drink. Some drinks may anticipate disturbances and hence avoid them. In expressing these reservations, one inevitably is also questioning whether we need to call the body-fluid level defended in the laboratory an 'ideal' value. It was suggested that the animal may allow body-fluids to fall somewhat (relative to laboratory values) when water is not so readily available, and this would reduce overall demand. Whether or not each of the fewer drinks that would occur under these conditions take body-fluids to the same value as observed at the end of a drink in the laboratory is mere speculation. Given the demands of life in the wild the size of each drink will depend on a wide variety of circumstances.

Having surveyed the thirst literature it is appropriate to consider the issue of set-points. Suppose there exist: (1) a neuron which shows a constant level of activity, i.e. it is insensitive to body-fluid state; and (2) a neuron which either increases or decreases its activity as a function of body-fluid volume (assume for the sake of argument a single detector cell). Suppose also that the activity of (2) is compared to (1); a synapse with excitatory and inhibitory inputs could perform this operation. When the activity levels differ significantly in the appropriate direction the animal drinks. This would be entirely analogous to the thermostatic room-heating system and activity of neuron (1) would form the set-point. Thirst may possibly work in this way but a simpler system is perhaps more likely to have been employed. Thus a neuron (assume again for this argument a single detector) whose activity is a monotonic function of body-fluid volume and which feeds into a behavioural controller could explain the evidence. If this is so then the expression 'set-point' seems somewhat redundant. Figure 5.15 shows a tentative model which arises from these considerations and refers to the cellular component of thirst. Cellular fluid volume is shown along with the degree of neural excitation arising from the osmoreceptor. Low levels of excitation, i.e. very hydrated cells, are associated with inhibition of water intake. An alternative

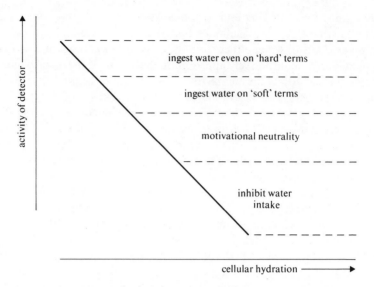

Figure 5.15 Tentative model showing the conditions under which the animal would ingest water

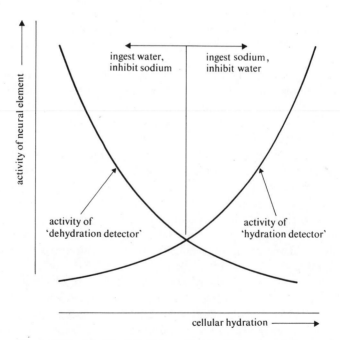

Figure 5.16 Model of fluid ingestion involving two detectors

(apart from the set-point model) is to employ two receptors, one of which detects cellular swelling and the other cellular shrinkage (Figure 5.16).

Perhaps the most demanding task for researchers in motivation is to develop understanding of how sensory information (for some purposes to be viewed as cue strength) and internal physiological state jointly hold responsibility for ingestion. This means examining the developmental history of individual organisms. In the present chapter 'cue strength' has been employed to mean that drinking tendency will be aroused when water is available. Thus an animal may drink even if body-fluids are not very depleted if, for example, it happens to pass water *en route* to obtaining food. Laboratory experiments which involve effort in obtaining relatively little water-reward may capture something of the ecological conditions of low availability.

The theory developed here attaches much importance to learning. This would serve to confirm the animal when dehydrated in its initial choices of water, i.e. a development role of learning. It would also play a role in timing initiation and termination of drinks. By drinking in association with a meal an animal may pre-empt a disturbing effect of the meal on body-fluids. Learning to terminate drinks is an alternative to the mechanism tacitly assumed in the model of Toates and Oatley (1970) which employed fixed inhibitory pathways running from the mouth and stomach to the thirst-controller.

Learning would be more flexible and allow the animal to recalibrate if the composition of available drinking fluid were to change. It would also be compatible with the synergistic interdependence of mouth, stomach, and post-ingestion stimuli in terminating ingestion. Such a proposal has a direct parallel with the model of feeding discussed in the last chapter. After a period of drinking the animal anticipates the subsequent systemic hydrational change and stops drinking at an appropriate point. Most, or at least a substantial amount, of the ingested water will still be in the gut at this point. Some will have been absorbed. The animal may use all available cues in terminating drinking.

At first sight a possibility for making the analogy between feeding and drinking termination nearly perfect seems within reach. The liver (or hepatic vein) appears to contain osmoreceptors which influence renal water excretion (Haberich, 1968; Adachi, Niijima, and Jacobs, 1976; Sawchenko and Friedman, 1979). Diuresis following stomach loading is remarkably prompt and there is good evidence that absorption of water is very quickly detected. One is tempted to argue that these same osmoreceptors could be involved in thirst satiety, but the evidence, though not addressed to this particular question, would not offer much support (Adachi, Niijima, and Jacobs, 1976; Blake and Lin, 1978).

CHAPTER 6

Interactions between hunger and thirst

6.1 INTRODUCTION

In the last two chapters, ingestion of food and water was discussed in a way which might suggest that these are two rather distinct and autonomous activities. As much as possible we deliberately examined situations where food intake was relatively unaffected by the fluid state of the animal and vice-versa. Initially such simplification is necessary and is in the traditions of scientific investigation. However, it is impossible fully to understand either activity without studying both simultaneously. Feeding–drinking interactions are so powerful and influential that unnecessarily limited information is available from studying just one of these activities and its physiological bases.

Although other behavioural systems show interactions, the need to understand these is probably not so important as in the case of hunger and thirst. Feeding and sexual behaviour may well influence each other; certainly sex hormones influence food intake. However, to the knowledge of the present writer, no serious fundamental problems in understanding either of these behaviours depend upon understanding the other. This is not the case with feeding and drinking.

6.2 LEVELS OF INTERACTIONS

Energy and water, both as commodities and as the roots of particular behaviour, show reciprocal interactions at various levels, and it may be useful to classify three such. Based upon the level of investigation most likely to be concerned with them, they can be termed ethological, psychological, and physiological. Consider again ethology. If an animal is feeding and suddenly a predator appears it may dive for cover. The animal's energy state has not changed; the set of causal factors for feeding are still present, but are overriden by the causal factors for escape. This is behavioural inhibition. An example of feeding–drinking interaction on the ethological level is a hungry–thirsty rat temporarily inhibiting feeding so as to drink.

The physiological/behavioural level of interaction has traditionally been studied in psychology laboratories. Here the level of causal factors for behaviour, the thirst or hunger signals, are adjusted by the physiological state appropriate to the companion motivation. For example, a thirsty animal will eat less food than a well-hydrated control. It would normally be argued that the energy signal for feeding arises in the same way in the two animals, but in the water-deprived animal an opposing signal is present which stops the animal feeding.

As an example of a non-behavioural and extraneural level of interaction, imagine that an animal previously adapted to a low protein intake eats a large meal of protein. It will need to excrete a lot of urine in order to lose the breakdown products of protein metabolism. A relatively large compensatory water intake would follow. As described here energy state does not directly influence thirst in this particular case. Rather it is a metabolic consequence of feeding which affects drinking, and acts through the medium of renal excretion. The distinction between the behavioural and non-behavioural level of interaction is brought into focus in the case of thirst and temperature, and a short digression is suggested. This anticipates the discussion of a later chapter, and shows an interesting parallel. If animals are put into surroundings of high temperature they increase their water intake. This could be because temperature detectors excite drinking—a behavioural interaction. Alternatively it could be due to high temperature causing loss of fluid, and this loss dictating increased fluid intake. Budgell (1970) performed an experiment to decide between these alternatives. He deprived rats of water at one temperature, but tested them at various temperatures. Following 24 h water deprivation at 20°C, rats given a brief drinking session at 0°C drank 5 ml, while at 40°C they drank 14 ml. Ambient temperature *per se* (or water temperature) appears to influence drinking.

First we need to examine ways in which food intake affects fluid balance and vice-versa on the extraneural level. Then we will be in a position to consider how energy and fluid intake directly interact at the behavioural level.

6.3 NON-BEHAVIOURAL INTERACTIONS BETWEEN FLUID AND ENERGY INTAKE AND PHYSIOLOGICAL STATES

Water is derived from food in two ways. Food usually contains a certain amount of water; laboratory chow contains only about 8% water, while lettuce contains about 90%. This means that by drying chow it will lose 8% of its weight. However, the potential, as opposed to actual, water in a diet such as chow is very much greater than this. As we saw in Chapter 4, metabolism of glucose yields water. Thus we must reckon with a substantial availability of water to the organism by energy metabolism. Indeed, in the case of some animals from desert environments the only source of water may be from the content and metabolism of food. In such cases, where plants, the staple diet, have a large water content it is perhaps somewhat arbitrary that we designate the animal's behaviour as feeding rather than drinking (see later).

Total lack of free water is one ecological extreme, but relative unavailability is perhaps more common. Chew (1965) goes so far as to argue that for most free-living mammals the major source of water is supplied via food. Some of Chew's conclusions dovetail nicely with those of the last chapter:

'When water is present ad libitum, probably considerable water is used only to bring about moment-to-moment optimum balances in the body—figuratively a "wasteful fine adjustment". On a restricted water intake, water balance is still maintained on a long-term basis, but probably not as satisfactorily from moment to moment.' (Reproduced by permission of Academic Press.)

Gerbils thrive on a diet of high water content but no free water. However, placed on a diet of laboratory chow, this desert-adapted mammal is unable to meet fluid needs by water of metabolism plus the small amount of water in chow, and dies unless free water is available.

Water of metabolism and water content of food are positive influences of feeding on the state of body-fluids. Most influences are negative. The amount of water lost as urine is a function of the quantity and quality of the diet. As food intake is increased then so renal loss increases. Protein diets demand relatively large quantities of water for the excretion of urea. The water-deprived organism is best served by fat and carbohydrate diets, which place less demand upon the kidney for water loss.

Since burning fat yields water, this raises the question of whether the large amounts of body fat maintained by some desert mammals is (in an evolutionary sense) to serve as a source of water during a drought (Chew, 1965). Chew argues that this is not the case that oxidation of fat causes more water to be lost in heat dissipation and urine than is gained. Seen in an evolutionary context the fat reserves are as a source of calories rather than water. The point may be a little academic; if water is unavailable then possibly food would also be unavailable, or at least moist food would be. Calories will be needed, and by burning body fat both calories and some water are made available. This may be as economical to fluid balance as eating food of low water content.

Food intake causes loss of water in the faeces. Rats excrete relatively wet faeces, in contrast to gerbils. But perhaps the most serious consequence of feeding on body fluids is caused by the osmotic content of food in the gut. If dehydration is such that the blood has lost water it can only make matters worse by ingesting food. Each gram of chow will pull about 1 ml of water into the gut (Oatley and Toates, 1969).

Although there are a number of ways in which fluid state is altered by intake of energy as food, it is not usually apparent where energy state is seriously altered by fluid ingestion. However, one such interaction is that if the animal drinks large quantities of water it will lose a substantial amount of heat in the urine. This could place metabolic demands on the organism. A failure of the ADH mechanism, and consequent excess urination accompanied by excessive drinking, would affect energy balance, and this is observed in some pathological states.

6.4 INTERACTION BETWEEN WATER AND ENERGY INTAKE UNDER *AD LIB* CONDITIONS

6.4.1 Minimum fluid demand

The minimum amount which an animal having food and water available *ad lib* needs to drink is dependent upon food intake amongst other factors. As food intake increases so does minimum fluid intake, since waste products must be excreted through the kidney. This raises the question of whether laboratory-housed rats generally drink in excess of the minimum need. That is to say, if rats are limited to less than their *ad lib* level of water intake, do they reduce food intake? Hsiao and Lloyd (1969) found that adding quinine in various concentrations to rats' drinking water always caused a reduction not only in water but also in food intake, and they argued that rats do not drink in excess of need. However, the rats studied by Dicker and Nunn (1957) and Rowland and Flamm (1977) exhibited more flexibility; reduction of fluid intake did not cause a reduction in food intake. Some individuals/strains seem to operate nearer to minimum levels of intake than others.

6.4.2 Circadian rhythms

In rats, both feeding and drinking show clear circadian rhythms. If rats are placed on a 24 h cycle of 12 h light–12 h dark then most feeding and drinking occurs in the dark phase, typically about 70% of food intake and 80% of water intake. Under conditions of continuous light or dark they still show a period of roughly 24 h in their ingestive activity (see Toates, 1979a for a review).

One or more intrinsic rhythms modulate the animal's ingestive behaviour. In the natural habitat the rhythm synchronizes with the 24 h rotations of the earth, while in the laboratory it synchronizes with the lighting cycle employed, provided it has a period of about 24 h. There could be intrinsic rhythms within the controllers of both food and water intake. Alternatively, there could be a rhythm within only the energy controller, and because the animal feeds mainly at night it tends also to drink then (the evidence does not suggest the possibility that the rhythm is within only the fluid intake controller). Oatley (1971, 1973) tested whether only feeding had an intrinsic rhythm. If the drinking rhythm is dependent upon the feeding rhythm then making a rat hungry and allowing it only to take meals at fixed and regular intervals during the day and night should flatten the rhythm of drinking. It did not; drinking was just as heavily nocturnal as before. Clearly, in some way the fluid controller or processes of fluid expenditure have their own rhythm. This is compatible with the observation that rats show a circadian drinking rhythm during starvation (Fitzsimons and Le Magnen, 1969).

A possible mode of interaction of a circadian rhythm and the drinking controller is a matter for speculation. I have argued that a biasing signal is added to the drinking command in the dark phase and subtracted from it in the light (Toates, 1978). For the laboratory rat with *ad lib* water, the effect is as though a body-fluid set-point rises during the dark and falls during the light. This would make drinking more likely during the period when the rat is normally active, and fluid regulation less demanding during the period of inactivity.

An analogy might help. Consider a well-insulated house, which needs a temperature of roughly 20 °C, but which should be heated mainly by night. If the set-point on the thermostat is moved from 17 to 24 during the night, and brought from 24 to 17 during the day this would give a temperature within a range of 20, and activity of the controller would be mainly nocturanl. Two points need making. First there is no proof for such a biasing signal in the drinking controller, but it does fit the evidence (Toates, 1978). Secondly it would not demand a set-point; a bias of the decision mechanism is all that is needed.

Other quite different explanations for the circadian rhythm can be given and are open to empirical check. Kissileff (1973) suggested that insensible loss is less in the light phase than in the dark, since the rat is less active. Peck (1979) argued that rats groom less in the inactive phase and hence oral expenditure is less. Both suggestions obviate biasing signals since they account for different intakes by different expenditures. Differences in water of metabolism and renal load need quantifying; nocturnal lipolysis may be relatively sympathetic to body-fluids. Of course, sleeping rats can't drink, and if body-fluids fall presumably ADH secretion is enhanced. A combination of such processes may provide the explanation. A developmental effect might also prove important and would relate to suckling experiences.

6.4.3 The timing of individual meals and drinks

This section looks at the timing of individual meals and drinks and the association between them. In rats, most drinks occur either just before or after a meal. Fitzsimons and Le Magnen (1969) adopted as a criterion of a meal-associated drink that occurring during, 10 min before, or 30 min after a meal. Under *ad lib* conditions meal-associated drinking is 70% of the total water intake in rats. Kissileff (1969) found the slightly higher association that 25% of water intake occurred in the 10 min before eating and 50% in the 10 min following eating.

One can attempt to explain the association in a number of ways. Drinking before eating could be because the fluids lift inhibition which thirst exerts upon feeding. This is why animals with *ad lib* food start to eat shortly after water is restored following deprivation. This process may be less evident under *ad lib* conditions. In the laboratory cage food and water are usually situated close together, and if the animal is aroused to satiate its thirst there is a fairly high chance that simply because it is aroused and active it will attend to any energy signal at the same time. Of course, this factor is not incompatible with water lifting inhibition; at times both factors may hold joint responsibility for the association.

Similarly a model simply involving arousal and activity could help to explain drinking following meals. However, here we have good reason to suspect a more specific interaction, as we have described already. Each gram of food in the gut pulls 1 ml of water from the extracellular fluid into the gut (Lepkovsky, Lyman, Fleming, Nagumo, and Dimick (1957); Oatley and Toates (1969)). If the animal does not drink in association with eating it will therefore be goaded to drink shortly afterwards by dehydration of the body-fluid compartments. In fact our measurements (Oatley and Toates (1969)) showed that drinking following meals is normally too rapid to be explained by such dehydration. By the time the animal would normally have taken 1 ml for every 1 g eaten there is insufficient time for water to migrate into the gut. In other words, instead of drinking in response to dehydration caused by eating, rats drink in anticipation of it and thereby avoid it.

It is possible to advance several reasons why it is more economical for animals to function in this way (Toates, 1974, 1978). First, in the natural habitat water and food may be at similar locations. If feeding introduces a demand for fluid it is probably best that the animal drinks as soon as possible after eating. Secondly, it is less strain on the organism to anticipate dehydration. If 1 gm of food were to dehydrate the organism by 1 ml and then 1 ml was drunk to correct this, the result would be 2 ml/g in the gut. It would appear more economical, especially when taking substantial meals, to thwart the osmotic shift in the beginning.

The gerbil behaves differently from the rat in this respect. Whereas rats drink mainly after feeding, gerbils have a strong tendency to drink before meals and a low probability of drinking immediately after meals (Toates and Ewart, 1977). Although it seems that eating dry food causes dehydration and drinking after a while, in so far as our experiences are concerned no anticipatory mechanism exists. This is perhaps not surprising when we recall that the gerbil is a species unadapted to drinking. Why does feeding so commonly follow drinking? It could possibly be due to the need to correct hypo-osmolarity in the gut caused by the unnatural stimulus of free water (Toates, 1979a), but no convincing answer is available.

In rats an interesting effect of diet on drinking was noted by Fitzsimons and Le Magnen (1969). When a carbohydrate-rich diet was replaced by one rich in protein, rats increased water intake considerably to match the new elevated demand for fluids. At first, additional drinks were taken at times remote from meals, but after a while meal-associated drinking increased to accommodate the new demand. Fitzsimons and LeMagnen proposed that the rat develops an association between the taste, smell, and texture of the protein diet and the relatively high fluid demand following its absorption. Increased drinking with meals removes the subsequent imbalance. When the animal is returned to a carbohydrate-rich diet, fluid intake does not suddenly fall with the reduced demand. Only after several days does intake match demand. As discussed in the last chapter, such anticipatory processes acting over a long delay were used by Toates (1979b) to advance a feedforward model of drinking.

6.5   EFFECTS OF FLUID REGULATION UPON FEEDING

6.5.1   Reduction of food intake by thirsty animals

When fed a dry laboratory diet, a variety of species show a reduction in food intake in response to water deprivation (see Toates, 1979a for a review). For instance, rats eat only about 25% of normal on day 2 of water deprivation (Cizek and Nocenti, 1965).

There is no reason to believe that energy needs fall during water deprivation and that this can explain reduced food intake, though as a *consequence* of food intake decrease there may be a reduction in resting metabolic rate. Considering in isolation the question of energy balance, the animal should carry on eating normally. Indeed, it may need to be more active than normal in searching for water, and therefore require more energy. It seems clear that the reduction in energy intake is imposed by factors associated with fluid regulation, and it is not difficult to see its adaptive significance.

Working with doves, McFarland and Wright (1969) found that rectal water loss is proportional to food intake. Following 4 days of water deprivation rectal water loss approached zero. For rats, Dicker and Nunn (1957) found that decreased feeding, amongst other factors, enabled rectal water loss to fall by 98% after 4 days of water deprivation. Voluntary food restriction may possibly allow body temperature to fall slightly in a species such as the Barbary dove, with a consequent saving of evaporative water loss (McFarland and Wright, 1969).

As we have mentioned already, one benefit from food intake reduction is that the extracellular space is not depleted of fluid by food in the gut. However, the animal's energy must be derived from somewhere, and if food intake fails to meet metabolic need then it must burn an intrinsic source. This means fat and lean mass. Burning of fat is fairly considerate to fluid regulation; the renal demand is not too high for the metabolic yield of water. Loss of lean mass is not so economical, but the advantage is that when fat and lean mass are being burned there is no shift of water into the alimentary tract. A combination of intrinsic fat and lean mass may be less costly in urine than taking a meal which could be rich in protein. It is significant that the water-deprived rat not only reduces food intake but shifts preference away from proteins towards fats (Overmann and Yang, 1973).

What is the mechanism by which water deprivation reduces food intake? Arguments in the literature often involve a dichotomy between central and peripheral factors. Rats could be unmotivated to feed or seek food; as a result of fluid displacements hunger is satiated despite the presence of an energy displacement that would normally cause feeding. Alternatively, a motivational command to feed is present, but lack of moisture in the mouth makes it difficult to swallow food. This is a somewhat different dichotomy from that between drives and appetites; the argument is not that the animal lacks an appetite, it is simply that it finds the mechanics of eating impossible. Evidence can be assembled for each view, and it has been argued that possibly both factors are jointly involved (Toates, 1979a). If we believe that motivation to ingest food is lowered, an additional signal needs to be added to a model of feeding so as to cancel the energy signal. Under a wide variety of conditions the strength of the inhibition is such that the equivalent of 1 ml of thirst inhibits $\frac{1}{3}$ g of food intake (Toates, 1979a).

Collier and Knarr (1966) argued for a central interaction on the grounds that saliva and gastrointestinal fluid dry up rather slowly following water deprivation, whereas reduction in food intake is immediate. McFarland and Rolls (1972) and

Rolls and McFarland (1973) found that the powerful central dipsogenic stimulus of intracranial angiotensin injection inhibits feeding and responding for food. This presumably reflected a purely central factor.

Van Hemel and Myer (1970) asked whether water restriction lowers the animal's motivation to feed and earn food. For animals deprived of food and water, and then loaded with water prior to being placed in a Skinner box, bar-pressing (VI 30 sec) increased with size of pre-load. The fact that lightly loaded animals ate less than those more heavily loaded could hardly be attributed to difficulty in swallowing, since even unloaded animals ate as much when food was freely available as did any of the loaded animals on the VI schedule. Body-fluid state clearly affected motivation, as reflected in working for food.

Hoarding by rats is a non-consummatory activity which seems to reflect degree of hunger (Herberg and Stephens, 1977). Prior food deprivation increased hoarding tendency, but water deprivation cancelled some of the effect of food deprivation. Herberg and Stephens suggested:

> 'The inhibition of hoarding by water deprivation therefore appears to be a subtractive operation, similar to that proposed for the inhibition of feeding. . . .'

The experimental evidence at present available does not allow us to decide whether or not learning is involved in the reduction of food intake during water deprivation. The effect soon appears following the first removal of water but it could be that the rat experiences a deterioration in fluid state by eating, and this rapidly gives rise to a learned reduction of food intake.

Ideally it would be necessary to attempt to demonstrate learned preference for food relatively sympathetic to fluid state and learned aversion to less sympathetic food (or its associated taste and odour). Water-deprived rats do indeed show a particular aversion to high-protein diets and preference for carbohydrate, but it cannot necessarily be concluded that this depends upon the diets' systemic hydrational consequences. Rats have difficulty in swallowing high-protein food (Corey, Walton, and Wiener, 1978), which confounds the issue.

Given that, in the wild, animals have to make the best of available foods and fluid, then learning about hydrational consequences might represent the optimum strategy (Toates, 1979b). To cut down indiscriminately on all foods when thirsty could be counter productive. This may be particularly so in the case of a species, such as the gerbil, living in an environment where free water is not normally available. In so far as animals have any choice of diet in such an environment it is between substances varying in degree of both water and energy content. Preferences may therefore arise in order to satisfy a compromise between energy and fluid needs. In the laboratory the gerbil satisfies this compromise by reducing intake of chow when water-deprived.

Some people are surprised that animals coming originally from an arid environment drink in the laboratory. Not all such animals do; some refuse water. But if we can consider that they have mechanisms for ingesting diets sympathetic to fluid needs then ingestion of pure water may be viewed as an extension of this.

6.5.2 Effects of deprivation on body structure

By lowering food intake over a length of time, the water-restricted animal loses lean body mass. Consider a rat deprived of water in the presence of food. Food intake does not match metabolic demand and so the body burns fat and lean tissue mass. Suppose we measure total body-water of the animal both at the start of deprivation and after, say, 48 h. We find it has lost x ml. The puzzle is why it does not drink x ml, but only 30% of x ml, when water is restored (Kutscher, 1972). I suggest that the reason is as follows. Some water is lost from the extracellular space. This represents a true deficit and would contribute to drinking. However, consider water lost from the cells. If the cells simply lose water but remain otherwise unchanged then cellular fluid loss represents a true deficit which would contribute to drinking. But the cells are not unchanged; the solid mass of the cells shrinks and loss of cellular electrolyte will occur. It is no longer the same animal we are dealing with, but a smaller one. To drink the amount of fluid lost would involve over-hydrating the now shrunken cells. Only after food intake has allowed lean mass to be re-built and cellular electrolyte to be restored would the fluid content of the body also return to normal. Remember, we believe that (apart from the extracellular stimulus) a cell in the brain is what activates drinking. We can lose mass, electrolyte, and water elsewhere, and the brain would not 'know'.

Following a given period of total water deprivation, animals will drink more if food was available during deprivation than if it was unavailable, drinking being repectively 30% and 10% of total body water loss (Kutscher, 1972). However, for any given period of total water deprivation, water loss is the same in the two cases. I believe that the explanation for this (30%–10%) difference may be seen by reference to Figure 6.1. If the animal is food-deprived then loss of lean mass and cellular electrolytes is greater than if it has food but voluntarily restricts intake because of thirst, though also in the latter case some lean mass will be sacrified if

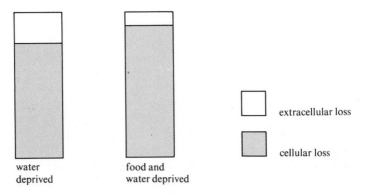

water
deprived

food and
water deprived

extracellular loss

cellular loss

Figure 6.1 Comparison of the site of water loss under the conditions of water deprivation and combined food–water deprivation

deprivation is prolonged. The rectangles represent total water loss, equal in the two cases. The distribution is different, though, in that the cellular loss is greater for total food deprivation. Extracellular loss is correspondingly smaller. Extracellular loss will 'demand' to be repaired. By contrast, cellular loss is, for a large part, a 'pseudo-deficit' which only becomes a true need when body mass and cellular electrolytes are repaired.

It may be surprising to learn that during water deprivation loss of water from the body is the same irrespective of whether food is available or not, apparently contradicting the assumption that it is beneficial in terms of reduced urine loss to stop eating. From Kutscher's study it appears that any such benefit is felt when a steady weight has been reached on a food-restriction schedule, but not in the dynamic phase of weight loss caused by total food deprivation (though remember that benefit derives from not having an osmotic load in the gut). If only a percentage of the animal's normal water intake and *ad lib* food are allowed over a period of time, then its food intake will be only a percentage of normal. After some days on such a schedule its weight will stabilize at a subnormal level. Under these conditions urine loss will be less than normal. Reduction in weight by decreased food intake over a period of water shortage may well be a not-uncommon experience in the natural habitat.

The effects of water restriction over a long period of time were studied by Collier and Knarr (1966). Rats were divided into groups and allowed either *ad lib* water (average 36 ml per day) or daily rations of 22, 13, or 8 ml. Food intake fell and body weight stabilized at lower than normal on the restricted schedules. A linear relationship between log (asymptotic body weight) and log (water intake) was obtained. They noted

'. . .the amount of metabolically active tissue which can be supported by an animal is determined by its water intake and is reduced when water intake is reduced.'

Further research by Collier and Levitsky (1967) ascertained the sites of loss of solid lean mass and water. Following water deprivation or restriction in the presence of food, a constant lean mass to water ratio in the various body organs was observed. This of course meant reduction in the solid mass and water content in the organs; the rat shrinks on a *pro rata* basis.

Adolph, Barker, and Hoy (1954) deprived rats of water for various periods of time and observed weight loss. The amount of water taken at the end of deprivation matched weight loss for short periods of deprivation, but was considerably less for long periods. To describe this, Adolph and associates coined the now common expression 'voluntary dehydration', but if we accept the arguments of the present chapter this is something of a misnomer (see also Wolf (1958, p. 113)). It seems to imply a behavioural tolerance to some detectable fluid deficit, an assumption we would not accept. Rats, rapidly dehydrated by being exposed to a high temperature and losing no solid body mass, precisely correct their water deficit at the termination of the test (Lewis, Rubini, and Beisel, 1960), but this appears not to be the case under some comparable conditions (Rabe, 1975).

It is in the present context that Bolles (1975, p. 164) refers to the 'shrinkage paradox':

> 'To put the paradox in the most stark terms, when the animal has been deprived so that it has lost, say, 25 per cent of its body solids and 25 per cent of its body water, and if either food or water deprivation can reduce it to this state, then how can it know if it is hungry or thirsty? It does know because it will eat in the one case and drink in the other; but what is the basis for this discrimination?' (Reproduced by permission of Harper and Row, Publishers, Inc.)

Probably in the case of starvation the 25% loss of water results largely from loss of whole cellular mass involving no depletion of brain fluid. In water deprivation not only is cellular mass lost but also extracellular osmolarity rises, extracellular fluid is lost and brain cells are dehydrated. This inhibits food intake. In the latter case the animal knows to drink because of cellular dehydration at the brain, and knows not to eat because this is actively inhibited by brain dehydration. In the former case, then, an energy signal is present but no detectable dehydration.

By now it should be apparent why there was reluctance in the last chapter to use the expression 'set-point' for body-fluids. Uncritical use of this term leads to the puzzle of why rats don't correct their set-point defined 'deficit' after a period of water–food deprivation. Set-point may be a useful metaphor for animals with *ad lib* access to food and water. The animal first deprived of water or water–food and then allowed water must be described in terms only of defence of extracellular volume and brain hydrational state. The assumption that brain hydrational state is a reflection of total cellular water volume will lead us astray when food is restricted. However, the assumption is valid at other times, for instance, the response of the rat to salt injections.

6.6 THE EFFECT OF FOOD DEPRIVATION ON DRINKING

The influence of body-fluid state on feeding and consequences for body-fluids has been considered. We now look at the reciprocal interaction, that of energy state on drinking. According to the argument that food intake largely dictates fluid loss which in turn determines drinking, if an animal is placed on restricted feeding or total food deprivation its water intake should fall. This is more or less true for most species, but it is certainly not true for some. The exceptions attract much attention.

Rats placed on restricted feeding or food deprivation show a gradual reduction in fluid intake (Wright, 1976; Toates, 1971). Some researchers have even found that on the first day of starvation rats increase their intake, but it always eventually falls (see Toates, 1979a for a review).

In response to food deprivation rabbits show a sustained increase in fluid intake (Cizek, 1961). There is an identifiable physiological cause for this. Rabbits lack a renal sodium concentration mechanism, and in the absence of a supply of sodium in their food rapidly suffer from sodium deficiency. Extracellular sodium concentration and hence osmolarity fall. The osmotic gradient following extracellular hypo-osmolarity pulls water into the cells and depletes the extracellular space. This extracellular volume deficit provokes drinking which lowers extracellular osmolarity still further and causes more shift into the cells. Only when restoration of food (containing sodium) allows extracellular osmolarity to rise can body fluids return to equilibrium and thirst be satiated. Evidence that specifically sodium deficiency is responsible for polydipsia is provided by the fact that fluid intake returns to normal if sodium is added to the water.

Gerbils also show polydipsia during starvation, but the cause is different from that in the case of rabbits. A sodium chloride supplement makes gerbils' polydipsia more severe. Further, they are known to have extremely efficient renal sodium conservation mechanisms (see Toates, 1979a for a review). No explanation for starvation polydipsia in gerbils is available. Weiss (1969) claims that for this species the evidence fits the interpretation that during food deprivation water is a source of bulk to fill the stomach. Presumably, in some way, this would generate a transient satiety signal for the energy controller. If non-nutritive solid bulk is made available gerbils eat this and starvation polydipsia is less in some cases (Weiss, 1969), though not all (Miller and Kutscher, 1975; Kutscher, 1979; VanderWeele, 1979).

Most gerbils do not become polydipsic on the first day or so of food deprivation (VanderWeele and Tellish, 1971; Toates, 1979a); the effect takes a day or so to appear. Toates (1979a) proposed the following explanation. The food-deprived gerbil has little physiological need for water: following a period of water–food deprivation it drinks very little. Most of the gerbil's daily water intake is dictated by the fact that it is eating. When food is removed there is a strong possibility that the gerbil will not drink at all for 24 h, since it suffers little or no water deficit. Only when physiological need for water provokes it to drink would water in the gut be able to exert any satiety upon hunger, or serve some other, as yet uncertain, role. Several such thirst-motivated journeys may be necessary before the association between drinking and its effect (transient satiety appropriate to food intake?) is established. Individual differences between animals in the speed of acquisition of the habit are enormous. We imagine this is due to differences in (1) the time at which thirst-motivated drinking occurs, and (2) speed of learning the association.

6.7 SCHEDULE-INDUCED POLYDIPSIA

This phenomenon caused considerable surprise when in 1961 it was discovered by John Falk. Rats, deprived of food but not water, and working for food pellets on a schedule where one pellet was delivered every, say, 1 min (FI 1 min) drank excessive amounts of water. At first Falk (1961, 1971) was led to believe that the drinking apparatus was leaking, but he later established that the water was passing through the body of the animal. The phenomenon is demonstrated as follows. Typically a rat is reduced to 80% of its *ad lib* body weight by food deprivation. Water is always available in the home cage. Each day the rat is placed in a Skinner-box for, say, 3 h, and obtains on average one 45 mg pellet per minute. It either has to work for it by pressing a bar, or simply obtains pellets freely at intervals. Either variable-interval or fixed-interval schedules are used, i.e. reward is not immediately available after the delivery of a previous reward, but a certain minimum interval separates rewards. In the case of bar-pressing, responding is not rewarded until the minimum interval has elapsed, which means of course that the animal can afford to waste time for a while after each pellet. Rats drink immediately after the delivery of each pellet. After several days, water intake in 3 h in the Skinner-box may reach as high as half of the animal's body weight, an excessive amount by any standards. The quantity of food obtained from the schedule is such that if it were available in the home cage as a normal meal the animal would drink about 8–12 ml in association with it, rather than the 90 ml or so which the schedule induces. Rats exhibiting such polydipsia drink little or nothing in the 21 h in the home cage. The food-restricted animal normally drinks rather less than when food is available *ad lib*. Further, it is not in the interests of a food-restricted animal to heat water to body temperature only to lose it as urine, risking water intoxication in the process. As Falk (1971) argues, the behaviour seems absurd. It does indeed when we think in terms of homeostasis and the paradigms of physiological regulation. We obviously need to escape from this model. Falk continues:

'But perhaps most absurd was not the lack of a metabolic or patho-regulatory reason for the polydipsia, but the lack of an acceptable behavioural account. That is the behaviour is absurd in the sense of philosophical existentialism. Now most animals are probably not philosophical existentialists, but perhaps we have run across a class of behaviours in animals which was imputed to man as his exclusive, if obscure, property.' (Reproduced by permission of Pergamon Press, Ltd.)

The response is not a reflex inevitably initiated by food reward; it takes several sessions to appear. It is not a form of superstitious behaviour, and appears in a variety of species (Falk, 1971, 1977). Falk coined the term 'adjunctive behaviour' to describe not only schedule-induced polydipsia but also other forms of behaviour maintained by similar circumstances. Falk (1977) defined adjunctive behaviour as:

'... behaviour that is maintained at a high probability by stimuli which derive their exaggerated reinforcing efficacy primarily as a function of schedule parameters governing the availability of another class of reinforcing event.' (Reproduced by permission of the Psychonomic Society, Inc.)

In this case water takes on a reinforcing property to a non-thirsty organism because of a schedule relevant to hunger. Falk (1977) reviews evidence showing that such behaviours as attack, escape, and wheel running, can be generated by a schedule of this kind using reinforcers other than food.

The relationship between the water intake of schedule-induced polydipsia and the interval between food rewards is bitonic, it is a maximum at a 120–180 sec interval, and falls off as the interval is either decreased below or increased above this. Similar bitonic functions apply to other adjunctive behaviours (Falk, 1977). The relationship between the magnitude of schedule-induced polydipsia and hunger is monotonic: as body weight is reduced so water intake increases. Falk (1977) reminds us that behavioural phenomena which are reliably observable and occur across several species usually serve an identifiable adaptive purpose (see also McFarland, 1966). He proceeds to develop an analysis of adjunctive and displacement activities in which it is argued that in the natural habitat adaptive value is served by switching to an alternative behaviour when the ongoing activity is in some way thwarted. Space prevents a full discussion of Falk's ideas, but basically the argument is that schedule-induced polydipsia is an exaggeration (in the unnaturally restrictive environment of the Skinner-box) of a natural switching mechanism. Similarly, Staddon and Simmelhag (1971) argue that it is of adaptive significance for animals to avoid situations at times when reward has been learned not to occur. They suggest that:

'. . . the means for ensuring that animals will not linger in the vicinity of food (or other reinforcers) at times when it is not available may be provided by the facilitation of drives other than the blocked one. . .'

In other words, subdominant motivational states are not just disinhibited but *elevated*. The authors continue:

'In the wild, such facilitation will usually ensure that the animal leaves the situation to seek other reinforcers.'

They remind us that in the laboratory study of adjunctive behaviour not only is the animal physically restrained by the Skinner-box but also restrained by the peculiar nature of the schedule.

Adjunctive behaviour is very often not irrelevant to the on-going activity, though it may be. Falk (1977) reviews evidence showing that the appearance of an adjunctive behaviour can be accelerated by the presence of stimuli normally relevant to this activity. In this context we should note that drinking immediately after feeding is entirely natural and a common activity in any case; it only becomes grossly exaggerated under these schedules. Feeding arouses drinking but the animal normally persists with feeding until a substantial meal has been taken. Meal termination is a cue for drinking. This persistence may partly mask the potential strength of the excitation from feeding to drinking. An interval schedule would have every opportunity to reveal this excitation, and by processes just described greatly enhance it. Water is close, i.e. high cue strength. Furthermore, the animal has little to inhibit it so long as ADH secretion is suppressed and its kidneys quickly excrete the load. This stands in contrast to the reciprocal interaction where drinking does not excite feeding, and in any case excess food could not be so easily accommodated.

Toates (1979a) argued that the reason gerbils do not show schedule-induced polydipsia is that there exists no natural tendency for this species to drink after feeding (Toates and Ewart, 1977). However, recent reports suggest that if both experimenter and gerbil persist for long enough polydipsia appears (Porter and Bryant, 1978). It may be slower to acquire due to a weaker excitation from feeding to drinking.

6.8 SUMMARY, CONCLUSIONS, AND DISCUSSION

As I have argued elsewhere (Toates, 1979a), if nature had not provided direct behavioural interactions between feeding and drinking, they would nonetheless arise indirectly from physiological and biochemical considerations. Three of these are immediately apparent:

1. Protein rich diets are unsympathetic to body-fluids; they demand a high urine rate but have a low metabolic yield of water.
2. Food in the gut pulls water from the body-fluid pool.
3. Reduction in food intake, for whatever reason, means a lower need for water (for a diet such as laboratory chow).

It appears that in (1) and (2) the physiological level of interaction can be pre-empted by acquired anticipatory action.

The living organism is a complex machine, whose morphological and behavioural characteristics serve the interests of survival. Motivation must be seen first and foremost in this context. Sometimes the gratification of a need associated with one motivational channel must be restrained in the interests of an associated physiological state.

The food-deprived organism has little to lose by drinking normally. Although hungry rats ultimately reduce their fluid intake, this appears to be due to a decreased need for water. There is no evidence that a thirst signal is inhibited by the presence of a hunger signal. In the case of the interaction of fluid balance on feeding, evolution seems to have arrived at a compromise between conflicting needs. Whereas it is in the interests of water balance for the animal to stop eating when water is unavailable, this is not in the best interests of energy balance. In fact a graded reduction in food intake occurs; food intake only reaches zero some days after water supply ends. The reduction of food intake which accompanies water restriction applies across species. We advanced several reasons why this is of adaptive significance to the animal. However, we must avoid the conclusion that indiscriminate reduction in food intake is useful to all species. It would be counter-productive for those which only obtain water through food and its metabolism. Here the interaction with thirst may serve to steer them to moist diets.

Cross-motivational excitation of ingestion as well as inhibition is apparent.

Intake of food necessitates water intake in fairly close temporal proximity so that the body fluids are not dehydrated by food in the gut. The timing of drinks reflects this to some extent. However, the fact that both water and food intake are heavily nocturnal cannot simply be attributed to such interaction. Even if meals can only be taken at regular intervals throughout the 24 h, drinking is still heavily nocturnal. Two influences working in concert appear responsible for drinking being mainly nocturnal in the rat: (1) an intrinsic rhythm either in the water intake control system/decision processes or in expenditure of water; and (2) excitation of drinking caused by the act of eating.

Quality and quantity of food intake set limits to the animal's minimum water intake. Some strains of rat seem to drink a minimum amount as dictated by renal concentrating ability and other sources of fluid loss. Other strains drink in excess of this and can afford to cut down on water intake. By changing diet or quantity of food allowed one can introduce a decrease in the animal's water need. The response to such a sudden change is very revealing as regards the way water intake is organized. Toates (1979a, 1979b) argued that various results, some described in this chapter, point to essentially the same conclusion: that in the steady state a rat's water intake develops its own momentum.

The evidence is as follows. When the diet is changed from one rich in protein to one rich in carbohydrate the water need drops abruptly, but water intake falls only slowly. It seems that the rat has learnt a strategy by which it abides even after demand is lowered. Experience of over-hydration presumably enables it to recalibrate its intake. Conversely, when the diet is changed from carbohydrate-rich to protein-rich there is a tendency for the strategy appropriate for carbohydrate to persist. Thus relatively small drinks are taken in association with each meal. Water intake therefore underestimates need and the animal is forced to drink between meals. After a time bigger drinks are taken with each meal and drinking between meals diminishes. To underestimate brings the organism into conflict with homeostasis more than to overestimate does; it is easy to excrete a surplus.

I believe that the animal's behaviour is showing the following form of adaptive control, closely related to that described by Rozin and associates (see Chapter 4). In some way, as yet very unclear, it appears to compute the effort involved in going to the spout and attempts to economize this cost. Imagine that on several occasions fluid state is far from optimal at, say, 50 min after taking a drink. It may be that the animal then either increases or decreases, as the case may be, its fluid intake at the next visit to the spout. The scope for further research here appears to be considerable.

When rats are food-deprived or—restricted, water intake often takes several days to fall to a steady state level, although demand is presumably reduced rather quickly (though individual differences are enormous). Again it seems that habit has its own momentum and only after hydration at a level incommensurate with its environment has been experienced does the organism recalibrate its ingestion .mechanism. It is interesting that on the first day or so of food deprivation the rat can easily be persuaded to drink much less than it otherwise would. For instance making it work for water by pressing a loaded bar causes it to reduce intake much more than is the case for the *ad-lib*-fed animal (Morrison, 1968). Similarly, adulteration of the drinking water with quinine seriously limits the food-deprived rat's intake (Rowland and Flamm, 1977). Momentum alone appears unable to sustain behaviour when there is a price to pay for drinking; under these conditions actual fluid state becomes more decisive as a determinant of intake.

A result in accordance with the theme being developed here was obtained by Milgram, Krames, and Thompson (1974). Rats were reared on dry food and lettuce but with no access to free water. Later it was found that with access to only dry food and water they were capable of regulating body-fluids. They drank normal amounts, but in response to food deprivation drank much less than a control group raised with water available. Milgram *et al.* proposed that the group raised with water were largely under the dictates of habit in their drinking. The reader is reminded at this stage of Figure 5.13 which showed the effect of momentum on bar-pressing for water.

Paradoxically the gerbil appears at times to behave more homeostatically in its water intake than the rat. Thus its usual immediate response to food deprivation, adipsia, can be explained homeostatically. Of course the adipsia is short-lived; the additional factor of drinking apparently motivated by hunger adds to the small thirst-motivated drinking factor.

For the gerbil to drink excessive quantities of water in response to food deprivation is maladaptive. However, this should not surprise us. It is certainly the case that in the natural habitat generations of gerbils have never been exposed to an abundance of water but no food. Therefore evolution could not have been influential in this aspect of their behaviour. We draw a distinction between the polydipsia of the gerbil and the rabbit. In the former case polydipsia is primary and excessive urination secondary. In the latter case polydipsia is caused by a primary excessive loss of water.

CHAPTER 7

Temperature control

7.1. BODY TEMPERATURE AND HEAT EXCHANGE

The body of a living organism contains an amount of heat. This heat quantity determines its body temperature, the relation between heat and temperature being defined by the composition of the body, i.e. its mean specific heat. An animal generates heat and also exchanges heat with the environment. According to circumstances it may experience either a net rate of heat gain or loss. Figure 7.1 illustrates this.

Body heat quantity is given by the integral of the net rate of heat gain. If gain equals loss over a period of time then temperature will be constant. When gain exceeds loss, temperature rises and when loss exceeds gain temperature falls. So far, this is pure physics. Regulatory physiology and psychology become of relevance when we consider the mechanisms by which an organism can *alter* its rate of heat exchange with the environment.

Vertebrates function efficiently only if body temperature is kept to within rather narrow limits (Heller, Crawshaw, and Hammel, 1978). Indeed, life is usually only possible at all if temperature is confined to within a certain range (the location of the range depending to some extent upon the species). The reason that animals can function efficiently only in a narrow temperature band is that physiological processes depend upon biochemical reactions which are a function of temperature. Regulation of body temperature can also be viewed in terms of its effect on the longer-term survival chances of the organism. For example, it has been suggested that tadpoles which seek the warmer regions of a pond have a higher metabolism and therefore reach metamorphosis earlier. There can be survival value in being a frog rather than a tadpole, particularly if climatic

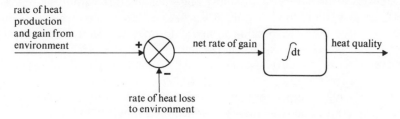

Figure 7.1 Relationship between heat quantity and rates of heat production and loss. Heat quantity is obtained by integrating net rate of heat gain. Net rate of gain is given by the difference between gains and losses

conditions either dry out the pond or wash it away in a flood (Brattstrom, 1970).

The terminology of temperature control is somewhat confusing, and expressions should not be applied uncritically. Thus, vertebrates are sometimes classified as warm-blooded (i.e. mammals and birds) or cold-blooded (i.e. fish, reptiles, amphibia). It is true that a lizard may have a relatively cold body temperature some of the time; but at other times its body temperature may be as high, or even higher, than that of a so-called warm-blooded animal. Technical expressions used are *homeotherm* for warm-blooded animals and *poikilotherm* for cold-blooded. These imply, respectively, constancy and variation in body temperatures. It is true that poikilotherms do in general fluctuate in body temperature more than homeotherms but poikilotherms in most cases also take corrective action to maintain temperature (e.g. basking) and so an alternative classification is in terms of the source of body heat when correction is being made. Biologists use the expressions *endotherm* (heat source from within) and *ectotherm* (heat source from without) for warm-blooded and cold-blooded animals respectively. However, amongst the so-called cold-blooded animals, the ectotherms, some can generate a relatively large amount of body heat and so might be considered to be slightly endothermic; an example is the Indian python.

Poikilotherm was orginally used to describe the situation where body temperature is the same as ambient temperature (Brattstrom, 1970). For example, aquatic amphibians have the same body temperature as the water. However, lizards absorb radiant heat by basking in the sun, and become much hotter than the temperature of the air around them. Panting in response to heat stress is shown by the Chuckwalla (*Sauromalus obesus*), a lizard found in the southwestern United States (Bogert, 1959).

Both endotherms and ectotherms have an optimal body temperature. This is the temperature which they will maintain if their circumstances allow them to do so. Deviation from this temperature causes corrective action. Given a choice, an environment will be found having a temperature which enables this optimum to be maintained. Both endotherms and ectotherms usually take action when body temperature departs from its optimum. Ectotherms in general show wider fluctuations in body temperature than endotherms, and so the more flexible term 'preferred zone' is often used to describe the temperature range which they maintain. Compared to endotherms, ectotherms have a limited number of mechanisms available for maintaining body temperature within the preferred zone. Since a cold-blooded animal such as a lizard performs regulatory behaviour (for example, basking to maintain body temperature), it might be called a *homeotherm,* an expression normally reserved for birds and mammals. However, since its source of heat is external it would have to be defined using the reservation that it is an *ectothermic homeotherm.*

Preferred temperature, or optimum temperature, in ectotherms may be defined not only in terms of body temperature which the animal maintains when placed in a thermal gradient, but also by such factors as the rate of its metabolic processes and its ability to move at speed as functions of temperature. In the case of endotherms, preferred temperature, as maintained in a gradient, coincides with that which involves minimum physiological cost on the part of the control system for body temperature. As Bligh (1973) expresses it, if free to do so animals elect to remain in a thermoneutral range which avoids both active heat gain such as shivering or active heat loss such as panting and saliva spreading.

It may appear to the reader that endothermic animals waste an enormous amount of energy in maintaining body temperature considerably above ambient temperature. Would it not have been wise of evolution to have programmed a lower set-point for body temperature? Obviously maintenance of a high temperature poses demands in winter, but of course the demands are less in summer. Crompton, Taylor, and Jagger (1978) speculate that if a lower temperature were defended then in summer the endotherm would need to lose an excessive amount of water to achieve this. Body-fluids might suffer as a result, since the cost in getting to water so often would be unreasonable.

There are basically two means by which an animal can alter its thermal state. First, some animals, the endotherms, can change their rate of heat production. Secondly, endotherms and ectotherms can, to various extents, adjust their rate of exchange with the environment. The second category of regulation may be further subdivided into two classes, as follows. The animal can move to another location. Alternatively, it can stay approximately where it is but change the physical property of its interface with the environment. Examples of the latter include changing feather orientation so as to increase insulation, sweating so as to increase heat loss, huddling with other animals, and nest-building.

We distinguish between *autonomic* and *behavioural* reactions to a temperature disturbance. Examples of autonomic reactions are sweating and shivering. These are normally the concern of physiologists, and are sometimes called *physiological* reactions. Examples of behavioural reactions are to huddle, build a nest, and in more recent years, to press a lever in a Skinner-box for heat reward. Bligh (1973) prefers the term 'autonomic' to 'physiological'. To describe a response such as wallowing as behavioural in no sense makes it 'non-physiological', whereas by exclusion it is not an autonomic reaction. Endothermic animals show a mixture of autonomic and behavioural temperature regulation (Bligh and Moore, 1972). Bligh (1973) draws an anatomical distinction in the organization of autonomic and behavioural temperature responses. The former depend upon the integrity of the hypothalamus but not the cortex. Behavioural temperature regulation depends upon motor reactions which, not surprisingly, are likely to be impaired by a damaged cortex. Some behavioural temperature responses appear to require an intact anterior hypothalamus, while others do not (Cabanac, 1972). For a detailed discussion of which hypothalamic regions are important in behavioural temperature control the reader can refer to Satinoff and Hendersen (1977) or Van Zoeren and Stricker (1977). The extent to which these two modes of control share common neural circuits is still a matter of discussion.

Whereas endotherms have a variety of autonomic as well as behavioural mechanisms to command in the interests of restoration of optimal temperature, ectotherms are more heavily dependent upon behavioural reactions. It can be misleading if we attempt to give weighting to the importance of autonomic and behavioural controls in endotherms. If a behavioural response such as moving to a new location is possible the animal will often do this, and spare some autonomic activity. Otherwise the autonomic mechanisms will take the full load.

An ectotherm has a very low rate of metabolic heat production when compared to an endotherm of the same body weight. By adjusting its rate of metabolic heat production the endotherm has a powerful mechanism for holding body temperature at a level above that of its surroundings. In addition, endotherms have good heat insulation in comparison to ectotherms. Thus the ectotherm is more at the mercy of the environment, and therefore dependent upon its behavioural temperature-regulation mechanism of moving to another environment. Having a high metabolic rate the endotherm normally needs to lose heat to the environment. Endotherms therefore favour an environment having a lower temperature than that of the body (Hart, 1970) so that heat may be lost down a gradient (the reader should compare his/her present room temperature with his/her body temperature). Ectotherms prefer an ambient temperature close to their optimal body temperature (Heller, Crawshaw, and Hammel, 1978).

In the present study we are primarily concerned with behavioural regulation but we will also need to consider autonomic regulation, since there is a concerted effort by the organism when a challenge occurs. The reader is probably reminded of behavioural (i. e. drinking) and non-behavioural (i. e. the kidney) regulation of body-fluid state, and the need there to consider both of these aspects in parallel.

From examining the literature, a number of differences in the way in which temperature and body fluids have been investigated become apparent. Researchers have shown more willingness to be familiar with both behavioural/autonomic temperature regulation than with behavioural/renal fluid regulation. Whereas attempts to apply control systems theory to fluid balance and drinking have been few in number, for a long time control systems theory has been at the heart of discussions on temperature (see, for example, Bligh and Moore, 1972). Sophisticated arguments and lucid discussions on the exact mode of operation of the control system abound in the literature on temperature. Also, with regard to temperature, a much wider spectrum of animals, including humans, has been studied in depth. Paradoxically though, amongst experimental psychologists, rather little regard is paid to temperature control. Even where it is mentioned in discussions of motivation, thirst and hunger invariably occupy very much more space. It should be noted that a convenient and simple technique for quantifying behavioural temperature regulation only appeared for the first time in 1957 (Satinoff and Hendersen (1977)). For an excellent recent discussion of behavioural temperature control with emphasis upon operant techniques the reader can consult Satinoff and Hendersen (1977).

7.2 THE CONTROL SYSTEM FOR BODY TEMPERATURE

Somewhere within the organism temperature detectors are located, and the signal which they generate is used to determine the autonomic and behavioural reactions. We believe that neurons, whose activity is dependent upon temperature, form the transducers.

Mitchell, Atkins, and Wyndam (1972) note that models of temperature regulation typically consider how two populations of neurons serve to supply the information for temperature control. These are in the core of the body and at the periphery. Mitchell *et al.* argue that although such neurons do provide information for the controller, temperature changes in almost any part of the body can initiate thermoregulatory responses. They claim that in some species, such as humans, average body temperature may be used to drive thermoregulatory responses.

Most emphasis in the literature is placed upon temperature detectors located in the core of the body, to be precise in the region of the brain known as the hypothalamus. Although there is no good reason to doubt that temperature control is exerted by such detectors the case made for the over-riding importance of this site is not always wholly sound. Mitchell *et al.* remind us of the common argument in the literature that since the core exhibits the most stable temperature in the body it is precisely here that detectors must be located and these provide the feedback signal. However, Figure 7.2 (derived from their paper) illustrates the reason why Mitchell *et al.* reject this logic. A vacuum flask (the contents of which are analogous to the core of the body) is surrounded by water (analogous to the outer part of the body). The temperature of the water bath is controlled by means of a thermostat and heater. Even though it is

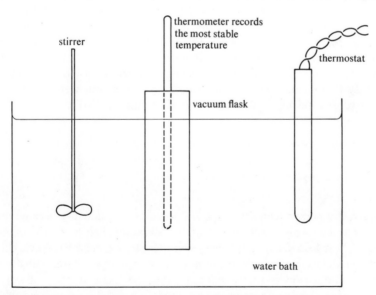

Figure 7.2 Water bath with thermostat located at one edge. When environmental temperature fluctuates the most stable temperature i.e. the least fluctuating, is recorded at the centre, even though this is not the site of control (Based upon Mitchell, Atkins and Wyndam, 1972. Reproduced by permission of North-Holland Publishing Co.)

peripheral temperature which is controlled, core temperature remains the most stable. In other words, wherever the detectors in the body are located, if the temperature stress arises from the environment then the core will show the least fluctuation.

The fact that temperature throughout the body is unlikely to be uniform for much of the time has, metaphorically speaking, posed problems for the evolution of temperature control systems. Were body temperature to be uniform then we might be in the fortunate position of analysing how a single detector determines thermoregulatory responses. Given that temperature gradients do exist within the body, where should we expect the site of detection to be located? It soon becomes apparent that a serious conflict of interests could arise here. Deep-core temperature is obviously of great importance, and if a departure from optimum occurs here then it needs to command a decisive thermoregulatory response. However, except in such cases as taking extremely hot or cold food as liquids (fever is a special case), temperature stress normally arises in the animals interaction with the environment, via the skin. In this case, the core is the last place to know of the disturbance. Usually, under temperature stress, the skin is the first location to be affected, and if the animal takes action in response to a temperature disturbance here then it may prevent core temperature being shifted from optimum. An animal whose defences against environmental temperature stress were under the exclusive control of core temperature would be at a disadvantage—it might freeze to death or overheat before reaching shelter.

However, in certain species serious problems for some aspects of temperature regulation could have arisen if too much weight were to have been placed upon particular peripheral signals. Take for example, in humans, the response of sweating following a rise in core temperature as a result of exercise. Sweating cools the skin relative to the core, which ultimately cools the core. It would obviously be maladaptive if a slight cooling at the skin were to inhibit the centrally driven response of sweating. As Benzinger (1964) argues:

'Temperature regulation in a home would not be improved if the warm sensor of the thermostat were sprinkled with water and cooled during periods of overheating.'

However, if a person were sweating and the weather suddenly became unduly cold it may be to his advantage to inhibit sweating slightly in *anticipation* of increased heat loss. Obviously a very delicate balance must be struck between peripheral and central signals, the animal giving appropriate weighting to each. We have good reason to believe that important species differences will appear in this weighting, depending upon the mechanism which the animal employs for heat loss/gain and its ecological background. Size of animal is an important factor. Small mammals such as chipmunks have higher hypothalamic temperature sensitivity than larger mammals such as seals (Heller, Crawshaw, and Hammel, 1978). Small body size means that the hypothalamus is more rapidly sensitive to normal temperature changes than is the case for a large body size.

Mitchell, Atkins, and Wyndam (1972) warn us of some of the dangers inherent in theorizing about temperature-sensitive neurons in the CNS. Just because a neuron can be shown to be temperature-sensitive does not mean that it normally serves as a temperature detector. Some temperature-sensitive neurons appear to have nothing to do with temperature regulation. Neurons in the core of the body may show temperature sensitivity and may indeed be shown to activate

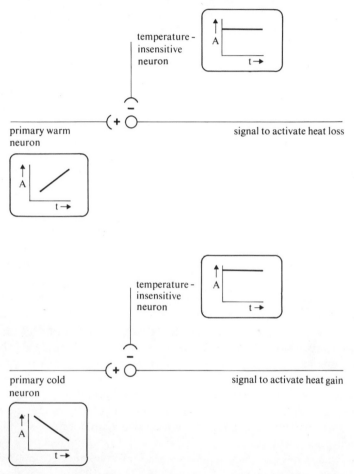

Figure 7.3 Control of body temperature by a set-point system. The activity (A) of a neuron is independent of temperature, i.e. the temperature-insensitive neuron provides a set-point. A primary warm neuron increases its activity as a function of temperature (*top*) and the activity of the temperature insensitive neuron is subtracted from this (i.e. comparison). Any disparity activates heat loss. A primary cold neuron increases its activity as temperature falls (*bottom*). By subtracting the activity of the temperature-insensitive neuron, i.e. the index of the set-point temperature, from this, the signal is derived for activating heat gain. (Based upon Bligh, 1973. Reproduced by permission of North-Holland Publishing Co.)

temperature-regulatory responses following local temperature changes. However, they may rarely be called upon to act that way since their immediate environment would be relatively well buffered against large temperature stresses. Some might claim that there is no conclusive evidence that temperature-sensitive neurons in the hypothalamus are involved in temperature regulation (Bligh, 1973). The argument rests on the foundation that this area is intimately involved in temperature control and that such neurons fit our ideas as to how the system works. So much for a few qualifying statements and reservations, the currently most acceptable view is that temperature-sensitive neurons at the core (hypothalamus and spinal cord), periphery (skin), and possibly other sites (e.g. blood vessels, etc.) supply the information upon which the temperature reaction is based.

Investigation of the activity of hypothalamic neurons has shown that, with regard to temperature, several different classes exist. By *primary warm sensors* we mean units which increase their level of activity as temperature is increased (Bligh, 1972). A number of models of temperature regulation have been proposed, and we will now examine some of these. One suggested model of temperature control, commonly advanced in the literature, is based upon a hypothalamic set-point and comparitor. It is assumed that the hypothalamus contains (1) neuron(s) whose rate of firing is a function of temperature, and (2) neuron(s) whose rate of firing is independent of temperature. The latter would discharge spontaneously and show a fixed rate of firing. It would provide a set-point for the system, and the difference between (1) and (2) is the cue for either active heat gain or loss, as the case may be. We could imagine a synaptic junction with the temperature-insensitive neuron exerting an inhibitory effect and the primary warm neuron an excitatory effect (Bligh, 1973, p. 176). When the post-synaptic neuron is active, this would mean temperature is above the set-point and would provide the stimulus for cooling. A second comparison, this time between a temperature-insensitive neuron and a primary cold neuron, would also be made, and the outcome would be the cue for heating. Figure 7.3 (based upon Bligh, 1973) shows these systems. If the set-point for cooling were to be at a slightly higher temperature than that for warming then a thermoneutral range in which no action is needed would exist between them. This may indeed be how the system functions, but other possibilities exist which do not involve comparison with a set-point.

In advancing our case against the necessity of involving set-points in body-weight regulation and feeding (Toates, 1975; Booth, 1976; Booth, Toates, and Platt, 1976), we overlooked very similar arguments in the literature on temperature control. Mitchell, Atkins, and Wyndam (1972) and Bligh (1972) revive and revise alternatives to the set-point model which was first advanced by H. C. Bazett in 1927. The essentials of one such model are shown in Figure 7.4. A population of neurons increases its activity as a function of temperature, and a representative signal is extracted from this. Heat loss mechanisms are dependent upon this signal; as it increases, so does the magnitude of the heat loss effort. As temperature falls another population of neurons increases its activity, and their

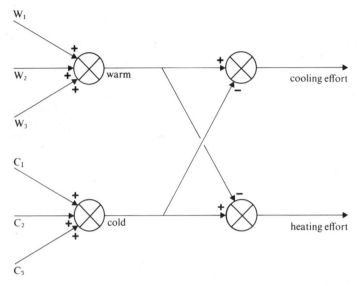

Figure 7.4 Model for the regulation of body temperature, not involving set-point control. W_1, W_2 and W_3 are neurons which increase their activity with increases in temperature. The warm signal is the outcome of the summation of these signals, and it promotes cooling and inhibits the heating effort. C_1, C_2 and C_3 are neurons which increase their activity with a fall in temperature. Their combined effort gives the cold signal. This promotes heating and inhibits cooling

integrated effort drives the mechanisms of heat gain. These two opposing processes are such that the body comes into equilibrium at a particular temperature. If temperature rises above this value, then heat loss dominates over heat gain until temperature returns to normal. Conversely, if temperature exceeds this value, then heat loss is activated until equilibrium is restored. Thus, stability is shown without the postulation of a set-point. Inhibitory pathways are also suggested, i.e. the signal from the warm detectors inhibits heat generation and that from cold detectors inhibits heat loss.

Figure 7.5 shows a diagram of temperature control in neurophysiological terms, and incorporates the essentials of Figure 7.4. It was suggested by Hammel (see Heller, Crawshaw, and Hammel, 1978). Four different hypothalamic neuron populations (1–4) are each represented by individual neurons, and this shorthand description will be employed here. Neuron 1 is a primary warm neuron. It increases the activity of neuron 3, which in turn increases the rate of heat loss. Neuron 2 is a primary cold neuron which facilitates the firing of neuron 4. This then determines the magnitude of heat conservation and production. Cross-connections within the hypothalamus are also shown. Activity of 1, the warm detector, not only excites 3 and therefore promotes heat loss, but also inhibits 4 and serves to inhibit heat conservation. The cold detector inhibits heat

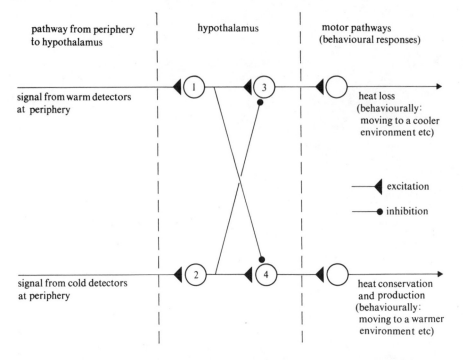

Figure 7.5 Simplified and modified version of a possible model of the neural basis of temperature control, proposed by Heller, Crawshaw, and Hammel (1978). Tentatively, behavioural control has been added. Neuron 1 is a warmth-sensitive neuron in the hypothalamus, which increases its activity as temperature increases. It serves to activate autonomic heat-loss mechanisms, and (tentatively) such behaviours as moving to a cooler location, saliva-spreading, wallowing, etc. Neuron 2 is a primary cold-sensitive neuron in the hypothalamus, activity of which promotes heat conservation and production. Also, tentatively, it is proposed that it is associated with behaviour such as moving to a warm environment, huddling, etc. Activity of neuron 1 is influenced by the reception of heat at the skin, i.e. an excitatory input derives from this source. Neuron 1 not only activates heat loss but it also inhibits heat conservation and production. This is represented by the inhibitory connection to neuron 4. Neuron 2 is excited by cold detection at the skin, and its output inhibits the activity of neuron 3. Thus detection of cold inhibits heat loss. Note that single neurons are shown, each representing possible populations of neurons serving a common function

loss. Information from heat sensors in the skin also increases activity of neuron 1. Activity of neuron 2 is facilitated by cold sensors in the skin.

For some purposes, particularly in a study concerned with the broad issues of motivation, it may suffice to treat the detector of temperature in terms of a 'black box'. In other words, we simply accept that certain neurons are able to change their rate of activity as a function of temperature. However it may help illuminate the fundamental processes involved if we look more closely at how a unit could change its activity with temperature.

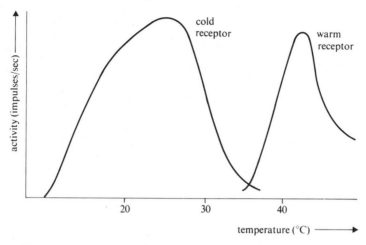

Figure 7.6 The steady-state response of temperature sensitive neurons. Cold and warm receptors

Figure 7.6 shows the kind of result which has been obtained from single-unit recording of afferent fibres which respond to peripheral temperature. Frequency of discharge as a function of temperature forms an inverted-U in each case. In this example the cold unit shows maximal sensitivity at about 24 °C and the warm unit at about 42 °C. In examining the published literature it is found that the maximal sensitivity of cold units is in the range 20–35 °C and that of warm units about 40 °C (Iggo, 1970). To meet the standards required of being a biological thermometer then a unit must show a sustained characteristic temperature-dependent activity, though an initial transient effect will also be apparent. Iggo reports that sensitivity remains constant over several hours of recording from such units. The transient response is complementary to the role of the detector, i.e. warm units are additionally sensitive to a rise in temperature and cold units to a fall. This is one of the criteria used for identification of cutaneous thermoreceptors (Hensel, 1970), and is illustrated in Figure 7.7.

It is worth distinguishing the temperature-sensitive units which play a role in normal temperature regulation from those which respond to sudden biologically harmful stimuli. The latter are sensitive to the extremes of temperature, and presumably give rise to the sensation associated in humans with burning (hot and cold).

Detectors which are described in the context of one particular sensory quality are often found also to be sensitive to other qualities. A favourite example is the pressure-sensitivity of the eye: in the absence of light energy falling on the retina flashes of light nonetheless can be made to appear by applying pressure to the eye. This is unlikely to matter very much under normal circumstances, since the situation in which pressure is applied is highly artificial. However, it is known that mechanoreceptors often show temperature-sensitivity (Iggo, 1970), and it could confound temperature control if temperature detectors were sensitive to

146

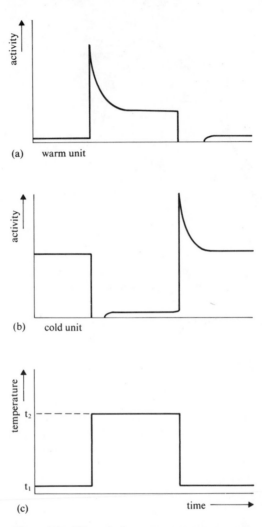

(a) warm unit

(b) cold unit

(c) time ⟶

Figure 7.7 The response of peripheral temperature detectors. Temperature changes from t_1 to t_2
and back to t_1 again (*bottom*). The warm unit
shows a transient response to the increase in
temperature as well as a sustained activity in
response to the high temperature t_2. Conversely
the cold unit gives a transient response to the fall
in temperature as well as a sustained response to
the cold temperature t_1

mechanical deformation. In fact, primary temperature-sensitive units are
insensitive to non-thermal stimuli. Parenthetically, it is worth noting that one
reason why poikilotherms need to exceed a certain minimum body temperature
for efficient locomotion (Houston and McFarland, 1976) could be the
temperature-dependence of muscle spindle characteristics (Iggo, 1970).

A change in temperature has a variety of effects upon neurons (such as to change potassium permeability of the membrane (see Sperelakis, 1970)), some of which tend to increase the frequency of discharge and others to decrease it. Primary temperature-sensitive neurons are distinguished by their particular sensitivity in one or the other direction. That an inverted-U shape relates frequency of action potentials and temperature is indicative of the dominance of first one and then the opposite temperature effect.

7.3 BEHAVIOURAL TEMPERATURE CONTROL

7.3.1 Introduction

After supporting a model of the kind shown in Figure 7.4 for autonomic temperature regulation, Hammel (1972) suggested that the basic features are applicable also to behavioural temperature regulation. It is a view also held by Cabanac (1972). This means that basically the same underlying causal basis for the activation of heat production and conservation applies to removing the animal from a cold environment. Similarly, the same signals which promote heat loss also goad the animal to remove itself from a hot environment. In the case of ectothermic vertebrates the behavioural reactions may be the only ones available, or certainly the most important. For endotherms both autonomic and behavioural responses would be aroused by similar stimuli, though that is not to say the mechanisms are identical in how stimulus information is integrated. Support for the idea that basically the same temperature-detection processes are involved in both behavioural and autonomic temperature responses may be found from examination of data on humans. Corbit (1970) noted that the magnitude of the autonomic reaction to either heat or cold exactly mirrors the magnitude of the subject's rating of thermal unpleasantness.

Animals exhibit a variety of behavioural regulations in response to temperature deviation. Like autonomic reactions, some of these involve unwanted costs as well as benefits, while others involve minimal cost. Any kind of locomotion or muscular activity involves expenditure of metabolic energy which could embarrass attempts to keep cool but would not harm attempts to keep warm. However, simply moving may increase the animal's surface area of exposure to a hostile environment. Bligh (1973) discusses the advantages and disadvantages of four mechanisms of heat loss in animals, i.e. wallowing, saliva-spreading, panting, and sweating. Wallowing does not involve loss of body water or minerals. Saliva-spreading and sweating involve loss of both water and minerals. Panting involves loss of water without minerals. These are crucial considerations when we remember that the animal under thermal stress may well find itself under hydrational stress a little later. Sweating involves no mechanical work, and wallowing may involve only little, in contrast to panting. Where an animal has a covering of fur then obviously evaporative cooling presents problems; rats lick only exposed surfaces. Wallowing and saliva-spreading are correctly regarded as *behavioural* reactions, in contrast to the autonomic functions of sweating and panting. Wallowing is particularly effective and commonly observed in a bare-skinned animal such as the pig, where autonomic mechanisms of heat loss are rather limited. Mud is an obvious medium for this behaviour.

Between 10 and 25% of rats fail to acquire the behaviour of saliva-spreading in response to elevated ambient temperature. Hainsworth and Stricker (1970) suggest that saliva-spreading may be acquired by the fortuitous reinforcement obtained from evaporative cooling following the association of normal grooming and profuse salivation.

When we consider mechanisms of heat production and conservation then although in the short term there is only gain in expenditure of metabolic energy this obviously involves long-term costs in terms of the need to seek extra food. Clearly the behaviour of huddling, common in many species (Hart, 1970), and, in the long term, nest-building are highly efficient with little cost.

7.3.2 Various species

In this section we will discuss in detail certain classes and species of animal in order to illuminate the different processes by which temperature is held constant.

7.3.2.1. Rats

Passengers waiting in winter for trains at Lancaster, in the north of England, and who wish to keep warm, are required to get up at regular intervals and press a button on the wall. In response, an electric heater comes on for a fixed time of a few minutes duration. It is clearly an operant situation, learned through imitation, insight or accident since no instructions accompany it. Very similar mechanisms have been successfully employed using rats in Skinner-boxes for the study of behavioural temperature regulation. In the absence of responding by the rat, the Skinner-box temperature is, say, a steady 0 °C, and in response to bar-pressing a heater comes on for a few seconds (Weiss and Laties, 1960; Carlisle, 1966).

Carlisle mentions one advantage this reinforcer has over the more conventional ones of food and water. Physiological state can be specified and monitored in terms of body temperature, whereas thirst and hunger are more distantly defined in terms of hours of deprivation. In fact, another advantage becomes apparent here. One cannot suddenly introduce a state of thirst drive except by a procedure such as salt or angiotensin injection or haemorrhage. In all probability, these are highly traumatic for the animal. To place an animal in a cold environment quite quickly motivates it to work, and is probably less traumatic. Carlisle (1966) found that, as temperature was lowered, so the rate of bar-pressing increased, as was also the case when reward size was lowered (lower-power heater or shorter 'on' duration). The inverse relation between rate of responding and magnitude of reward suggests that activity is dictated by the rate of reinforcement needed to prevent body temperature falling (Weiss and Laties, 1960). However, schedules which yield relatively little profit, i.e. a low rate of reward, are sometimes abandoned by the animal, which then suffers hypothermia.

In terms of the so-called natural history approach or 'preparedness' (Seligman, 1970) to learn, it is perhaps surprising how well rats do perform on this task. Whereas bar-pressing for food reward arguably bears some similarity to the kind of manipulation and gnawing that would naturally accompany the gain of food, bar-pressing for heat reward bears no relation to any natural reinforcement. Of course no consummatory behaviour is involved in heat reward. Carlisle (1968) reported that he had been unable to obtain thermoregulatory bar-pressing in a cold environment unless the rats had been shaved prior to the test. Attempts at explaining this were made, but to the present author it suggests that perception of heat reward at the skin is crucial to maintain responding. Fur would be expected to prevent an immediate impact of the reward. Carlisle found that if the reinforcement conditions were changed the animal's rate of responding changed accordingly, and this occurred before central temperature changed. Again a peripheral source of information is implicated. Hypothalamic temperature was found to be normal during these operant sessions. In other words, it appears that in this experiment rats are able to use peripheral temperature information in order to prevent hypothalamic temperature falling. Rats which are very competent at regulating temperature in a Skinner-box when on a continuous reinforcement schedule fail to maintain temperature on fixed ratio schedules. Carlisle (1969) notes the similarity of heat and electrical stimulation of the brain in this respect.

In a study by Murgatroyd and Hardy (1970), made on rats in a $-5°C$ environment, hypothalamic temperature did fall. Rate of bar-pressing was roughly a linear function of the deviation from normal of hypothalamic temperature. These researchers estimated the rate at which the rat would need to bar-press in order to maintain thermal balance, and found that it worked at only about one-quarter of this rate. In other words, it placed a major share of responsibility upon autonomic regulation, even though in theory it could have worked at a rate such as to avoid this. However, it must be noted that the rats were shaved and to have maintained thermal balance by behavioural means alone might have necessitated excessive radiant heat on their backs. The authors make the cautionary comment that under other circumstances the rat might give more weight to behavioural mechanisms.

We saw earlier that if food or water are made 'expensive', by requiring a relatively large effort on the part of the animal to obtain them, then daily ingested quantity falls. It is possible to compare this with responding for heat reward. If the length of time for which the heater comes on with each bar-press is halved does the rat respond twice as hard? The answer is no, though as we have already seen response-rate increases. The total quantity of heat obtained falls as it is made more expensive (Corbit, 1970). This applies to a situation where the rat has control over when the heat comes on but not when it is switched off. If, alternatively, the heater comes on for as long as the lever is depressed then when the power of the heater is halved the rat spends twice as long holding the lever down. Monkeys shift reliance slightly towards autonomic and away from behavioural regulation when operant behaviour is inefficient in terms of yield for effort exerted (Adair and Wright, 1976).

In a series of experiments using various species it has been shown that in response to a displacement of either skin or hypothalamic temperature animals will work for a corrective change in skin temperature. When skin or hypothalamic temperature is increased rats work to turn on a draught of cool air. Rats can also be persuaded to work for a change in hypothalamic temperature (see Corbit, 1969b). If either skin or hypothalamic temperature is displaced upwards rats will work to reduce hypothalamic temperature, and if both are displaced the effect on operant behaviour is additive. In the experiment of Corbit the skin was exposed to cold air and the hypothalamus was cooled by a water-perfused thermode when the animal responded. Since Corbit's data fitted an additive model he concluded that:

'. . .signals arising from cutaneous and hypothalamic temperature receptors are integrated additively before the final common pathway for thermoregulatory behaviour.'

An equation of the kind was proposed:

cooling effort $= (K_1 \times$ displacement above normal in hypothalamic temperature)

$+ (K_2 \times$ displacement above normal in cutaneous temperature)

where K_1, K_2 are constants

In other words effort or 'drive' is proportional to the temperature deviation. Corbit (1970) shows that response rate for hot/cold reward in an operant test is proportional to the deviation of air temperature from thermo-neutral. Although, as Corbit notes, skin temperature does not equal air temperature, it is some indication of the extent of the temperature stress. An additivity model of temperature control as opposed to a multiplicative model is supported by the fact that responding begins when either detection site departs from normal (Corbit, 1970). Hypothalamic temperature displacements in the absence of peripheral stimuli are also able to motivate behavioural temperature responses in baboons (Gale, Mathews, and Young, 1970). It should be noted though that other investigators favour a multiplicative relationship (Murgatroyd and Hardy, 1970).

Corbit argued that in the case of autonomic thermoregulation more weight is placed upon hypothalamic than on peripheral temperature signals. In other words K_1 is greater than K_2. The opposite is the case for behavioural temperature control: K_1 is less than K_2. As Corbit notes, the extra weighting given to cutaneous temperature sensations in behavioural control is particularly useful. As far as the animal in the natural habitat is concerned thermoregulatory behaviour usually means changing its immediate environment. The temperature stress usually arises from the surroundings, and is of course felt first by the skin. If a thermally acceptable environment has been chosen and yet body temperature is not at its optimum then autonomic reactions would be appropriate, these being driven mainly by core temperature. Bligh (1973) discusses the neurophysiological embodiment of such a theory. We could imagine the dominant synaptic influence on a neuron being exerted by one or the other set of afferent signals. The post-synaptic unit would thereby reflect a suitably weighted version of the incoming information.

So far we have discussed the rat in a highly artificial laboratory setting. In the laboratory it is possible to study situations which are in some respects similar to those encountered by rats in the natural habitat. The behavioural response of moving to a warmer environment in a gradient test is apparent in rats at an age of about 5 days, this appearing earlier than the autonomic components of temperature regulation (Fowler and Kellogg, 1975). A rat placed in a cold environment without nest materials will shiver, increase food intake and metabolic rate. If materials are available it will construct a nest and hence rely less upon autonomic regulation (Mogenson, 1977).

Richter (1943) describes an experiment in which rats were given access to a paper roll (1.3cm wide). The amount of paper taken was measured using a cyclometer, and each day the paper was removed. The length of paper taken each day for nest-building increased from 500 to 6000 cm as temperature was decreased from 80 to 45°F.

7.3.2.2 Birds

Budgell (1971) observed Barbary doves which pecked a key for heat reward. From a mathematical analysis of the time-dependent relationship between environmental temperature, hypothalamic temperature, and responses, Budgell concluded that behavioural thermoregulation is under the control of brain temperature rather than peripheral temperature. However, others have doubted that the hypothalamus of birds is sensitive to temperature, it being claimed that core temperature transducers are located in the spinal cord (see Heller, Crawshaw, and Hammel, 1978 for a discussion of this). No-one doubts that the hypothalamus plays an integrating role in avian thermoregulatory behaviour. Possibly both hypothalamic and spinal temperature-sensitive units are implicated (Whittow, 1976). Bligh (1973) takes the view that temperature responses in birds are activated by peripheral, hypothalamic, and spinal sensors, but that extra weighting is given to the spinal units. Behavioural and autonomic components may depend upon different populations of sensors (see Laudenslager and Hammel, 1977). Although in Barbary doves the mathematical relationship between temperature and behavioural effort is compatible with the hypothalamus being the site of detection, this in no way establishes the precise location since in all probability spinal cord temperature would have shown a similar time-dependence.

In Barbary doves the technique for bringing behaviour under operant control by heat reward is to give food reward first. Then, in a cold environment, the experimenter gives combined food–heat reward, and slowly phases out food (Budgell, 1971). In the case of rats, operant behaviour for heat reward develops without the need for prior food reward (Carlisle, 1966). The findings with Barbary doves prompted McFarland (1973) to form some general conclusions regarding behavioural and non-behavioural aspects of homeostasis:

'In my experience it is easier to train an animal to obtain a reward by operant means when the reward is relevant to an homeostatic system in which appetitive behaviour normally plays an important role. For example, Barbary doves (*Streptopelia risoria*) seem to have poor energy conservation, and are very easy to train to work for food rewards. They have quite good water conservation mechanisms, and are harder to train to peck for water rewards. They have extremely effective temperature regulation [McFarland and Budgell, 1970], and it is very difficult indeed to train them to regulate environmental temperature by operant means [Budgell, 1971]. Thus there may be constraints on learning corresponding to the weighting given to behavioural aspects of homeostasis within a particular system. In addition, it seems likely that the relationship between the animal's normal behavioural repertoire and the operant response required by the experimenter may constrain learning. Pecking is presumably a more relevant response for a dove in a feeding situation than in a temperature regulation situation.'
(Reproduced by permission of D. McFarland.)

Although, as will be described shortly, the response of key-pecking for heat in the cold is a difficult one to teach birds, it may have proven easier to teach Barbary doves to work for cold reward in a warm environment. The reason will be given shortly when we consider other bird species. Using the Barbary dove, an interesting interaction between motivational systems was investigated by McFarland and Budgell (1970). Drinking appeared to be aroused by increases in core temperature, which forms an additional ingestion mechanism to body-fluid controlled drinking, and would serve to anticipate water loss following an increase in temperature.

Domestic fowl exposed to a temperature of 40°C rapidly learn the response of pecking a disc for a 30 sec burst of air at 22°C. Indeed, Richards (1976) reported that after training had been completed the operant response even preceded or replaced the physiological responses and postural changes which would normally accompany high temperature. By contrast it was found that birds exposed to a temperature of − 5°C did not perform this same instrumental response for the reward of air at 22°C or radiant heat. When food and heat were given together the birds would peck for food but pecking rate dropped by 75% when food was omitted. Under these conditions birds take a crouched posture which minimizes skin exposure. Richards concluded that:

'... in birds behavioural thermoregulation is directed chiefly toward supplementation of the relatively limited physiological defences against heat.'

Chickens can be persuaded to peck a key for heat reinforcement in a cold environment but consideration must be given to the competing activities which appear. As temperature is lowered they first exhibit what Horowitz, Scott, Hillman, and Van Tienhoven (1978) recognize as species specific reactions as well as autonomic responses. Only if these fail will an instrumental response (originally learned in association with food reward) be performed. By species specific reactions, Horowitz *et al.* mean fluffing feathers, tucking the head under a wing and squatting, which has the effect of protecting the legs and feet. When the bird's feathers were removed (such that species specific and autonomic reactions were inadequate) it readily performed the instrumental response. Given adequate insulation there will be a tendency to remain still in the cold which would interfere with any behavioural reactions. Schmidt and Rautenberg (1975) found that whereas it was relatively easy to train pigeons the response of breaking a photocell beam for the reward of cooling in a hot environment it was difficult to teach them to perform in any sustained manner a response for heat reward in a cold environment.

An interesting study of behavioural temperature regulation in partridges was performed by Laudenslager and Hammel (1977). By interrupting a light beam at a corner of its cage the bird was able to cause the cage to move into a different temperature environment. Two beams were present at opposite sides of the cage and interruption of one moved the cage from warm to cold and the other from cold to warm. As a result of the bird's behaviour, the cage was observed to shuttle back and forth between the two environments. Frequency of alternation increased as the temperatures were made more extreme on each side of thermoneutrality. The authors note the biological appropriateness of an exploratory response to gain a new environmental temperature, in distinction to such responses as bar-pressing or head-nodding.

In the natural habitat heat stress is alleviated in many species of bird by withdrawal to a shaded area (see Dawson and Hudson (1970) for a review). The Rock wren moves into deep crevices in the rocks when the summer heat is most intense. Cactus plants offer protection to the Elf owl in the south-west of the United States and Mexico. Soaring to great heights, bathing, urinating on the exposed legs, and splashing water over the feathers in the process of drinking (Hafez, 1964) are other methods used, according to the species.

7.3.2.3 Reptiles

Most reptiles show the capacity for body temperature regulation, mainly by behavioural means. There are apparent exceptions, such as certain marine turtles (Templeton, 1970) and iguanas (Bogert, 1959). The exceptions appear to have little need for temperature regulation since their natural thermal environment is very stable. Reptiles produce very little metabolic heat in comparison to birds and mammals, and body surface insulation is very poor (Templeton, 1970).

The preferred temperature of a reptile appears to vary as a function of the animal's other activities. A circadian rhythm in temperature is often present. Lizards commonly prefer a lower ambient temperature by night than by day. A relatively high temperature is sometimes sought following a meal, and it is argued that this aids digestion (see Templeton, 1970). Reptiles show relatively flexible behaviour, and although it is possible to define a preferred or optimum temperature they are able to depart widely from this. For example, if defence of territory is called for, thermoregulation takes low priority.

Lizards were once thought to exhibit no regulatory abilities, and were observed to come into equilibrium with the laboratory temperature. As Bogert (1959) notes, this is like asking a man with a weight attached to his foot to demonstrate his ability at running. In the natural environment lizards may be observed to shuttle back and forth between sun and shade, and in so doing maintain body temperature at a value intermediate between the equilibrium temperature at the two sites (see Templeton, 1970). The boundary limits which result from such behaviour are just a few degrees apart. Similarly close boundaries are associated with the thermoregulatory responses of orienting positively and negatively with respect to the sun's rays (Bogert, 1959). When body temperature is low the lizard positions itself at right angles to the sun's rays so as to maximize the surface exposed. Lizards also seek slanting surfaces so as to derive the maximum absorption from the early morning sun. It is possible to devise control models which would account for behaviour such as shuttling.

Templeton, for example, suggests two thermostats, one controlling the behavioural 'heater' and the other the behavioural 'cooler'. The settings would be a few degrees apart and this would account for the range where the lizard is not motivated to move. There is evidence to show that the hypothalamus of lizards contains temperature-sensitive neurons which, acting together with peripheral detectors, instigate thermoregulatory behaviour (see Bligh, 1973, p. 308 for a review). It should be noted that in a small animal with low thermal insulation at the skin, such as the lizard, core temperature would rather quickly come into equilibrium with that of the more peripheral regions. Given standard environmental conditions, of sun and shade in close proximity, it may be possible to predict the movements of a lizard between two environments in terms of two temperature limits. However, if the environment were different, these limits would probably not apply. Lizard's movements are a function not only of temperature state but of *cost* involved in changing from one location to another (Huey, 1974). Where the distance between suitable hot and cool sites is high then alternation is infrequent. In the extreme, the lizard *Anolis cristatellus* appears virtually to abandon active regulation and passively accept the ambient temperature of the forest (Huey, 1974). The lizard must, like other species, be viewed in terms of a survival machine which performs optimal behaviour sequences. There is a cost involved in not being at optimal body temperature in terms of locomotion, but the cost involved in maintaining optimal body temperature may outweigh the advantages.

7.4 SUMMARY, CONCLUSIONS, AND DISCUSSION

Despite the relatively sparse consideration which is often given to this subject in studies of animal behaviour, some fundamentally important and interesting questions arise when we examine behavioural temperature control. It shares features in common with other motivational channels. For example, the question of species-specific behaviour arises in that animals can more easily earn heat or cold if the required task is compatible with natural sequences of behaviour appropriate to temperature control.

Consideration of the site of temperature-detection in the body introduces problems common to the study of hunger. In both cases explanatory weight has traditionally been placed upon central detectors of physiological state, to be precise hypothalamic detectors. In both cases a disadvantage in exclusive reliance upon central detection becomes apparent, i.e. the hypothalamus is cushioned against disturbances. This means in both cases it would be relatively slow in instigating corrective responses. In the case of energy, the liver is at a location which 'knows' first that the gut is empty. By analogy, the skin is the first location to feel a change in environmental temperature. However, the evidence strongly suggests that both central and peripheral detectors are involved in temperature control. Peripheral temperature detectors give advance warning. If behavioural and/or autonomic action is taken, then the core temperature may be little affected by an environmental temperature stress. But if hypothalamic temperature does shift from optimum then powerful corrective reactions are instigated. Similar arguments concerning the necessity of fixed set-points for body weight and temperature exist. In both cases it is possible to argue that stability is attained without such a fixed reference. We have discussed several candidates for explanation of temperature control which are in the form of models. The evidence available at present does not allow us firmly to decide between them.

It is suggested that basically the processes of detection and integration have the same characteristics in the case of autonomic and behavioural temperature control. This is not to say that identical neural machinery is employed for both purposes, since lesions can selectively disrupt one or the other sub-system. It is probable that the same temperature-sensitive units are employed in both cases but that the integration operations are quite distinct. Evidence suggests that different weighting is given to central and peripheral information by the behavioural and autonomic controllers.

The most simple model of temperature control is one in which information from a population of temperature-sensitive units feeds into a neuron with appropriate weighting given to the information components. The output side of the neuron then determines both autonomic and behavioural temperature reactions. We have been forced to consider a more complex model in which behavioural and autonomic component reactions are separately computed in this way. However, it may be that this is still too simple, even though it provides a good provisional model of the processes and behaves like the biological system. For behaviour, there may be no unitary temperature command, but rather a series of commands each appropriate for one or more particular response. In the case of both the animal's autonomic and behavioural repertoire, lesions may disrupt one component temperature response but leave others intact (Roberts and

Martin, 1977). This is quite apart from the disruption of either the whole autonomic or behavioural set. In rats, extension and lowering of the body in response to heat appears to be mediated purely by temperature units in the hypothalamus, whereas grooming and other components are also under the control of peripheral units (Roberts and Martin, 1974). It may be possible to see adaptive significance in some responses being activated early, such as locomotion to a new environment, whereas others are 'last-hope' reactions, such as lying extended with minimal metabolic rate.

CHAPTER 8

Sex

8.1 INTRODUCTION

The behaviour described in this chapter is fundamentally different from the previous chapters. Up to now we have discussed the internal physiological state of the organism, i.e. fluid environment, temperature and energy state. The behavioural mechanisms described all serve to maintain the internal environment at an optimal state for survival. Sexual behaviour, by contrast, primarily serves not individual survival but gene perpetuation. That is not to say individual survival chances are unaffected by the organism participating in sexual activity. Pair-bonds and status may be confirmed by sexual behaviour. Conversely, the animal may place itself in a vulnerable position by such behaviour.

The primary purpose of sexual behaviour is of course fertilization. This is mainly a question of bringing the male sperm and the female ovum into contact, but for some species genital friction is also necessary for bringing the female into a hormonal state which allows fertilization. Some species, such as the rabbit, are what is known as reflex ovulators, and genital stimulation is necessary to cause ovulation. In the case of rats, repeated intromissions are needed in order (1) to facilitate sperm transport through the cervix, and (2) initiate a neuroendocrine reflex in the female so that she enters the so-called progestational state (Adler, 1969). Secretion of progesterone by this reflex induces implantation and allows pregnancy. If six or more intromissions precede ejaculation the possibility of pregnancy is very much higher than if only three or fewer do.

In this chapter several aspects of sexual behaviour are examined. We consider the effect which early hormonal environment has on the future course of the development of sexual behaviour, and also the way in which hormonal condition of the adult animal affects sexual potential. We will also take into account the sensory information which impinges upon the organism and arouses sexual activity. The actual mechanics of copulation will occupy our attention in so far as it reveals underlying processes, and finally we will need to examine the factors which terminate sexual behaviour.

8.2 THE PATTERN OF COPULATION IN RATS

Although the discussion will not exclusively concern itself with rats this species will occupy most of our attention. It is therefore necessary to describe in some detail the copulatory pattern of the rat. For a more detailed discussion the reader should consult the excellent review by Sachs and Barfield (1976).

Immediately following introduction of a sexually receptive female to a male there is a stage of reciprocal investigation. After a period of time, the length of which varies according to the state of each animal, the male mounts the female from the rear. He then makes a series of rapid pelvic thrusts, spaced some 10–40 msec apart. If, in the course of these thrusts, the penis does not achieve insertion into the vagina the behaviour is simply classed as a *mount*. The animal then dismounts, may very likely groom its penis, and then makes another mount attempt. *Intromission* is characterized by a relatively long deep thrust of duration between 200 and 400 msec, in contrast to the rapid shallow thrusts of a mount. In addition, other distinctive characteristics of intromission may be seen (Sachs and Barfield, 1976). *Mount latency* is the time from introduction of male and female to the first mount. This may or may not be the same thing as the intromission latency, depending upon whether intromission was attained on the first mount. *Intromission* latency is the time to the first intromission.

The rat executes a series of intromissions of between three and ten. The time-interval between intromissions is known as the interintromission interval. The series of intromissions is terminated by a particular kind of intromission, *ejaculation,* where sperm and a seminal plug are deposited in the vagina. Ejaculation is characterized by, amongst other things, a slower and deeper pelvic thrust than an intromission which is not accompanied by ejaculation.

We sometimes speak of mounting, intromission, and ejaculation in female rats. It hardly needs saying that by this we mean something rather different from what we have just described for the male. Females are sometimes observed to mount other females and perform pelvic thrusts characteristic of male mounting, intromission, and even ejaculation. To be more precise we usually employ the expression intromission response or ejaculatory response.

The normal female rat heterosexual role is not passive, contrary to what some earlier writers have tended to suggest or imply. In fact some investigators seem to have given no more attention to the female than researchers in thirst motivation have given to the water spout. Doty (1974) was motivated to call his paper 'A cry for the liberation of the female rodent: courtship and copulation in *Rodentia*'. After introduction of male and female, the female, if sexually receptive, makes responses such as ear-wiggling and darting around the cage. She finally arches her back concavely, a posture known as *lordosis,* to facilitate the male's intromissions.

8.3 HORMONES

8.3.1 General

A hormone is a specific chemical which is secreted into the blood at one site, carried by the blood to a distant site, and there affects a physiological process. The site affected may be, for example, the kidney, as in the case of anti-diuretic hormone (ADH), which we discussed earlier. In this chapter we will be concerned with how hormones affect sexual behaviour. There can be few, if any, areas of scientific investigation more intellectually demanding, and yet frustrating and irritating for student and lecturer alike. I will attempt, as far as is possible, to summarize the literature in an economical and logical fashion, but the student must be warned that God and/or evolution has left this area very effectively booby-trapped for the investigator!

Oestrus (a noun and spelt without an 'o' after the 'r') refers to hormonal and behavioural state of the female at the time when she is able to be fertilized. Not surprisingly, it is also the time when she is most active sexually, and indeed the name means gadfly and refers to hyperactive behaviour. In the case of rats, oestrus is a period lasting about 15 h, and occurs every 4–5 days. Heat is another expression for oestrus. *Oestrous* is the adjectival form of the word. The oestrous cycle refers to the cycle of hormonal activity of 4–5 days period. An *anoestrous* female is one at a stage of the oestrous cycle other than oestrus. The female is unable to be fertilized in this condition, and in a species such as the rat is behaviourally unreceptive to any advance by a male.

The term *oestrogen(s)* refers to a class of hormones secreted cyclically by the female ovary and which are responsible for inducing sexual receptivity. The word is sometimes in the plural because it refers to several distinct chemicals having very similar properties and secreted by the ovary. The word oestrogen(s) is often used either as though these different chemicals were really one, or to refer to one specific kind of oestrogen. *Oestradiol* is an example of an oestrogen secreted by the female. We will also need to refer to exogenous oestradiol and its effect on behaviour. *Progesterone* is another hormone secreted by the ovary (and adrenal) and exerting an effect on sexual behaviour.

The hormones secreted by the testes of the male and which influence sexual behaviour are known collectively as *androgen(s)*. An example of a specific androgen, the one with which we will be mainly occupied, is testosterone. Sometimes androgens are called the male sex hormones and oestrogens the female sex hormones. These expressions are somewhat unfortunate since they seem to imply that male and female sexual behaviour is under the control of androgens and oestrogens respectively, with no overlap. As we will see shortly the situation is very much more involved than this.

The processes which cause the adult female ovary to secrete oestrogen and progesterone are complex. Their secretion is determined by other hormones whose secretion is in turn controlled by yet another hormonal level. Secretion of oestrogen and progesterone is under the control of luteinizing hormone (LH) and follicle-stimulating hormone (FSH). These two hormones, known as gonado-tropins, are secreted by the pituitary gland at the base of the brain. We now believe that gonadotropin secretion is itself under hormonal control. Originally investigators spoke merely of a 'factor' which caused LH and FSH release. A substance present in the hypothalamus has been found to stimulate release of both LH and FSH from the pituitary; it is therefore called LH–RF/FSH–RF, or more simply, LH–RF because it is a more potent stimulus for the release of LH than of FSH. Although this releasing factor may be the physiological stimulus for the secretion of both gonadotropins, it is possible that another, as

160

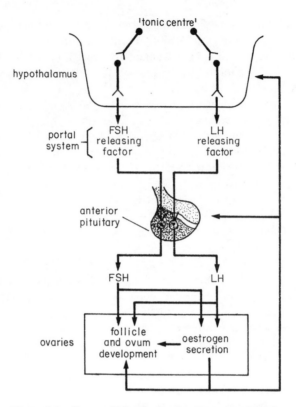

Figure 8.1 Some of the feedback effects involved in oestrogen secretion. FSH and LH secretion is under the control of FSH and LH-releasing factors. There is believed to be a 'tonic' effect which tends to excite release of the releasing factors. Acting to counter this excitatory effect is inhibition from oestrogen, hence one of the negative feedback effects. These interactions give rise to rising and falling levels of oestrogen, i.e. the oestrous cycle (From *Human Physiology—The Mechanisms of Body Function* by A. J. Vander, J. H. Sherman, and D. S. Luciano, Copyright © 1975. Used with permission of McGraw-Hill Book Company.)

yet uncharacterized, factor is responsible for the finer control of FSH secretion.

The secretion of oestrogen in the female is then under the control of the hypothalamus acting via intermediate hormones. Growth of the ovarian follicle is stimulated by FSH. Oestrogen secretion by the ovarian follicle is stimulated by LH and FSH. In turn, feedback effects exist such that oestrogen secretion influences LH secretion and also secretion of LH–RH/FSH–RH. The details of these feedback effects need not concern us; suffice it to say that both positive and negative feedback effects from oestrogen arise. This generates cycles of rising and falling oestrogen level, the oestrous cycle. Figure 8.1 shows these interactions.

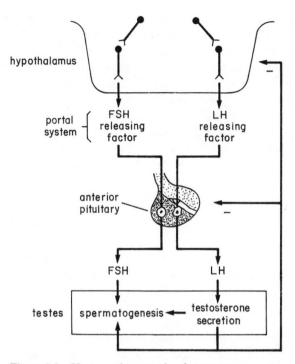

Figure 8.2 Hormonal control of testosterone secretion. Note the similarities with Figure 8.1, in that feedback pathways are present. (From *Human Physiology—The Mechanisms of Body Function* by A. J. Vander, J. H. Sherman, and D. S. Luciano, Copyright © 1975. Used with permission of McGraw-Hill Book Company.)

Figure 8.2 shows the interactions leading to testosterone secretion in the male (Vander, Sherman, and Luciano, 1975). Spermatogenesis is under the control of FSH and testosterone. Testosterone production is controlled by LH. If the anterior pituitary is removed testosterone secretion almost ceases. A negative feedback pathway between testosterone and LH exists. When testosterone concentration in the blood reaches a certain level it inhibits secretion of LH, both at the level of the pituitary and the hypothalamus. Although hormone levels (LH, FSH and testosterone) show some slight fluctuations (notably seasonal and circadian), compared to the female the whole system is relatively static during the course of adult life.

8.3.2 Hormonal control of the development of mechanisms underlying sexual behaviour

We can draw a distinction between two primary effects of the sex hormones on the organism. In the adult they are responsible for maintaining the sensitivity of neural structures so that external sexual stimuli are able to arouse copulation. In this context we are dealing with structures having a relatively fixed capacity for displaying certain forms of behaviour depending upon whether or not an adequate supply of hormone is present. In many cases if the hormonal environment is not appropriate in the adult, sexual behaviour cannot be aroused, but it reappears when the hormonal environment is restored. The sensitization of sexual behaviour by the hormone usually lasts only a very short time after the supply of hormone has been terminated (as in the case of non-human females), or at least shows a decline over a period of days or weeks (as in most males).

The other primary role of hormones is developmental, and, depending upon the species, occurs at around the time of birth; just before or just after. These effects are sometimes known as 'critical' or 'sensitive' period effects (Brain, 1979). By developmental we mean that the early hormonal environment gives a shape or bias to the future sexual potential and proclivity of the animal.

Masculine sexual behaviour includes mounting, pelvic thrusting, penile intromission, and ejaculation. Feminine sexual behaviour includes behaviour patterns which apparently attract or solicit the male's attention. When this is secured, the female places her body in a position which facilitates penile intromission. In the rat, soliciting behaviours include darting and hopping movements and also rapid vibration of the ears, known as 'ear-wiggling'. When these behaviours have attracted the male's attention, the receptive female responds to his mount by arching her back concavely, to facilitate intromission. As Beach (1975) notes, intact and untreated females of a variety of species occasionally attempt mounting of both homosexual and heterosexual partners. Just as in the case of the male, pelvic thrusts are made. When in oestrus these same females show normal receptivity to males. The reciprocal situation, presenting for mounting, is less common in normal males, but those male rats which have been observed to take the lordosis position also prove to be active copulators in the presence of receptive females (Beach, 1975). These results demonstrate that the potential for both masculine and feminine sexual behaviour is present in both sexes. We now believe that the early hormonal environment normally gives a bias towards masculine sexual behaviour in males and feminine sexual behaviour in females. Despite this bias it is still possible to observe behaviour associated with the opposite sex both in males and females, though of course it is usually less common than that associated with the genetic sex of the individual concerned.

We believe that the early hormonal environment of the male serves to masculinize and defeminize his sexual potential. These two processes are caused by the small amount of testosterone secreted by the prenatal or infant male. If these processes do not occur then a feminine bias results, which occurs in normal genetic females (McEwen, 1976). If a new-born male rat is castrated, thus preventing an early androgen influence, when adult it will exhibit a less strong masculine sexual bias than an intact control. In response to testosterone administered in adulthood it will mount, but intromission is less common than normal and the ejaculation response is not shown. Relatively strong feminine behaviour occurs in response to a combination of oestrogen and progesterone. The feminine response which occurs following such injection and in response to a stud male is less strong if the animal was castrated at 14 days of age rather than at birth, indicating that the masculine bias had already been partly established by then (see Beach and Buehler, 1977 for discussion). Given testosterone on day 4 of life, when adult a female rat will behave like a normal male in response to testosterone and in the presence of a receptive female. Beach (1975) investigated the response of sexual presentation by adult female dogs to active male dogs. It was almost totally abolished by androgen treatment at birth, even if oestrogen and progesterone were given just before the response test. Whereas no intact males showed sexual presentation following oestrogen and progesterone treatment, a significant percentage of males castrated at birth did so.

If male rats are castrated on the day of birth and, when adult, given daily testosterone injections the capacity to ejaculate is absent (Södersten and Hansen, 1978). The capacity to intromit is still present (but to a reduced extent) while mounting is still very strongly present. Thus an androgen source is not necessary in neonatal male rats for the future performance of mounting. Presumably a prenatal source is sufficient. If the same hormone treatment in adulthood is given to male rats which were castrated at 10 days of age, not only is mounting present, but intromissions occur in a large percentage of animals and the ejaculation response can be obtained in a large percentage of the sample. A source of androgen in the neonatal stage is normally necessary for the ejaculatory response to be shown by the animal. Above-normal doses of testosterone shortly after birth strengthen the ejaculatory potential over and above the level of the untreated male. Rats castrated on the day of birth and given large doses of testosterone until day 10, showed a greater ejaculatory capacity in response to testosterone when adult than animals castrated on day 10 and untreated up till then (Södersten and Hansen, 1978).

Beach (1975) collected a considerable amount of evidence to show that structures responsible for the organization of both masculine and feminine sexual behaviour are present in the brain of both sexes. Beach draws a very important distinction, which is commonly passed over; on a dimension of masculinity–femininity, loss of one aspect of behaviour is not necessarily associated with gain of the other. Androgenic stimulation at around the time of birth can cause a reduction in feminine behaviour, not necessarily associated with an increase in masculine behaviour. For example, female dogs given androgen treatment at birth, but not in utero, were strongly defeminized but masculinized little if at all. Females given androgen both in utero and at birth were strongly masculinized and defeminized. Tentatively Beach extrapolated to the belief that, by analogy, in genetic males feminization and demasculinization are two distinct processes. At one stage of prenatal development in female guinea-pigs injections of testosterone serve to inhibit ultimate expression of feminine behaviour without enhancing masculine behaviour, while the reverse is the case if injections are made at a slightly different stage in time (Goy and Goldfoot, 1975).

What is the effect at a neural level of the early hormonal environment which later reveals itself in the character of sexual behaviour? One is hesitant to volunteer suggestions, since as Beach (1975) warns us, all we know is the relationship between early hormonal gain or loss and the ultimate expression of behaviour. Two possibilities are to be found discussed in the literature. One is that hormones actually organize the underlying structure, forming circuits appropriate for later behaviour. The other alternative, and the more plausible, is that the neural circuits are already formed but their sensitivity to a future hormonal environment is predetermined by exposure or lack of exposure to an early hormonal environment. This is compatible with the evidence that circuits serving both aspects of behaviour are so clearly present in many animals. Brain lesions can be employed selectively to impair masculine sexual behaviour and leave feminine sexual behaviour intact, and vice versa (Beach, 1975; Singer, 1968).

The theme which is emerging in this chapter is that androgens enhance the potential for future masculine behaviour and inhibit potential future feminine behaviour. This is known as *androgenization*. Unfortunately at this point we must introduce what to many students is the most perplexing aspect of this subject. Injection of a so-called feminine sex hormone, oestradiol, is sometimes as potent as testosterone in giving a masculine bias to the behaviour potential. Indeed, under normal circumstances it appears to be oestradiol which bears responsibility for masculinization of the brain. Although this should not be regarded as absolutely established, it is reasonable on the basis of evidence at present available to give the following account of the process involved. Brain cells convert testosterone, secreted by the male testes, into oestradiol (a process known as *aromatization*) and dihydrotestosterone (McEwen, 1976; Plapinger and McEwen, 1978). In the case of the male rat either oestradiol alone or oestradiol and dihydrotestosterone together masculinize the brain shortly after birth. Why then does the female brain normally remain immune to the oestradiol which is naturally present to a certain extent in females? The answer may be as follows, McEwen (1976) describes a protection mechanism employing alpha-fetoprotein. This is an oestrogen-binding blood protein which is present in substantial quantities only during the critical period following birth. The oestradiol present in the blood of the female is therefore prevented from reaching the critical brain regions. In the case of testosterone this is not bound by alpha-fetoprotein, so in the male testosterone reaches the brain, is converted there to oestrogen and masculinizes the brain. Early androgen treatment of females, in a variety of species, increases the chances that they will perform mounting as adults (see Brain, 1979, for a review). Neonatal oestrogen treatment of females also tends to masculinize their adult behaviour. The ability of alpha-fetoprotein to protect the brain of the female is rather limited in the face of large exogenous doses of oestrogen. Confirmation of the masculinizing and defeminizing action of neonatal oestrogen was revealed by J. Booth (1977b). She observed the effect of neonatal injections of the anti-oestrogen MER-25 on adult male sexual behaviour. Rats were less likely to ejaculate in the presence of receptive females, and more likely to show lordosis in the presence of an active male. Booth tentatively suggested that neonatal MER-25 raised the threshold of sensory stimulation necessary to reach ejaculation.

J. Booth (1977b) found that injection of neonatal male rats with the anti-oestrogen MER-25 sometimes prevented suppression of feminine behaviour without affecting masculine behaviour. She argued that in the developing rat brain more oestrogen is needed to suppress feminine characteristics than to organize masculine characteristics. This is supported by the fact that the androgen androstenedione, given to male rats castrated on the day of birth, can prevent loss of adult masculine behaviour without suppressing lordosis (Stern, 1969).

Södersten and Hansen (1978) castrated a group of rats on the day of birth and gave them dihydrotestosterone (DHT) for the first 10 days of life. When adult, the rats were given daily injections of testosterone and compared with animals castrated on the day of birth and given no early treatment. Early DHT treatment allowed an increase in penis weight and development of cornified papillae on the glans penis in response to testosterone injections when adult. However, these rats did not show enhanced ejaculatory capacity when adult as a result of early DHT treatment. They did show an increased tendency to mount and intromit in comparison to untreated neonates (when adult both experimentals and controls were given testosterone). Another group of male rats were castrated on the day of birth, and given injections of oestradiol for the first 10 days of life. In response to injections of testosterone when adult they behaved differently to postnatally DHT-treated rats. An increase in ejaculatory capacity in response to testosterone injections when adult resulted. No penis sensitivity increase in adulthood in response to testosterone was shown by these rats over and above those untreated in the first 10 days of life.

Treatment of the neonate with a combination of oestradiol and DHT was more effective in allowing the ejaculatory response to be revealed in the adult by testosterone than was oestradiol treatment alone. Södersten and Hansen argued that the inhibition of the lordosis potential, as well as facilitation of the mounting potential in the brain of young male rats, are both dependent upon oestradiol formed in the brain from testosterone. This occurs in essentially the first 10 days of life. In fact neonatal treatment with oestradiol was found to be more potent than treatment with testosterone, in so far as the revelation of ejaculatory behaviour in the adult was concerned. These authors reinforce their argument by noting that neonatal anti-oestrogen treatment reduces ejaculatory capacity in adult male rats. They claim that those studies which found a disruption of mounting behaviour in males used abnormally large doses of oestrogen at the neonatal stage. J. Booth (1977a) found that the female sexual response of lordosis could be elicited easily in male rats castrated on the day of birth, if oestrogen and progesterone injections were given when the male was adult.

If male rats, castrated at birth, were injected on days 1–5 of life with testosterone, or especially if injected with synthetic oestrogen (RU 2858), lordosis in response to later doses of oestrogen and progesterone was suppressed. Booth found that for males castrated at birth to show ejaculation as adults when given testosterone and presented with a receptive female it was necessary to inject them at birth either with testosterone or a combination of oestrogen and DHT.

Either of the latter two substances on its own was inadequate, though the number of mounts observed was roughly the same for all treatments. As Booth notes, either of two possible explanation may be advanced. Ejaculation could depend upon a combination of the peripheral action of DHT and the central action of oestrogen. Alternatively the peripheral action could be solely the result of DHT, while centrally there could be a synergistic action of oestrogen and DHT (see also Larsson, Södersten and Beyer, 1973).

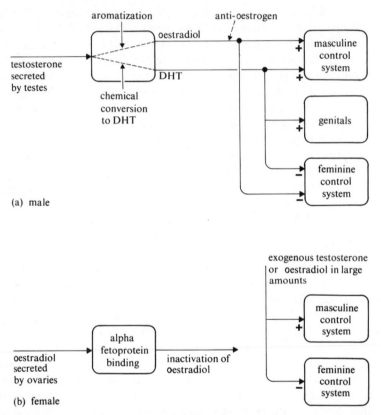

Figure 8.3 Summary diagram of hormonal factors relevant during developmental stages. (a) In the male, testosterone is converted to oestradiol and DHT. Either oestradiol alone, or in combination with DHT, sensitizes part of the brain's control system for masculine behaviour and desensitizes the feminine control system. If oestradiol is absent during this period, the feminine control system emerges as dominant at the expense of the masculine. DHT is shown to sensitize the genital surface. Injection of anti-oestrogens prevents oestradiol from performing its masculinizing and defeminizing role (or at least hinders it). (b) In the female, oestradiol secreted by the ovaries is inactivated by alpha-fetoprotein. This allows the feminine control system to assume dominance. If androgens (or excessive amounts of oestradiol) are given at around this time the female may be masculinized or defeminized, or both

In response to long-term treatment with oestrogen or by treatment with testosterone and electric shock, adult female rats, untreated when young, can be induced to show not only mounting of a receptive female but also intromission and the ejaculatory response (Barfield and Krieger, 1977). In many respects the intromission and ejaculation pattern of males and females appeared identical. The authors concluded that:

> '...the substrate for the execution of the ejaculatory pattern is developed in normal females, but that the responsiveness of the system is too low to allow expression under normal circumstances.'

Untreated females commonly show mounting and some intromission but, it was argued, fail to emit the ejaculatory pattern because intromissions are too few and infrequent. In response to hormone treatment intromissions become sufficiently regular to permit the ejaculatory threshold to be reached. Barfield and Krieger tentatively propose that perinatal androgenization which occurs normally in the male has the effect of strengthening the mechanism underlying intromissions.

I fully appreciate that all of this is a bombardment of facts which are not at all easy to assimilate, and therefore Figure 8.3 is a summary diagram of what is believed to be occurring at around the time of birth. The effect of hormones on the adult animal is discussed later, after a systems model has been developed.

8.4 THE ROLE OF EARLY EXPERIENCE

Sexual differentiation serves to establish a potential for future sexual expression, but experience up to the time of sexual maturity also plays an important role in determining behaviour. In some species mate-selection in the adult animal may be partly determined by an imprinting process shortly after birth (see Bateson, 1978). This allows a certain amount of flexibility since the animal need not be rigidly 'pre-wired' in terms of the stimulus which will ultimately elicit mating.

In many cases sexual expression is abnormal if the animal is raised from birth cut off from contact with other animals. A male rat raised in isolation may, instead of mounting from the rear, initially attempt to mount by climbing over a female from the side (see Diakow, 1974). This subject was recently reviewed by Larsson (1978), and the reader should consult this paper for further details. The present section is based upon Larsson's account. In both rat and rhesus monkey early social deprivation impairs the later appearance of sexual behaviour. However, whereas in the case of the rat the impairment can be overcome by later exposure to a receptive female the effect on masculine behaviour in the rhesus monkey is permanent. As Larsson notes, this failure to re-adapt is surprising in light of the greater brain capacity of the monkey. The failure of the male rhesus monkey raised in isolation to perform normal copulation is not a result of lowered sexual interest. Masturbation frequency is high in such animals. It is rather the case that the perceptual and motor coordination needed for successful copulation is absent.

In the case of the rat Larsson arrived at several conclusions:

1. Adult sexual activity is related to infant play activity. Social isolation deprives the animal of an important source of stimulation, and this impairs subsequent masculine activity.
2. Lordosis is not impaired by isolation, but the masculine response is affected in both sexes.
3. Castration is particularly destructive as far as sexual behaviour is concerned if the animal is sexually inexperienced at the time of the operation. It is much less harmful if the animal already has obtained sexual experience.

8.5 SEXUALLY MOTIVATED BEHAVIOUR IN THE MALE

8.5.1 Introduction

Superficially at least, sexual behaviour in the male would seem to be a candidate for analysis in terms of homeostasis. An analogy with feeding and drinking seems not unrealistic. One could imagine that a physiological displacement arises during a period of sexual abstinence and this is corrected by sexual behaviour. Ejaculation, or a series of ejaculations, leaves the male refractory to sexual stimuli. Following an adequate period of recovery the male becomes arousable to sexual stimuli again. What could possibly form a physiological displacement holding responsibility for the state of sexual arousability of the male? The view has sometimes been expressed that the accumulation of seminal fluids leads to a local build-up of pressure at the genitals and this arouses drive. Ejaculation lowers pressure, and with it drive also. However, it has been shown that removal of the seminal vesicles, the chamber holding seminal fluids prior to ejaculation, does not cause loss of sexual drive in animals (Beach and Wilson, 1963; Beach and Jordan, 1956). Further evidence also fits the interpretation that drive is not simply a function of the seminal fluid pressure. Thus, artificial ejaculation caused by electric shock applied either systemically or at brain sites appears to leave the animal motivationally unchanged (Vaughan and Fisher, 1962; Arvidson and Larsson, 1967; Beach, Westbrook, and Clemens 1966; Beach, Goldstein, and Jacoby, 1955). As Beach, Westbrook, and Clemens (1966) note, it emerges from the Kinsey survey that stimulation can cause orgasm in the absence of fluid discharge in pre-pubertal human males. Orgasm in the absence of seminal fluids appears to be rather similar to normal ejaculation in some cases (Singer, 1973, p.140). Drive can sometimes be very easily re-aroused in an otherwise sated male animal by a change of partner, a process which could hardly operate through a seminal fluid change.

So far no blood hormone or chemical has been found which fluctuates with the state of sexual arousability/exhaustion of the male animal. At present we are not in a position to explain sexual motivation in terms of chemical changes, though it would seem a reasonable guess that there does exist in the brain a biochemical correlate of the animal's state of arousability/exhaustion.

Some of the evidence which we do have on the nature of the system underlying sexual motivation can be shown in the form of a model, adapted by Toates and O'Rourke (1978) from an earlier model of Freeman and McFarland (1974). It is convenient to use such a model as a means of organizing the experimental literature, and this is the use to which it will be put here.

8.5.2 AROUSAL

The model is based upon the fact that stimuli arising from the female have the capacity to arouse sexual behaviour in the male. Whether or not the male is persuaded by these stimuli to mount and copulate depends upon the state of his nervous system. I find it useful to consider a term called *arousability*, which refers to the state of the male's nervous system in the context of sexual motivation.

Figure 8.4 illustrates the meaning of this. *Actual arousal* refers to a hypothesized state of neural activity somewhere in the male's nervous system which is the stimulus to copulate: it goads him to mount and intromit. *Potential arousal* refers to the summated total within the male of incoming sensory information from the female appropriate to sexual arousal. This information enters the male's nervous system through the tactile, visual, olfactory, taste, and auditory channels. Whether actual arousal occurs in response to potential arousal depends upon arousability. If arousability is high, for instance the male has had a long time to recover from previous sexual activity, then the chances that copulation will occur are high. Conversely, if the male is sexually exhausted then even a high level of potential arousal will not permit actual arousal. At times it is possible to discuss a trade-off between arousability and potential arousal. Some female animals have more power to arouse than others, and this is reflected in the male's recovery time.

This subject has been discussed by Konrad Lorenz (Lorenz, 1966). He describes an observation of Wallace Craig on ring doves. The female was removed from the male and the sexual arousability of the male tested following various deprivation intervals. A previously ignored white dove was courted after several days deprivation. Given a few days longer a stuffed pigeon aroused the male's interest, and then even a rolled-up cloth. Lorenz argued that deprivation has the effect of lowering the sexual arousal threshold. In the terms of Lorenz, action specific energy builds up as a function of deprivation and so a lowering of discrimination is shown, as almost any stimulus becomes adequate. Although the model developed here would not use the metaphor of energy, increasing

Figure 8.4 Actual arousal depends upon potential arousal and arousability

arousability with deprivation gives rise to very similar predictions (see also Walton, 1950).

What is the physiological embodiment of the term arousability? We have seen that it is not necessarily reflected in the state of seminal fluid volume or pressure. At present we have no answer, but an intelligent guess might be that a neural circuit or pathway changes its sensitivity. This change in transmission characteristic is reflected in arousability to sexual stimuli. Possibly synaptic transmission at certain sites is blocked immediately following ejaculation(s), and slowly recovers with time. Seward (1956) uses the expression 'short-circuiting' of the sexual arousal pathways to describe the sexual fatigue which occurs following ejaculation(s). This accords with the theoretical approach being adopted here.

8.5.3 Sexual arousal and the pattern of copulation

At this stage it is useful to take another look at the pattern of copulation of the male rat. Earlier we discussed what happened up to the time of the first ejaculation. Following this ejaculation there is what is known as a postejaculatory refractory period of about 5min. The male then starts to intromit again. The second ejaculation is typically obtained with fewer intromissions than the first. There then follow up to seven or so further ejaculations, the number depending upon how long the male has had to recover from the previous sexual encounter. The male then shows satiety; by this is meant that it does not attempt further intromissions.

Although individual quantitative differences between animals are enormous, a number of qualitatively regular characteristics can be used as indices of arousability. This is based upon the observation that there is a correlation between these various measures. A rat deprived of sexual contact for a relatively

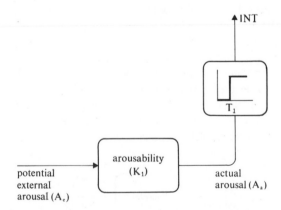

Figure 8.5 Block diagram representation of the dependence of intromission (INT) upon actual arousal (A_a). Actual arousal depends upon potential external arousal (A_e) and arousability (K_1)

long period of time will, apart from showing relatively little discrimination, exhibit the following characteristics:

1. the latency to intromission is short;
2. relatively few intromissions are needed to obtain the first ejaculation;
3. relatively many ejaculations are needed before satiety appears.

With reference to (1), Toates and O'Rourke (1978) proposed the following explanation. Only if the summation of the incoming signals from the female exceeds a copulatory threshold will the male initiate an intromission. In Figure 8.5 T_1 represents the copulatory threshold with INT as its output; INT takes a non-zero value when the animal performs an intromission. At all other times it is of course zero. If the arousability is high then even a low value of potential external arousal may be adequate to reach T_1. If arousability is not so high then only a relatively high value of potential external arousal (given the label A_e) will enable T_1 to be exceeded. Toates and O'Rourke argued that A_e is dependent upon the behaviour of the female. If the male is very arousable then even a static female may hold sufficient arousal value to cause a mounting attempt and possible intromission. By contrast if the male is fairly exhausted then only a very arousing response on the part of the female will provoke mounting. Some responses of the female such as ear-wiggling, darting, etc. have a high arousal value for the male. At any point in time there is a probability of the female emitting such a response and the male being at an optimal location to take advantage of this. If the male is sexually fatigued a long period of time may need to elapse before the combination of female response/location and male location is adequate. Beach (1956) found that intromission latency varied from 18sec (following a 15-day recovery from sexual exhaustion) to 3600sec (1 day of recovery).

Figure 8.6 shows a typical male rat sexual response. It is characterized by a number of intromissions and ejaculations. As the length of time allowed for recovery from the previous sexual response increases so does the number of ejaculations shown before satiety appears. Given 10 days of recovery then, say, 8 ejaculations may result, whereas only, say, 3 would follow 2 days of recovery. By the measure of number of ejaculations the male only recovers fully by about 10–14 days (Beach and Jordan, 1956). A fully recovered rat may need only 4 intromissions to obtain the first ejaculation whereas one with only 1 day of recovery might take 10.

Toates and O'Rourke (1978) argued that sexual exhaustion which follows a series of ejaculations is represented by a loss of central arousability. This loss occurs progressively throughout the series. In terms of Figure 8.5, arousability is lowered slightly by each ejaculation. The term K_1 will be used to represent arousability.

Use of the expression *sexual exhaustion* is common in the literature, but is unfortunately used at times almost synonymously with general physical exhaustion. This is misleading since in terms of general fatigue an animal could hardly take 14 days to recover fully from copulation. It must be a specific sexual

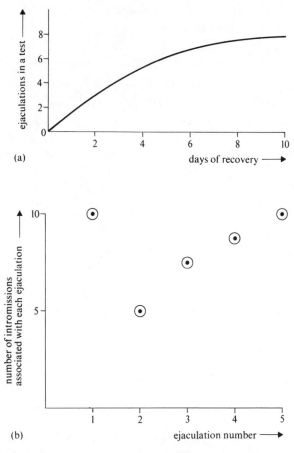

Figure 8.6 (a) Number of ejaculations shown in a test as function of the time allowed for recovery before the test. (b) A test in which five ejaculations were shown. The number of intromissions associated with each of the five ejaculations in the series is shown. In this case the first ejaculation needed ten intromissions. If a longer recovery period had been allowed then, say, only eight intromissions may have been needed for the first ejaculation, two for the second and so on. Also, more ejaculations would have been shown before the animal satiated

arousability mechanism that is 'exhausted'. McGill (1965) notes that in various strains of mice, length of recovery bears no relation to effort exerted in copulation. Some strains take 1 hr while others take 4 days to reach the same criterion of recovery. In rats, sleep following ejaculation has different electrical characteristics from sleep following general exhaustion (Boland and Dewsbury, 1971).

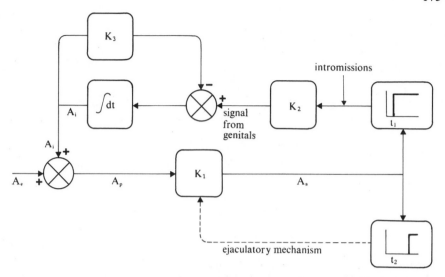

Figure 8.7 Model of the controller of male rat sexual behaviour. Potential arousal (A_p) is multiplied by arousability (K_1) to give actual arousal (A_a). If this exceeds t_1 the animal intromits, i.e. INT takes the form of a series of pulses. K_2 relates INT to the signal generated by genital friction (N). N is integrated to give internal potential arousal (A_i). A_i adds to A_e to give total potential arousal (A_p). Internal arousal dissipates unless the animal continues to make intromissions. The rate of this loss of internal arousal is determined by the term K_3, hence the negative feedback effect around the integrator. When A_a exceeds t_2 the animal ejaculates, and K_1 is reduced as a consequence of ejaculation (the dotted line). (Based upon Toates and O'Rourke, 1978)

We believe that each intromission sets up a neural signal at the male's genitals which travels to the brain. Figure 8.7 shows the full model. K_2 relates the intromission to the genital signal generated (its magnitude reflects genital sensitivity). It is an assumption throughout the literature in this area that each intromission contributes a certain increment to the state of activity of some internal neural process. After a sufficient number of intromissions have been performed this excitation is at a level which triggers ejaculation. However, if intromissions are very widely spaced it has been noticed that animals are unable to ejaculate. This may be shown by the experimenter removing one animal for a period of time, say 15min, after each intromission. It is believed that the stored internal excitation which accumulates with each intromission has a natural tendency to dissipate with time. Only fairly closely spaced intromissions enable the *ejaculatory threshold* to be reached. However, if intromissions are slightly less frequent (e.g. one per 90 sec) than the animal's natural *ad lib* rate (e.g. one per 60 sec) then ejaculation is obtained with a minimum of intromissions (Sachs and Barfield, 1976). For instance ejaculation may require 8 intromissions when the animal spaces them 60 sec apart, but only 5 if the experimenter forces them to be 90 sec apart. Figure 8.8 illustrates what we believe to be happening. The internal level of neural excitation is shown as a function of time. Each intromission contributes

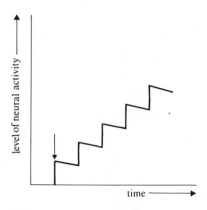

Figure 8.8 Level of internal arousal or excitation (A_i in Figure 8.7) as a function of time. Arrow indicates first intromission. Each intromission adds an increment to internal arousal. Between intromissions there is a fall of arousal

a component of excitation, and between intromissions there is a slight loss of excitation.

The neural impulse arising from each intromission is integrated to give a level of excitation (see Figure 8.7). The negative feedback pathway around the integrator means that in time the level of neural excitation will decay unless further excitation is provided. Following Freeman and McFarland (1974), Toates and O'Rourke (1978) argued that the internal neural excitation which is generated by intromissions adds to the level arising from external stimulation to give a total potential arousal. Thus we have an example of positive feedback: external stimuli arouse the animal to copulate, copulation increases arousal which potentiates still further the tendency of the animal to copulate. In Figure 8.7 it may be seen that internal potential arousal adds to external potential arousal. The strength of this combination initiates intromissions when it exceeds T_1 and ejaculation when it exceeds T_2.

Why does enforced spacing of the interintromission interval to a value a little longer than the animal's natural interval mean that fewer intromissions precede ejaculation? Several possible explanations may be discussed (Sachs and Barfield, 1976), but we (Toates and O'Rourke, 1978) prefer the following one. The sensitivity of the system (K_1 and K_2) may be lowered by each intromission and fully recover in a minute or two after the intromission. It could be the case that recovery is not complete at the time when the animal voluntarily resumes intromission. If a little more time is allowed the signal arising from intromission may be greater than under *ad lib* copulation conditions.

In Figure 8.6 it may be seen that the second and third ejaculation in the series is obtained more easily than the first, i.e. with fewer intromissions. This may seem to contradict the claim made earlier that each ejaculation lowers central arousability. Clearly if the second and third ejaculation are obtained with lower arousability there must be an additional facilitatory factor at work. It seems that there is a carry-over of stored excitation from the first ejaculation, some of which is available 'gratis' to the next ejaculation. That arousal is high following the first ejaculation is suggested by the observation that other males are likely to be mounted at this time, if the test male is returned to the home-cage (Ford and Beach, 1952, p. 148). It is reasonable to suppose that by the third or so ejaculation loss of arousability is so great that even the carry-over of stored excitation is inadequate to prevent ejaculation from becoming more and more difficult. An increasing number of intromissions is needed and hence the U-shaped curve.

8.5.4 Hormones and copulation in the male

Toates and O'Rourke (1978) argued that with the help of a model it is useful to focus upon two primary sites of hormone action, though this by no means excludes other sites. The hormonal environment of the infant animal (described earlier) determines whether or not androgens are able to exert the effects which we are about to describe. We will be working on the assumption that a supply of androgen was present at the neonatal stage so that structures were given the potential to be responsive to androgens in the adult. First, the strength of the central arousability of the nervous system to sexual stimuli is dependent upon an androgen supply in the adult. As Beach (1971) so clearly expresses it, androgens serve to lower the threshold for the elicitation of sexual behaviour. The preference of the intact male for the odour of an oestrous female is lost following castration, this being a motivational change and not a loss of sensory discrimination capacity (Carr and Caul, 1962). Castrated males spend less time orientated towards females (Hetta and Meyerson, 1978). In some cases it has been found more difficult to arouse mating attempts in the castrated animal, but there are also reports that frequency of mounting in castrated rats is as high as the intact ones (Sachs and Barfield, 1976). What is beyond dispute is that it is much more difficult to obtain intromission and particularly the ejaculation response from castrated male rats. Following castration in rats the ejaculation response is the first aspect of sexual behaviour to be lost, followed by intromission, and lastly, if at all, loss of mounting (Beach and Holz, 1946). Conversely, when androgens are restored to castrated male rats, at low doses only mounting is observed but increasing the dose allows intromission. Still higher doses permit ejaculation (Beach, 1948, p. 272). Failure of the ejaculatory response could be due to a loss of central arousability which means arousal is never able to reach the ejaculatory threshold. As we shall see in a moment the two sensitivity parameters K_1 and K_2 are androgen-dependent, so it would not be surprising if arousal could never reach the ejaculatory threshold. In addition failure to achieve intromission as reliably as in intact animals could be due to androgen-sensitive spinal reflexes associated with erection (Hart, 1978).

In some strains of mice adult castration has a hardly noticeable effect (McGill and Manning, 1976; Thompson, McGill, McIntosh and Manning, 1976), and we can only conclude that the pre-castration supply of androgen served as a catalyst and this resulted in a lasting effect on central arousability.

In general, the so-called higher animals such as man and apes suffer less from castration than the so-called lower animals such as the rat (Beach, 1947), though this should not be used as too broad a generalization.

Restoration of androgen in appropriate doses to castrated animals leads to a return of sexual arousability, indicating that no irreversible structural change took place following hormonal withdrawal. In seasonal breeders drive level can be correlated with blood androgen level.

Individual differences in the intensity of sexual activity in adult male rats cannot be correlated with differences in testosterone levels (Larsson, 1966). Above a certain minimum level of hormone sexual activity seems to be unrelated to adult hormone level. Male rats showing various degrees of sexual activity were castrated and post-castration copulation compared to pre-castration performance. 'High-active' males remained active following castration, whereas 'low-active' males ceased sexual activity rather soon. Subsequently they were all injected with large doses of testosterone, and all returned to their pre-castration level of sexual activity. Persistence of individual differences in the strength of sexual behaviour despite castration and subsequent injection of various doses of testosterone was demonstrated for the guinea-pig by Grunt and Young (1953). Södersten and Hansen (1978) suggest that differences in the strength of sexual behaviour between adult males depends upon different magnitudes of early sexual differentiation caused by testosterone levels. One might say that in the adult animal hormones reveal a predetermined behaviour potential. However, it is also the case that graded effects of androgen may be observed under some conditions. Beach (1948) reports that if males are castrated and given relatively low androgen doses sexual behaviour is directed only at an oestrous female of the same species. Larger doses cause discrimination to fall, and attempts may be made to copulate with animals of other species.

Peripherally, androgen influences the surface structure of the penis (Aronson and Cooper, 1968). Following castration in rats the penis loses its rough surface structure because of involution of penile spines. This loss of surface texture means that genital friction is less easy to generate, even if the male achieves an intromission. Both the central and peripheral effects of castration, i.e. reduction in central arousability (gain K_1) and genital sensitivity (gain K_2) would be expected to make ejaculation more difficult to attain.

As we discussed earlier, testosterone is converted to oestradiol and dihydrotestosterone (DHT) in the organism. It appears that oestradiol is responsible for maintaining central arousability, while DHT maintains genital· sensitivity (though see Brain (1979) for some words of reservation). Species differences are important. It is possible to arouse copulation in castrated males of some species by DHT administration (see Hutchison, 1978 and Feder, 1978 for a review). It is possible that oestrogen and dihydrotestosterone act synergistically to maintain central arousability in the adult animal, as well as to masculinize the infant. Some support for the idea that oestradiol and DHT have a synergistic effect on central mechanisms was provided by Baum and Vreeburg (1973). Male rats castrated when adult and injected with both substances showed considerably more mounts with pelvic thrusts than animals injected with only oestradiol. Baum and Vreeburg argued that this was unlikely to be due to a peripheral effect of DHT, though it could have been, since mounting may have been more rewarding if the genitals were more sensitive.

What exactly is the hormone(s) responsible for central arousability doing? At one level it provides us with a description to say that the threshold for the elicitation of sexual behaviour is lowered by the hormone. However, it may be worth looking slightly closer than this. Certain neural circuits in the brain relate incoming sensory information to the executive brain regions responsible for initiating pursuit of the female and mounting, etc. It appears that the neurons in these brain regions have their excitability changed by hormones taken up from the blood. We imagine that neural circuits which would otherwise be relatively difficult to activate become excitable because of the influence of hormones. We will have more to say on this subject in the section on female sexual behaviour.

8.5.5 Novelty and the Coolidge effect

Although definitions differ slightly, the Coolidge effect refers to the observation that animals may often be particularly aroused sexually by changing their partner. The name derives from the former U.S. president, and the remarks he is supposed to have made on a visit to a poultry station. It is said that the president and first lady were taken around the farm in two different groups. When being shown a particularly sexually active male, Mrs. Coolidge asked 'does he perform like that all day?' The technician replied that this particular cockerel did indeed perform for long periods at this intensity, and Mrs. Coolidge asked that this should be pointed out to President Coolidge when he came to this point on the visit. When the president arrived the technician mentioned the instructions from Mrs. Coolidge, and pointed out the particular male. President Coolidge asked, 'but do they regularly change the female?' The technician answered that this was necessary. 'Well, don't forget to tell that to Mrs. Coolidge!', replied the president. The Coolidge effect usually is discussed in terms of males, but this does not imply a male chauvinistic bias on the part of the investigator, though the story of the origin of the name may also suggest this.

The Coolidge effect specifically refers to arousal through novelty, but it is not always easy to devise a sufficiently well controlled experiment to reveal an uncontaminated effect. If one removes a female, replaces her with a novel female, and finds that an apparently sated male is now rearoused, this does not prove it is novelty which was responsible. Being sexually fresh, the novel female possible behaves in a more stimulating way than the original female. In order to control for this factor experimenters have tried using novel females which had recently been mated with other males, and such experiments showed a slight Coolidge effect in rats (Wilson, Kuehn, and Beach, 1963). Tiefer (1969) found that if a male rat has simultaneous access to 5 oestrous females it will divide its intromissions and ejaculations amongst them. However, it shows exhaustion after the same number of ejaculations as a control rat having access to only one oestrous female. It seems then that in the male the speed of onset of sexual fatigue is not dependent upon the female(s) present, but once fatigue has occurred the novelty of a subsequent partner plays a role in recovery.

An inevitable complication bedevils interpretation of experimental results in this area. If we give symbols to the contributing factors it enables the issue to be brought into clear focus. Let A represent the arousal induced in the male by the female at the end of copulation. Let a fresh and novel female have arousal value A + B + C, where B is the contribution to the male's arousal due to novelty *per se* and C is the contribution attributable to the fact that the new female has not recently mated. C represents the effect of her arousing behaviour over and above A. We wish to eliminate factor C from this particular study, in order to isolate novelty, and so we present a novel female which has recently mated with another male. We imagine that any rearousal of the male represents a pure Coolidge effect. It could well be the case that it is the novelty of the new female, but it could be that the female is particularly aroused by the novel male and behaves in a particularly arousing way towards him. Conversely, suppose that this procedure elicited no more rearousal than simply removing the original female and replacing her. Could we conclude that no Coolidge effect exists in the male? The evidence would be compatible with such an interpretation, but it would also equally fit two others. There could be a positive Coolidge effect in the male which is cancelled by a negative Coolidge effect in the female. A partner-change may have a negative effect upon her arousal. Alternatively, the fact that she has recently mated with another male may be detectable to the new male and may be distracting. Admittedly, these possibilities seem remote, but in rats we are dealing with a rather small effect.

Some evidence that males may be more susceptible to the effect of novelty than females was given by Krames (1970). Male rats with a polygamous sex history showed a preference for the odour of a novel female (compared to that of a female with which they had recently copulated, the odour being taken before copulation). Female rats having a polygamous history did not show a preference for a novel male's odour.

When we consider certain other species the Coolidge effect is very apparent, and the results less open to alternative explanations. The Coolidge effect is less strong in species such as the rat, guinea-pig and mouse than it is in ungulates such as the bull and ram (Cherney and Bermant, 1970; Schein and Hale, 1965). The effect appears to be most strong in species where the male normally associates with several females, and may have adaptive value in spreading the male's genes more widely.

In an experiment by Michael and Zumpe (1978) it was found that male rhesus monkeys having regular sexual contact with four different females declined in potency over a two-year period. Number of ejaculations in a test session declined and latency to the first mount increased. This was reversed when novel females were presented. However, the revival of potency did not last when the original females were returned. One cannot, though, attribute the effect entirely to the male since it is of course male/female interaction, but Michael and Zumpe could see no difference in the behaviour of the old and novel females towards the males.

It is necessary to consider how we might relate the Coolidge effect to the model shown in Figure 8.7. Possibly the decrease in gain K_1 which accompanies ejaculation is, to some extent, specific to the partner in whose presence the male was desensitized, though as we have seen, speed of desensitization is not a function of partner(s). If incoming sensory information from the partner is different to some extent from the partner in whose presence the animal was sexually exhausted then this evokes a more powerful reaction. We are reminded of the ubiquitous phenomenon of *habituation*.

8.6 FEMALE SEXUAL BEHAVIOUR

8.6.1 Introduction

The female has an active role to play in the initiation of copulation. In the case of the rat this ranges from making responses which excite the male, such as ear-wiggling and darting movements, to the more obvious responses of investigating the males' genitals. As Beach (1976) reminds us, the oestrous female rat is not only most attracted to the male but is at the same time most attractive to him. In some species the hyperactivity of oestrous females makes them more conspicuous to males and increases the probability of making contact with males. In the case of rats little male–female interaction occurs outside oestrus, but when the female is in oestrus mutual investigation is maximal. Beach uses the term *proceptivity* to describe the 'appetitive activities shown by females in response to males'. In the case of some responses there is not a clear distinction between appetitive and consummatory behaviour. Lordosis may serve to arouse an inactive male or it may facilitate intromission. In a variety of species when the female is in oestrus, tactile stimulation of the perineum or of the flanks and back causes her to take a position for mating. Neural signals coming from the animal's body surface and arising from contact with the male are able to influence motor reactions which put the female in a lordosis position. Intermediate neurons are sensitized by hormones. Thus these hormone-sensitive neurons are, in the present context, a switching mechanism, allowing or not allowing transmission. The switch is on when hormones have influenced it (McEwen, 1976). It is possible to trade-off oestrogen level and magnitude of sensory input in so far as the lordosis response is concerned (Diakow, 1974). In ovariectomized rats, given low doses of oestrogen, the probability of lordosis occurring increases with graded increases in the level of sensory input, e.g. light scratching of the back, flanks and perineum (10% of females responded), pressure applied to the same regions (40% responded), and pressure applied there and to the cervix (100% responded). With constant pressure the percentage of females showing lordosis increased as a function of oestrogen dose.

In summary, then, the hormonal environment of the female at oestrus arouses her to make solicitous advances at the male and respond positively to mounting by, in the case of the rat, performing lordosis. What terminates this sexually active phase of behaviour? Ultimately the sex hormones will induce a state whereby the female is either refractory to stimulation or actively hostile to the advances of the male. Hormones will have this effect irrespective of whether the female has mated or not. However, if the female mates then the effect of genital stimulation is often to make the female refractory. For at least some species there is evidence that genital stimulation and the hormonal environment towards the end of oestrus may act synergistically to inhibit further sexual behaviour. So we have either a termination of oestrus caused purely by the action of hormones on the nervous system or a somewhat quicker termination brought about by the additional factor of genital stimulation.

8.6.2 SEXUAL BEHAVIOUR IN NON-PRIMATE FEMALES

8.6.2.1 Hormones

Figure 8.9 shows the concentration of oestrogen and progesterone during the course of the oestrous cycle in female rats. It may be seen that oestrogen peaks some 12 h before the onset of behavioural oestrus. The peak in progesterone is much nearer to behavioural oestrus. In this species receptivity lasts roughly 14 h. It is generally believed that oestrogen serves to condition the tissues underlying sexual behaviour and that progesterone further acts upon this conditioned tissue to induce receptivity (Joslyn, Feder, and Goy, 1971). We speak of oestrogen and progesterone acting synergistically to induce receptivity. In the case of ovariecto-mized guinea-pigs to induce lordosis the optimal interval by which progesterone injection should follow oestrogen injection is 24–48 h (see Joslyn, Feder, and Goy, 1971; Feder, Landau, Marrone, and Walker, 1977). A combination of progesterone and oestrogen is more powerful in stimulating lordosis than oestrogen alone. However, extended treatment with oestrogen alone is capable of inducing receptivity in ovariectomized rats (Powers, 1970). It is easier to obtain, though, with the assistance of progesterone, and without it then above-physiological levels of oestrogen must be given to induce lordosis (Powers, 1970). Androgens are also able to facilitate lordosis, presumably by conversion in the brain to oestrogen (Feder, 1978).

What hormonal event or hormonally induced change terminates the period of sexual receptivity? Powers (1970) presents two possible alternative answers. Progesterone could have a bi-phasic effect in that it at first stimulates receptivity but some hours later reverses its role and inhibits receptivity. Alternatively the end of sexual receptivity could be due to loss of the oestrogen conditioning effect. It was argued that the latter alternative better fits the evidence. However, Morin (1977) reviews evidence which suggests a bi-phasic action of progesterone, excitatory and then inhibitory. By means of either loss of a hormonal effect or the appearance of a different hormonal effect receptivity will be lost irrespective of

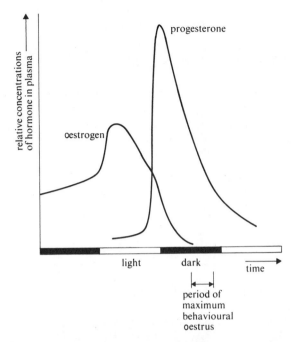

Figure 8.9 Diagram showing very roughly the concentrations of oestrogen and progesterone in the plasma of a female rat during 48 h. It may be seen that oestrogen peaks before progesterone, and behavioural oestrus follows soon after the progesterone peak. (Reprinted with permission from *Physiology and Behaviour*, **5**, J. B. Powers, 'Hormonal control of sexual receptivity during the estrous cycle of the rat', Copyright 1970, Pergamon Press, Ltd.)

whether the animal copulates or not. In addition, copulation possibly acting via progesterone-sensitive pathways can advance the onset of the sexually un-receptive state.

In the case of the guinea-pig, it was argued by Feder, Landau, Marrone, and Walker (1977) that progesterone has two quite distinct effects, first an excitatory effect in synergy with oestrogen and later an inhibitory effect on lordosis. They mention the possibility that distinct excitatory and inhibitory brain regions may be involved, though at present we have no firm evidence for this. Whatever the exact nature of the mechanism, its likely effect may be illustrated as in Figure 8.10. The excitatory effect is very rapid after the rise in blood progesterone level. Following progesterone injections to oestrogen-primed guinea-pigs receptivity appears in 4–5 h (Feder *et al.*, 1977). The inhibitory effect is slower and of longer duration.

Parts of the brain, notably the hypothalamus, are selectively able to take up oestradiol from the blood (McEwen, 1976). Behavioural receptivity occurs some 20 h after an oestradiol injection in ovariectomized female rats, which is roughly the time between the peak of oestradiol secretion and receptivity in normal

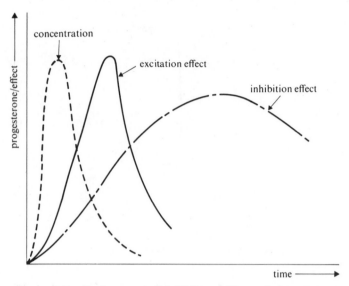

Figure 8.10 Excitatory and inhibitory effects of progesterone. Note the excitatory effect following soon after the peak of progesterone concentration, and the later rise and fall of the inhibitory effect

females. It appears that soon after peaking/injection (0–2h) maximum uptake of oestradiol occurs. Oestradiol need no longer be present in these brain regions at the time of receptivity. Rather, it appears to initiate actions—which in synergy with progesterone provoke receptivity (McEwen, 1976).

8.6.2.2 The effects of genital stimulation

Pierce and Nuttal (1961) developed a piece of apparatus in which the female rat was able to pace her own copulation. She could move freely between a compartment containing males and one where no males were present. It was observed that following every ejaculation the female entered the escape compartment, and that 95 per cent of intromissions were followed by escape from the compartment where males were housed. Evidence from Bermant and Westbrook (1966) also suggests that neural activity derived from genital contact briefly inhibits further sexual activity in the female rat. A female had been trained to bar-press for the reward of re-establishing contact with a male. The male was removed following either mounting, intromission, or ejaculation. Latency to initiate bar-pressing increased in the ascending order of (1) mount, (2) intromission, and (3) ejaculation. The latency following ejaculation could have been longest because of either the distinctive dynamics of this response on the part of the male, or because of the arrival in the female of the seminal discharge. Males were therefore injected with a substance that prevented seminal discharge.

The female's bar-pressing latency was still longer than that following intromission but not quite as long as following normal ejaculation. It may be the case then that genital contact and the presence of the seminal discharge both serve to inhibit temporarily further sexual contact.

Hardy and DeBold (1971) established that a large number of intromissions is followed by a diminution in the probability of further lordosis. This diminution was not observed in female rats wearing vaginal masks although numerous attempts at mounting were made by males. Hardy and DeBold (1972) report that stimulation of the vagina causes firing of units in the hypothalamus and a sleep-like EEG after-reaction. Towards the end of oestrus it is harder to elicit lordosis by manual stimulation if mating has occurred at the beginning of oestrus. Further, the period of sexually receptive behaviour has a relatively long duration if intromission is prevented by the female wearing a vaginal mask.

Female rats can be observed to offer violent resistance to further advances by males following a period of sexual activity. Tissue damage can follow an extensive period of copulation (Hardy and DeBold, 1972). Hardy and DeBold argued that hormonal change and the effects of intromission acting in concert terminate behavioural oestrus. Following a series of intromissions, both intact females in oestrus and ovariectomized females brought into heat by injections of oestrogen and progesterone show rejection of further intromissions. It is unlikely that nature would normally employ pathological means to achieve a behavioural effect. Vaginal tissue damage probably represents an extreme case. Before damage occurs, it would seem that information derived from genital contact serves to desensitize the female and makes her unwilling to respond to the advances of males. If contact has not occurred then receptivity and proceptivity will be ended by reversal of hormonal sensitization.

Goldfoot and Goy (1970) working with guinea-pigs found that behavioural oestrus, as measured by the tendency of the female to exhibit lordosis, is shortened by the mechanical stimulation of coitus. The abbreviation of receptivity was not prevented by administration of excess oestrogen and progesterone or by ovariectomy. Exogenous pituitary hormones also had no effect. The authors argue for an afferent neural inhibitory process, independent of hormonal mediation. If this source of inhibition is not present, that is to say the female remains unmated, then a hormonally mediated termination of receptivity is seen. Goldfoot and Goy note that male guinea-pigs perform a single ejaculation and then show a refractory period of an hour or more. Unlike rats, guinea-pigs do not require multiple intromissions to secure fertility. It was suggested that termination of the highly conspicuous oestrous running pattern of the female guinea-pig after probable fertilization could be of survival value.

In a study carried out by Carter, Landauer, Tierney, and Jones (1976) on Golden hamsters the influence of copulation and progesterone on the termination of behavioural receptivity was examined. Ovariectomized females brought into sexual receptivity by injection of oestrogen alone showed an inhibition of receptivity following mating, but a partial recovery appeared within 24 h. If oestrogen was followed by progesterone and the same exposure to mating allowed, then the inhibition of receptivity was considerably longer; it lasted at least nine days after copulation. Carter *et al.* suggest that:

> 'Prolonged exposure to progesterone may mediate the long-term storage of mating effects in the female hamster.'

Progesterone and copulation seem to act synergistically to terminate receptivity in this species, and probably others as well.

8.6.3 Sexual behaviour in primate females

Construction of this section has been particularly assisted by the review of Keverne (1976) to which the reader should turn for further particulars.

Whereas rodent females are critically dependent upon ovarian hormones, primate females appear to show less dependence. This is evidenced by the fact that they show less clearly defined cycles of receptivity than rodents. Although there are periods of maximal receptivity, primate females engage in coitus to some extent throughout the ovarian cycle (though both within and between-species differences are considerable). Furthermore, their sexual behaviour is not so sharply attenuated by ovariectomy as in non-primates. This led Beach (1947), amongst others, to propose that females of the so-called higher species had achieved emancipation from the immediate control of hormones. It was argued that such emancipation was one aspect of behaviour associated with a more sophisticated brain, in which higher regions have taken over more responsibility for behaviour from the hormonally sensitive lower regions.

In the case of laboratory-housed animals, the mounts, intromissions, and ejaculations made by the male chimpanzee, rhesus monkey, and pigtail macaque show a maximum frequency at the time of the female's ovulation. Again, as in the case of the rat, we are dealing with a very complex interaction between two organisms in which hormonal state of either male or female can affect both male and female. The terms *attractiveness* and *receptivity* were introduced by investigators in an attempt to give some demarcation to this area. The sexual motivation of the female may be measured by the number of sexual invitations that she makes and by her willingness to respond to male mounts. Rate of lever-pressing by the female to gain access to the male is another indication. These are generally spoken of in terms of receptivity and, as Keverne notes, they also fit Beach's (1976) term *proceptivity* — active rather than passive involvement. Attractiveness can be defined in terms of the ability of the female to stimulate the male to mount. These terms serve a certain descriptive purpose but of course should not be thought of as referring to distinct categories. Thus, the female which shows proceptive behaviour is presumably also attractive to the male because of her behaviour. Conversely, the unreceptive female loses some of her attractiveness even if her hormonal condition is optimal from the male's point of view. In the case of caged rhesus monkeys, Keverne (1976) found that the bar-pressing activity of the female for the reward of access to a male peaked at the time of ovulation. Correspondingly, so did the number of ejaculations delivered by the male. Ovariectomy slowed up the rate of bar-pressing to gain access to males. Some females stopped pressing altogether within intervals of from 14 to 200 days after the operation. As Keverne notes, at long intervals after ovariectomy a decline in receptivity is likely to be indicative of a painful consequence of intromission involving a genital tract long deprived of the influence of ovarian hormones. Oestradiol replacement caused a return of bar-pressing on the part of the female, though not to such high intensity as was demonstrated by the intact animal at the ovulatory stage. It did, however, increase female attractiveness to the extent shown at the ovulatory stage. Keverne then investigated the effect of giving testosterone in addition to oestrogen to the ovariectomized females. Access to the male was faster following the addition of testosterone. In so far as the active sexual role of the female is concerned it would appear that both androgen and oestrogens play a role. The adrenals provide a background level of androgen. In the intact animal the ovary also seems to provide a source of androgen, which could explain the mid-cycle peak in plasma androgen. Keverne suggests that during the menstrual cycle the lowering of proceptive behaviour in the luteal phase can be attributed to secretion of progesterone, although the mode of action of this hormone is not clear.

Adrenal androgen is indispensable for the performance of receptive behaviour by the ovariectomized female, even if oestradiol is given. Adrenalectomized female monkeys given oestrogen show low sexual receptivity. That is to say, there is a relatively large number of times at which they refuse the male's mating advances.

In the case of the rhesus monkey oestradiol acts peripherally in stimulating lubrication to the female's vagina. Alone it is ineffective in arousing sexually proceptive behaviour, but appears to act synergistically with androgen in this respect. However, as Keverne observes, any such conclusions risk being based upon what is inevitably confounded experimental data. The female certainly exhibits more sexually proceptive behaviour at the ovulatory stage, but since she is more attractive to males at this stage then their attention *per se* may partly stimulate heightened proceptivity. Even brain implants of oestrogen appear to have been able to induce peripheral effects upon attractiveness.

As we saw in the earlier discussion, in the case of rats, testosterone is converted to oestradiol in the brain. Could it be that testosterone is effective in female monkeys because it is converted to oestradiol? Keverne regards this as unlikely since adrenalectomized females given oestrogen are unreceptive.

In rhesus monkeys the male is apparently sexually aroused by a combination of both the behaviour of the female and the state of her sexual regions, particularly the emitted odour. Giving an ovariectomized female testosterone increases the frequency of sexual invitations which she makes, but does not stimulate male sexual behaviour. The importance of the olfactory cue to the male was emphasized in the following study. Male rhesus monkeys were required to press a lever 250 times in order to gain access to a female. Through the partition dividing them the male could smell and see the female. Males would respond to ovariectomized females which had been treated with oestrogen but not to untreated females. Familiar females still elicited the bar-pressing response in such a situation even if the male was temporarily without his sense of smell. Keverne suggests that earlier sexual experience with these females was the crucial factor in determining that the male would bar-press, since unfamiliar ovariectomized oestrogen-treated females did not arouse bar-pressing in the absence of the male's sense of smell.

In the case of normal males, application of vaginal secretion taken from oestrogenized attractive females to unoestrogenized, and therefore unattractive, ovariectomized females made them attractive to males. Very many mounting attempts by the male resulted from this treatment, though since the females were unreceptive, ejaculations were few. However, it must be added that the response of the male to unreceptive females treated in this way varied very much from female to female (similarly the female will sometimes show unreceptivity to particular males but obvious receptivity to others). Some females seem to have very much greater power to arouse the interest of males than others, and such differences are sustained over repeated tests. The male's sexual response exhibits idiosyncrasy, such that Keverne is reluctant to employ the term *pheromone* to describe the channel of communication by which the male is aroused. Strictly speaking a chemical means of communication between organisms is described as pheromonal, but Keverne is uneasy about some of the associations of this word since it generally is used in the context of the stereotyped responses of insects. In the case of monkeys the pheromonel is only one of many sources of information impinging upon the male which is acted upon with a considerable degree of plasticity.

8.7 HOMOTYPICAL AND HETEROTYPICAL SEXUAL RESPONSES

In earlier sections, we reviewed evidence showing that each genetic sex contains the potential to exhibit both homotypical and heterotypical behaviour of the sex concerned. To what extent each potential is revealed in adult behaviour depends upon the early hormonal environment. In this section we look at the adult animal and consider both homotypical and heterotypical behaviour in the light of the evidence we have now collected on the effects of sex hormones on the adult animal.

Beach (1968) reviews evidence indicating that mounting by females occurs in at least 13 species, which represent carnivores, primates, and rodents, amongst other orders. If spayed heifers are injected with small doses of oestrogen then sexual interest is shown in that they place themselves near to males. Larger doses induce willingness to copulate, while still larger doses provoke attempts at mounting other females (Beach 1948, p. 272). In the case of cattle, the animal mounted is normally a second female in oestrus (Beach, 1968). The same is true of female lions and domestic cats, but the hormonal state of the stimulus animal is apparently of little importance in guinea-pigs. In no way does mounting imply a reversal of sex roles. Mounting of other females is commonly observed in hormonally normal females, and they show normal receptivity to male animals.

As we mentioned before, it was once believed that hormones were sex-role specific and, by implication, for most normal individuals gender specific also. Therefore testosterone was the natural hormone to investigate in the context of mounting behaviour by females (Beach, 1968). Early studies were indeed able to stimulate mounting by means of androgen injections. Beach places the subject in its historical context by noting that:

> 'It was only after mounting by females was accepted as something other than "abnormal" or "sex-reversed" behaviour that it became reasonable to consider the possibility that such behaviour might be related to homologous gonadal secretions.'

That oestrogen may stimulate mounting was suggested by its timing in, amongst other species, guinea-pigs. W. C. Young (see Beach, 1968) was one of the first to make systematic observations in this area. The period of most intense mounting was usually seen in the pro-oestrus stage, i.e. before the onset of behavioural oestrus as measured by receptive behaviour. The mounting phase may begin up to 53 h before behavioural oestrus and would sometimes cease at about 1–3 h following the beginning of receptivity. In later studies, it was found that as well as preceding oestrus, mounting overlapped in time with it.

In the ovariectomized female guinea-pig mounting can be stimulated by injections of oestrogen followed by progesterone. The duration of heat and mounting are independent in both the case of intact and ovariectomized injected females. From these early studies emerged the conclusion that although the same hormone combination, oestrogen and progesterone, was responsible for both mounting and receptivity, different mechanisms were involved.

In an experiment reported by Beach (1968), it was found that female rats would mount irrespective of their hormonal condition. Mounting was not abolished by ovariectomy, though receptivity was lost. Injection of oestrogen and progesterone to ovariectomized females resulted in restoration of lordosis, but no change in mounting frequency. However, according to Kow, Malsbury, and Pfaff (1974) ovariectomy decreases but does not destroy the capacity of female rats to perform the masculine mating response. It may be strengthened in such females by exogenous oestrogen as well as androgen. In rats, but not in guinea-pigs, the masculine response on the part of the female seems to have achieved some degree of autonomy from strict hormonal determination. One is reminded of the fact that in some species/strains masculine behaviour in the male can survive for some time after castration, whereas this persistence is almost never shown in the feminine response of females, except the higher primates. In the case of the female dog mounting does not show a peak at oestrus (Beach, Rogers, and LeBoeuf, 1968). Hormonal condition does not seem to affect this aspect of their behaviour, but females are more likely to mount oestrous than anestrous females.

Beach (1968) reminds us that the reason females rarely show the ejaculatory pattern even though performing mounting at a high frequency is that, lacking an intromittent organ, there is insufficient sensory feedback as a consequence of mounting. In terms of the model developed earlier, this means that the high intensity of arousal needed to trigger the ejaculatory reaction cannot be attained.

Beach (1947) summarized homotypical and heterotypical behaviour in the following way:

'... one becomes aware of the possibility of dealing with courtship and copulatory behaviour in terms of stimulus—response relationships which are relatively constant in all individuals regardless of sex; and it is apparent that the neuromuscular mechanisms responsible for many such relationships are present in both male and female mammals. In either sex the application of pressure to the dorsolumbar region tends to elicit lordosis and opisthotonus; and in either sex the multi-sensory pattern of stimulation provided by the receptive female tends to evoke pursuit, mounting and palpitation with pelvic thrusts.' (Reproduced by permission of The American Physiological Society.)

We will now direct the discussion to the male animal. Lordosis is commonly shown by male hamsters treated with oestrogen and progesterone when adult. Even without hormone treatment it is sometimes shown in this species. In a survey of the literature, Kow, Malsbury, and Pfaff (1974) assembled evidence showing that although lordosis may be obtained from normal intact adult male rats, large doses of oestrogen are needed relative to the dose which elicits lordosis in the female. Some investigators have failed entirely to obtain lordosis from male rats. In comparison to females, where mounting is easily provoked, lordosis in normal male rats is relatively uncommon and difficult to obtain. Given, though, that the necessary potential for lordosis is present in both sexes, why is it more evident in the female than in the male? It could merely be a reflection of a difference in sensitivity of essentially similar mechanisms. In other words, androgenization means that it is much more difficult to activate these same lordosis mechanisms in males than in females. In addition to this another factor may be involved, as is discussed by Kow et al. Progesterone acts synergistically with oestrogen to potentiate lordosis in intact females and oestrogen-primed ovariectomized females. Enhancement of oestrogen-induced lordosis by progesterone is not present in male rats castrated when adult. The fact that the male is insensitive to progesterone is attributed by Kow et al. to neonatal androgenization.

Baum and Vreeburg (1973) found the oestradiol injections caused a small but significant increase in the lordosis quotient of castrated male rats placed in the presence of stud males. Untreated castrates and those injected with testosterone never exhibited lordosis. Lordosis quotient is the ratio of lordoses made in response to mounting attempts by a stud male. Baum and Vreeburg note that there is normally very little oestradiol in the blood of male rats, and this could explain why lordosis is seldom shown. Although oestradiol is needed for the masculine response of mounting it is converted from testosterone in the neural tissue underlying this behaviour. The lordosis quotient was lower in animals given both oestradiol and dihydrotestosterone. This could possibly mean that the synergistic effect of these two substances both excites masculine behaviour and inhibits feminine behaviour, but that would be to speculate beyond what is permitted by the available evidence. Thus Pfaff (1970) found that whereas oestradiol stimulated lordosis in females it did not do so in males. In fact no hormone treatment caused any increase in receptivity in male rats as compared to oil-injected controls.

In summary and conclusion then, it is clear that the behaviour of adult animals supports the notion of bi-sexuality. This is much more evident in the case of the female than the male. Species differences are considerable, and in connection with masculine behaviour in females, vary from dependence on oestrogen to independence. Although feminine behaviour may be exhibited by males it is not common. It appears that early androgenization very considerably raises the threshold for the elicitation of this behaviour.

8.8 SUMMARY AND GENERAL CONCLUSIONS

In line with the philosophy of the present study, the issues discussed in this chapter on sexual motivation concern the kinds of mechanism which generate the observed behaviour. Thus we have discussed such issues as the hormonal state of the animal, genital feedback, and the effect of ejaculation on arousability.

In the female hormonal determinants are crucial in spacing sexual behaviour. Also in the male, where hormonal state fluctuates little if at all, it may be seen that sexual activity occurs at intervals with satiety or 'sexual fatigue' between bouts of activity. This behaviour is not unlike that to which homeostatic theories have been proposed in the case of feeding and drinking. It is important to note that systems employing different modes of organization and serving quite different ends can generate similar behaviour sequences.

It is sometimes argued that a necessary condition for the arousal of hunger, thirst, and sexual behaviour is in each case a specific biochemical event in the blood, respectively low glucose, high osmolarity, and an adequate amount of hormone. Let us ignore for the moment the discussion about what constitutes a signal for hunger and thirst, and whether a biochemical displacement is necessary for their arousal. Certainly (given the presence of food and water) low energy supply and dehydration are *sufficient* if not necessary conditions to provoke feeding and drinking, respectively. Post-ingestional satiety corresponds to a reversal of these conditions. Here it is important to draw a distinction with respect to sexual behaviour. Testosterone may be necessary for the arousal of sexual behaviour in the male, but normal satiety does not appear to be accompanied by a low blood level of testosterone. Rather, testosterone at a fairly constant level provides long-term sensitization, both central and peripheral.

In the case of the male it was argued that somewhere within the nervous system a pathway essential for sexual arousal loses its sensitivity with ejaculation(s).

Such a statement is somewhat tautological, but it directs attention to the kind of change which is involved. Thus we no longer believe (at least in mammals) that loss of seminal fluids *per se* causes loss of sexual arousability. Rather, ejaculation(s) in all probability causes a chemical change at a synapse, or alternatively a change in a neuron's transmission characteristics. Ejaculation not only lowers the ability of the animal to be aroused to initiate copulation but by the same process appears able to make further ejaculation more difficult. However, is was argued that in the case of rats a tendency exists for the first ejaculation or two to facilitate further ejaculation by a carry-over of excitations. These two effects work in opposite directions and an inverted-U curve relating intromission number to its corresponding ejaculation is apparent. A model of sexual behaviour was presented, and it was shown where an explanation of these effects in systems terms was possible.

Arousability was the term used to denote the ability of the organism to be sexually aroused. Arousability is to some extent dependent upon the stimuli arising from the partner. A change of partner appears to increase arousability, a phenomenon known as the Coolidge effect. Also it is possible to observe a trade-off between arousability and sensory stimulation. If the male has been without sexual contact for a long period of time arousability will increase to a high value. A previously inadequate stimulus, such as a member of another species, may then be sufficient to arouse mating attempts. If androgen level is low extra androgen may increase the arousability of the system. However, provided a certain minimum level is present it is often the case that arousability is not dependent upon androgen level. Exceptions to this are to be found though; in some cases discrimination is lowered with high androgen doses.

Some useful explanations may be obtained if we examine the system in terms of copulatory and ejaculatory thresholds. Thus an individual animal may have a low copulatory threshold and therefore be induced to copulate frequently. However it may take considerable effort to reach ejaculation. The reason is that little time would have been allowed for recovery before copulation occurred again.

In the case of the female the situation is somewhat more complex than for the male. Although female rats can sometimes be observed to exhibit a reaction analogous to that of the ejaculating male this is in the context of mounting another female. In the course of normal heterosexual contact in rats and some other species it appears that a cumulative effect of genital friction is sufficient to terminate sexual receptivity and interest. Another sufficient condition, even if the female has not mated, is a reversal of the hormonal conditions present at receptivity. If mating occurs it seems that genital stimulation and a hormonal change act synergistically to terminate receptivity.

In the discussion of hormones in the female we considered both rats and primates in some detail. In this respect, as in others, it is wrong to treat primates as sophisticated versions of rodents. The hormonal basis of sexual proceptivity and receptivity in primate females is organized rather differently than in rodents.

It is only fair, though perhaps not entirely necessary, to point out that my own biases are evident in the choice of material for this chapter. It is by no means an exhaustive nor representative account of the processes involved in mating and the possible underlying mechanisms. A particular mechanism, loss and gain of arousability, was given prime importance. Alternatively, active mechanisms of neural inhibition may play a role in satiety (Freeman and McFarland, 1974). In rats an active inhibitory process may also be involved in the refractory period following each ejaculation and even after each intromission. Space prevents a detailed and balanced account of these issues.

Another bias in the study is that, although some interspecific comparisons were made, the laboratory rat has dominated the discussion. Although I am committed to the comparative approach, it is my belief that, in terms of formal theory building, a certain minimum level of understanding is needed before we can take interspecific comparison too far. In this area we have a large amount of data on behaviour and the components of the physiology underlying behaviour in various species, but the theoretical level is still weak. Once a reasonably convincing theory on the behaviour of the laboratory rat is available, interspecific comparison will inevitably follow. As Beach (1956) reminds us, in this context:

'To attempt a multispecies explanation from the beginning would be fruitless and ultimately frustrating.'

It would seem reasonable to start on the basis that for all species that employ internal fertilization, a certain minimum number of design features must underlie the motivational system. For the male, a minimal level of appropriate stimulation must be present to initiate a mount and penile insertion. Then a certain amount of genital friction is necessary to trigger ejaculation. A variety of evolutionary influences will have shaped the length of time that is needed for ejaculation. Finally, ejaculation or a series of ejaculations leaves the male refractory to further stimulation. Within these general constraints there exists considerable scope for species differences in the pattern of copulation. Some species attain ejaculation with a single, brief, penile insertion. Others, such as the rat, require a number of insertions. In a functional sense the male rat's behaviour is related to the amount of stimulation needed to bring the female into a hormonal state permitting pregnancy. Some species perform a long series of penile thrusts before withdrawal. Ultimately these species differences will need to be reviewed in terms of a model of sexual behaviour. It is a matter of speculation as to how far we could get by simply modifying the parameters of the model proposed here.

Another bias, not just with this chapter but with the available theoretical literature, is that it is male orientated. This is not (I hope!) a sexist bias on the part of male scientists, but is dictated by pragmatic considerations of how best to tackle a hideously complex behavioural interaction. The next logical step will involve formal model building in terms of (1) the female's oestrous cycle and (2) male–female interactions in terms of, amongst other things, ultra-sounds

(Sachs and Barfield, 1976). The latter would involve somewhat more attention being paid to the post-ejaculatory refractory period than has been the case in the present review.

With regard to the arousal of mating, the model developed here was for the laboratory rat. The natural habitat may involve much more complex factors even in the case of rats. When formal theory building is applied to some other species the problems appear to be even greater. In some cases a very long period of courtship must elapse before any recognizable consummatory behaviour occurs. Indeed, the consummatory behaviour of sex may appear to be only one of several goals of such courtship. In nature the extremes of coyness and promiscuity are to be found, and it must be admitted that, at the extremes, in terms of sexual arousal mechanisms and species differences there is little we can say at this stage.

CHAPTER 9

Aggression and fear

9.1 INTRODUCTION

This chapter might alternatively have been called attack and escape, since escape is usually associated in the literature with fear, and attack associated with some underlying 'aggressive' motivation. By aggression I mean fighting and initiating attack. By fear I mean escaping from a situation either by fleeing or freezing. I do not intend to pursue considerations of terminology, but will use rather liberally what seem to be appropriate expressions, borrowing from the researchers in this area.

Aggression and fear share important characteristics in common with, on the one hand, feeding and drinking, and, on the other, sexual behaviour. They serve the survival of the individual animal, and therefore allow perpetuation of its genes. Aggression also directly affects reproductive success. The animal which either flees or freezes in response to a threat increases its own chances of survival. Attack may also increase individual survival chances since an animal which is a potential threat may retreat in response to a display of aggression. In the case of some species, the male which can secure territory and a female(s) for himself obviously stands a good chance of perpetuating his genes. Any offspring would be in a good position to survive. The female with young who attacks an intruder can increase the survival chances of her young.

All of us must have experienced intense fear, or at least moderate fear, at some time. Furthermore we usually take evasive action. Interestingly, most of us living in the affluent and temperate Western world have never experienced either serious thirst or hunger. Perhaps more than in other areas of behaviour our scientific view of fear is coloured by our own private experiences. Whether or not animals experience conscious sensations of fear in any way similar to our own is a subject of endless speculation, but one which never yields useful explanations. It may prove to be a philosophically unsound question to ask. Animals exhibit external behaviour and internal physiological responses which, if it were a human, would undoubtedly be called fear, and accompanied by subjective reports of fear. Most of us in the comfortable world of academia never commit explicit acts of aggression, though the subjective feeling of anger is one that seems to be aroused in most of us on some occasions. In the present study we are concerned not with the feelings but with external behaviour, i.e. freezing, fleeing, and attack. Archer (1979) warns of the dangers involved in placing too much faith in subjective experience, particularly if we then generalize to animals. It is perhaps for this reason that some writers prefer to speak of escape or avoidance rather than fear, and attack rather than aggression.

Fear and aggression usually occupy a less prominent place than feeding and drinking in discussions of motivated behaviour. This may reflect some ambivalence on the part of the author concerned as to whether they really do constitute examples of motivation or drive. They are included here because they literally *move* the animal to flee, freeze, or attack. They can completely dominate the animal's behaviour, with feeding and drinking suppressed.

As Marler and Hamilton (1967, p. 166) remind us, some types of behaviour show neither regular periodicity nor anything resembling an appetitive or searching phase. Wallace Craig in 1918 (see Marler and Hamilton) had noted that they are more closely associated with aversion than with appetite, and continue until the appropriate external stimulation is terminated. Examples of responses which fall into this category include the wiping reflex of frogs and the scratch reflex of dogs. In the category of more complex behaviour, escape and aggression are two examples.

9.2 THE BASES OF AGGRESSION AND FEAR

In a recent review of the literature, Archer (1976) argued that basically the same type of stimuli serve to elicit both fear and aggression. In a review of the literature on fear behaviour a broad evolutionary perspective was introduced by Archer (1979), the essence of which is as follows.

Every living organism is liable to suffer physical damage during the course of its life. Mechanisms of repair and regeneration of living tissue represent a homeostatic type of response to such damage. But prevention is, of course, better than cure. In the case of species such as Hydra and the Sea Anemone, withdrawal into a protective enclosure is the only defence available. For other species the organism can physically move itself away from the noxious stimulus. *Paramaecium* perform an avoidance response to such stimuli as excessively hot or cold water, noxious chemical/osmotic stimuli. Archer (1976) argues:

'A later refinement in evolutionary development would have involved the ability to decide in advance of tactile or chemical stimulation whether danger was likely.' is this what we do?

Such mechanisms served as the prototype for the more sophisticated fear mechanisms we now study. Animals subject to predation have evolved mechanisms whereby the sight, sound, or smell of a predator (or even an innocuous intrusion) evokes avoidance behaviour. In the evolutionary development of such behavioural traits it has, for obvious reasons, proven useful for the species to inherit such anticipatory mechanisms. It would not reflect good evolutionary design for an animal only to take avoidance action in response to direct contact with a predator or conspecific attacker.

Archer (1979) argues that, with more sophisticated sensory equipment, animals would monitor the environment for specific dangers. In Archer's terms what constitutes danger is '. . . any large discrepancy between observed and expected stimuli . . .'. In other words, in the course of its activities the animal acquires a model or hypothesis of the expected state of certain features of the environment, and fear is evoked when the actual state of these environmental features differs significantly from the expected state. Archer considers this also to form the basis of aggression:

'. . . any large discrepancy between observed and expected stimuli would induce escape or withdrawal responses (or attack if the appropriate stimulus was present).'

In an early study by Hebb (see Hebb, 1966, p. 241) fear was evoked in chimpanzees by presenting a familiar stimulus in an unfamiliar context. A different coat worn by a familiar experimenter was one such example. A more extreme case was a dummy chimpanzee head without a body. For such stimuli to be effective the animal must have some experience of other normal and intact chimpanzees. In chimpanzees it appears that the bizarre is only fear-evoking when seen in terms of a history of exposure to natural stimuli.

Archer (1976) classifies stimuli which can elicit either fear or attack from the simple to the most complex, and argues that they may be understood in terms of a disparity model. Thus, sudden tissue damage or pain is the most simple stimulus and lies at one extreme. The so-called frustration of non-reward in an operant situation lies towards the opposite extreme. I will now list those situations which Archer (1976) describes as evoking either attack or fear in vertebrates. Often I omit the name of the author whose work it is that Archer cites; the reader who wishes to pursue these should consult the original article. The list is as follows:

(a) *Pain*
In a wide variety of species pain-induced fighting has been noted. Electric shock elicits attack in rats. For example, it is claimed that attack is directed at either an inanimate object or member of the same species that happens to be standing nearby. Behaviour associated with fear, i.e. continuous vocalizations, freezing, and defecation, is also elicited by electric shock.

Johnson (1972, p. 38), however, doubts whether shock-elicited fighting between two rats can really be classed as aggression for any useful purposes. Shock is a totally unnatural stimulus. It tends to elicit only certain components of the normal fighting behaviour of rats, most obviously the 'boxing' posture of standing on the hind legs. Johnson raises the point that this posture may simply minimize contact with the electrified grid floor. This view is echoed in the observations of Blanchard, Blanchard, and Takahashi (1978). They argue that on close examination the rat subjected to shock does not react in a way characteristic of attack, but rather characteristic of *defence*. In terms of the site to which it directs its bites the shocked rat behaves like an intruder to an established colony who has come under attack. It makes bites on the snout of the other animal, a reaction never shown by colony rats towards an intruder. This behaviour is not surprising. Sharp sudden pain would normally be associated with predation, which calls for somewhat different reactions than those exhibited in a situation which we could unambiguously class as attack. However, the distinction between attack and defence is perhaps not absolutely clear, even though certain differences emerge. Take for instance the cornered animal which attempts to flee, finds its way blocked, and then attacks. Is this aggression or defence? Is the savageness of wild rats towards the prospective handler defence or aggression? We will continue to consider defence a form of aggression but will remember to qualify our arguments. Of course the possible political implications are enormous; the strategic Air Command of the U.S. Air Force used the motto 'Peace is our profession'.

(b) *Individual distance intrusion*
In a variety of species attack has been noted when an intrusion is made into an animal's 'personal space', or as it is sometimes called, 'individual distance'. The intruder may be another animal or even an inanimate object. Defence of individual distance may be the evolutionary precursor of specific territory defence. Fear, in the form of escape, is of course also associated with individual distance intrusion.

(c) *Territory*
Closely associated with (b) above is attack elicited by intrusion of another animal of the same species into the territory of the attacker. Animals are likely to show attack more readily when on familiar territory than when it is unfamiliar. If the intruding animal is completely unfamiliar to the resident of the territory then the probability of attack is particularly high. Male mice, for example, secrete a characteristic odour which increases the probability of them being attacked by another male. However, males which are familiar with each other are unlikely to show attack, despite the presence of the potential attack-eliciting odour. Female mice secrete an odour which decreases the probability of them being attacked.

A wide variety of inanimate but novel objects have been reported to evoke attack when they suddenly invade an animal's territory. Thus, Archer concludes that novelty is an important feature in the causation of attack. This is particularly manifest in the case of an unfamiliar conspecific animal, but is also shown by novel objects. Archer suggests that:

'. . . whereas individual stimulus properties are undoubtedly important in inducing attack, and may be crucial for some species, one common factor in "territorial" and similar forms of fighting is the presence of an unfamiliar object in an area with which the animal is familiar'. (Reproduced by permission of Plenum Press.)

It was found that wild rats of the third generation raised in captivity show savageness towards humans who attempt to pick them up (Galef, 1970a). If they are handled daily from a very early age this savageness is eliminated. However, handled rats are no less aggressive towards mice and other rats, indicating that the quite specific stimulus of human contact is involved in their later behaviour. Similarly, mouse-killing is inhibited if rats are raised in the presence of a mouse. It should be noted, though, that strictly speaking mouse-killing may be closer to predation than attack.

Archer recognizes that specific features of the intruder such as a red breast or an odour may be particularly strong stimuli for attack, but sees them within the context of novelty. Novelty forms the basic plan, but specific stimuli are important in some species as a result of specific evolutionary adaptations for that type of animal. Adaptations would be built into the sensory side and superimposed on the novelty mechanism. Learned biases would also be possible. The disparity detector must obviously be heavily biased, a point which Archer (1976) acknowledges. He expresses it in the terms that certain species-typical stimuli may be particularly resistant to incorporation in the expectation copy.

Under this heading examples may also be given of where basically the same stimuli may elicit attack or fear. Sometimes the animal seems to show ambivalence in its behaviour, switching from one to the other. Sometimes the resident animal is observed to flee from its own territory. The fear response evoked by intrusion of an inanimate object into the territory of an animal has in several cases been shown to habituate with repeated presentation, which Archer sees as support for an interpretation in terms of novelty. As in the case of aggression, it is acknowledged that specific features give a bias towards fear.

(d) *An unfamiliar environment*

In some cases animals which have previously been familiar with each other, and which did not fight, started to do so when placed in an unfamiliar environment.

Fear is commonly observed when a single animal is placed in a strange environment. Rats are prepared to cross an electrified grid in order to escape from an unfamiliar environment back to the home cage. Freezing, crouching, and jumping at the sides of the unfamiliar cage are also labelled as fear responses. Chicks give distress calls under these conditions.

(e) *A familiar object in an unfamiliar location*

In the case of the three-spined stickleback an increase in frequency of attack was shown when a familiar stickleback was presented in a different location to normal. Rats show more shock-induced attack towards an object when it is in an unfamiliar place than in a familiar place.

(f) *Extinction-induced aggression*

This situation is probably familiar to the reader; the empty dispensing machine is frequently the victim of a violent blow. The hungry animal, normally obtaining reward on a continuous reinforcement schedule, responds with aggressive behaviour when reward is omitted. The objects of attack observed in the laboratory include a conspecific (or a model of one), a mirror, or a food dish. In the case of rats, when changing from continuous to partial reinforcement schedules it has been found that attack of a conspecific occurs on the unrewarded bar-presses but not on the rewarded. Archer argues that non-reinforcement is an example of a change in the stimulus situation from what is expected. Attack appears appropriate when judged in terms of the emotion which accompanies non-reward in humans. It may therefore come as some surprise to the reader to learn that, in animals, fear can also arise under these circumstances. Escape has been observed in rats, defaecation, jumping, and distress calls in chicks, and freezing in mice.

(g) *Thwarting*

This is where a previously rewarded response can no longer be performed to completion because of the imposition of some barrier. It has been reported to produce attack of a conspecific in the case of male mice and both sexes of domestic fowl. Escape behaviour, as well as alarm calling, has been reported in hens. Archer argues that competition for food may be classed as thwarting in that a rival animal provides a physical barrier. This is a stimulus which provokes fighting in a number of species.

(h) *Low reinforcement schedules*

In the case of male pigeons a target animal (or model of one) is the object of attack when the pigeon is faced with fixed-ratio schedules which yield low return for a high rate of bar-pressing. The attacks usually occur during the post-reinforcement pause.

This represents Archer's basic categorization of stimuli which elicit fear and/or aggression. Additivity of stimuli may be observed, and this increases the probability of fear/attack. Within limits, in the case of rats and mice, a combination of pain and novelty produces a higher probability of attack than either component on its own. The frustration of a low fixed-ratio food schedule is able to add to the effect of pain in the causation of attack. Perhaps the word *additivity* can be a little misleading in some cases, implying simple linear addition of two information components. Thus shock combined with introduction of a novel object into the cage of a wild rat is very likely to evoke attack of the novel object (Galef, 1970b). Although shock alone evoked attack of a familiar inanimate object it was very much less than that directed at a novel object. Presentation of the novel object but without shock caused no attacks. Thus I would prefer to call this a synergistic or multiplicative relationship. In the natural setting a sudden pain on its own would be unlikely to occur. Sudden pain accompanied by the appearance of a novel object might be a common occurrence in the history of a species.

In attempting to define the process by which attack and fear are instigated, Archer (1976) proposes:

> '. . . we would assume that the animal maintains a continuous complex representation of expectancies based on:

(a) the sum total of its experiences, i.e. what types of relations particular stimuli generally have to one another, what types of relationships commonly occur, what types occur less often, and what types never occur;

(b) more precise spatial representations of particular habitually used areas of the environment, so that certain habitually experienced stimuli become expected;

(c) temporal representations of the expected outcome of particular sequences of previously rewarded responses—thus a sequence culminating in a rewarding event would come to evoke a fairly precise expectation when the animal again found itself in a similar situation. It is also suggested that just as rewarding events become easily incorporated into the animal's expectation model, the model is particularly resistant to incorporating intense or painful stimuli.' (Reproduced by permission of Plenum Press.)

It is assumed that the animal's nervous system constantly compares incoming information with such stored 'hypotheses', and a discrepancy causes fear or attack. The fact that pain stimuli show less tendency to lose their power to stimulate fear/attack is an obvious qualification to be added to the model. Even the response to pain stimuli seems, though, to show some habituation if the intensity is not too great and the stimulus is applied at frequent intervals.

As Archer notes, his model bears a strong similarity to the theory advanced by Sokolov to account for the orientation reaction. This theory involves a comparison of incoming sensory information with established neuronal representations. For instance, people exposed to the ticking of a clock quickly form a neuronal representation of this. When this sound subsequently impinges upon us it finds a match with an internal representation, and does not lead to arousal and turning the head towards the direction of sound. Throughout the animal kingdom cases may be given of where clear signs of arousal, e.g. turning the head, ears pricking up, increased activity of the autonomic nervous functions associated with mobility, are caused by unfamiliar stimuli. Archer suggests that whereas relatively small disparity signals cause orientation, somewhat larger disparities are, in addition, associated with fear and aggression.

The theoretical position represented by Archer dovetails neatly with that of Bolles (1970). Bolles notes that much of the theory of avoidance learning (reacting so as to avoid a noxious stimulus) which has developed in the laboratory tacitly assumes that avoidance in the wild is largely due to classical conditioning. We can put it slightly facetiously as follows. An owl hoots just before pouncing. The mouse feels the pain of the tallons but escapes. Next time the mouse hears the hoot it escapes before the owl has time to pounce, and hence lives to pass on its genes to another generation. But Bolles argues that the chances for learning in such situations are absolutely minimal, and that animals such as mice are equipped with innate species-specific defence reactions (SSDR). Rather than avoidance being the result of conditioning, the SSDR occur when animals '. . . encounter any new or sudden stimulus'.

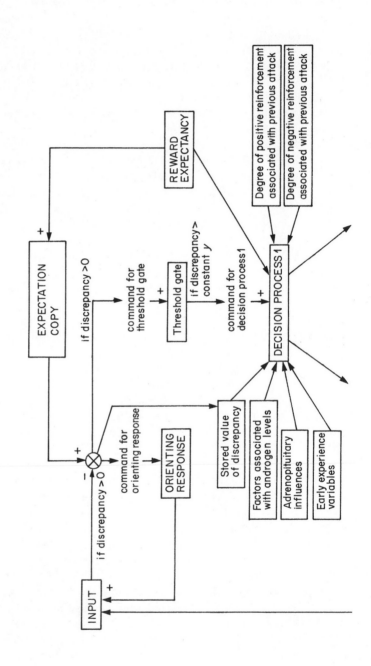

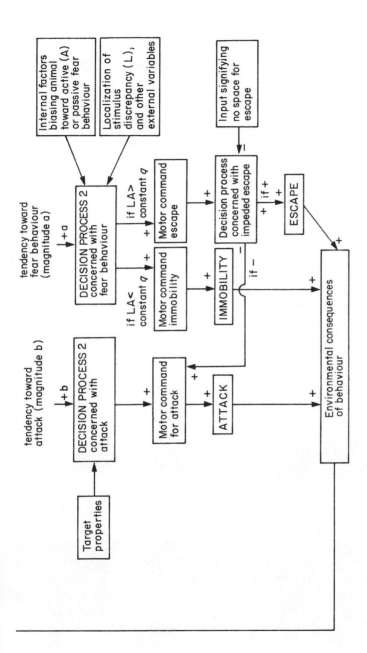

Figure 9.1 Factors influencing fear and attack, as proposed by Archer. (Source: Archer, 1976. Reproduced by permission of Plenum Press)

The congruence of Archer's and Bolles' arguments may be appreciated by the following quotation from Bolles (1970):

'These defensive reactions vary somewhat from species to species, but they generally take one of three forms: animals generally run or fly away, freeze or adopt some type of threat, that is, pseudo-aggressive behaviour. These defensive reactions are elicited by the appearance of the predator and by the sudden appearance of innocuous objects. These responses are always near threshold so that the animal will take flight, freeze or threaten whenever any novel stimulus event occurs. It is not necessary that the stimulus event be paired with shock, or pain, or some other unconditioned stimulus. The mouse does not scamper away from the owl because it has learned to escape the painful claws of the enemy; it scampers away from anything happening in its environment, and it does so merely because it is a mouse. The gazelle does not flee from an approaching lion because it has been bitten by lions; it runs away from any large object that approaches it, and does so because this is one of its species-specific defence reactions. Neither the mouse nor the gazelle can afford to learn to avoid; survival is too urgent, the opportunity to learn is too limited, and the parameters of the situation make the necessary learning impossible. The animal which survives is one which comes into its environment with defensive reactions already a prominent part of its repertoire.'

Figure 9.1 shows a summary of what Archer (1976) considers to be the most important features of the orientation/fear/attack decision mechanism. Although the diagram may at first appear formidable it can be used merely as a descriptive aid. We will consider each stage of the model.

At the top left INPUT means incoming information via the animal's nervous sytem. EXPECTATION COPY is a neuronal representation of the expected information arriving. The plus and minus signs at the top-middle accompanying the circle indicate comparison of actual and expected. The statement 'if discrepancy > 0' in fact refers to the output of this comparison and indicates, in the tradition of Sokolov, that disparity is the cue for the orientating response. The statement 'if discrepancy > 0' represents the assumption that large discrepancies are the cue for fear or aggression. DECISION PROCESS 1 represents the operation of deciding whether to emit a fear or attack response. This decision is influenced by a number of factors which are represented by boxes with arrows touching DECISION PROCESS 1. According to Archer one of the most decisive factors is the magnitude of disparity. Small disparity only causes orientation, larger disparity aggression, while very large disparity causes fear. Archer echoes Hebb in this aspect of the model (Hebb, 1966, p. 257):

'It is also possible that milder degrees of disruption innately lead to aggression, stronger ones to avoidance.'

Archer uses for his model the fact that as electric shock intensity increases aggression occurs at first, but in the higher ranges this is replaced by fear. However, as we have already seen, experiments in this area should not be interpreted uncritically. Electric shock of high intensity may work through means which should be regarded more as pathological rather than behavioural in the strict sense. Aggression may be more demanding in terms of sensory-motor co-ordination than is the case for what we call fear responses—in the extreme, unconsciousness is hardly a fear response. However, Archer also cites a study by Bateson (1964) to which this criticism would probably not be relevant. Young chicks were exposed to a moving object and the initial response was avoidance. Subsequently, fear gave way to vigorous pecking of the moving object. Similarly, in a study carried out by Kuo (1960) it was found that in familiar territory quail would attack a conspecific but when in unfamiliar territory would respond by undirected running and jumping, which Archer labels as fear.

Hormones are assumed to influence the fear/attack decision, and this explains the input termed 'factors associated with androgen levels'. Many species show annual variation in the tendency to aggression which are associated with variations in androgen levels. In the extreme, territorial behaviour is only shown during the breeding season in some species. The fact that males of most species are more likely to attack than females (though there is a significant number of exceptions to this) is usually attributed to differences in neonatal and adult androgen levels. The box labelled 'adrenopituitary axis' reminds us that pituitary and adrenocortical hormones may also influence the decision to attack or flee. Leshner (1975) also proposed a model of agonistic behaviour, e.g. attack and fear. In this model, based on rats and mice, the organism's baseline hormonal state gives it a predisposition to be more or less aggressive and more or less submissive. Increasing androgen level up to a point increases tendency to attack. Hormones of the pituitary–adrenocortical axis are also implicated, according to this model. Increase in ACTH levels increases the tendency to show fear.

The animal's motivational states can play a role in determining what is the necessary stimulus to evoke attack. As Hinde (1970, p. 340) reminds us, when in the winter flock, species such as chaffinch and yellowhammer show aggression towards conspecifics which are a few inches or feet distant. Such conflict is associated with feeding and has a rather clear adaptive significance. In the breeding season an area of as much as one or two acres is defended against conspecific males.

In the theoretical model which we are developing experience is given an important role in biasing the decision mechanism. At present it is difficult to give a convenient summary of the role of early experience, except to say that interactions with the mother and with siblings appear to influence the tendency to aggression in the adult animal. In the case of both rats and mice isolation increases the chances that a conspecific will be attacked upon its introduction.

An animal's success or failure in attack is a particularly powerful influence biasing the attack/fear decision mechanism. Winning a fight increases the chances that on subsequent encounters, in response to a similar stimulus, the animal will fight. Conversely, stimuli which in the past evoked fighting will cause fear responses if the fights were unsuccessful. This is the reason why Archer shows inputs called 'degree of positive reinforcement associated with previous attack' and 'degree of negative reinforcement associated with previous attack'. Such feedback also plays a crucial role in the model of Leshner (1975). Defeated animals show increased adrenocortical activity and lowered androgen levels. This gives a bias away from attack and towards fear. Feedback also seems to involve interaction with an actual or potential rival in the case of mice. The aggression-promoting pheromone released by male mice appears dependent upon androgen level, and so defeat may be followed by lowered androgen level which decreases both attack tendency on the part of the animal concerned and that shown towards the animal by a rival.

At this stage, we are now at the output side of DECISION PROCESS 1. On the basis of the various inputs a 'tendency towards attack (magnitude b)' and a 'tendency towards fear behaviour (magnitude a)' arise. Thus, there is a stimulus for both behavioural responses present. One or the other will be stronger and this will be the behaviour which is exhibited. DECISION PROCESS 2 is responsible for the motor command for attack. ATTACK then has certain environmental consequences; hopefully the intruder flees! This, of course, then changes the INPUT to the system. The reader will recognize this as an example of negative feedback; disparity causes the animal to take action such as to correct disparity. In this context, Archer quotes a comment of Craig made in 1928:

> 'When an animal does fight he aims, not to destroy the enemy, but only to get rid of his presence and his interference.' (Reproduced by permission of Plenum Press.)

'DECISION PROCESS 2 concerned with fear behaviour' has at least two options to decide between. It may either execute a 'motor command immobility' or a 'motor command escape'. Archer would class both reactions as fear, as would Hinde (1970, p. 349), though in searching the literature the reader would find some differences in terminology. The decision as to which form of fear behaviour to implement will be biased by internal and external factors which are shown entering the decision process. If IMMOBILITY occurs then the favourable environmental consequence is that the predator loses sight of the immobile animal and moves elsewhere. This of course returns the system to a safe equilibrium. ESCAPE, if successful, means that the animal is out of the reach of the predator. Even at this late stage further biasing may occur. The information may indicate the appropriateness of fear and indeed the animal may be executing the motor reaction of escape, but it finds its exit blocked. Freezing is inappropriate since the animal has been spotted. The only alternative is attack. This is indicated by the cross-connection 'decision process concerned with impeded escape' which, despite all other factors being appropriate for escape, nonetheless instigates attack when the exit is blocked.

Other factors which influence implementation of an attack decision must be mentioned. One such is dependent upon the specific characteristics of the target.

If the target is localized then a motor command for attack is able to be put into action. In Archer's terms DECISION PROCESS 2 must receive 'clearance' from 'Target properties' before a motor command for attack occurs. If this clearance is not obtained then according to Archer's diagram the animal is in a somewhat ambivalent position. Presumably by the animal doing nothing the discrepancy increases to the extent necessary to evoke fear. From the evidence which Archer assembles it may be concluded that if the object of the discrepancy itself is of a form such as to make attack impossible or improbable a nearby object, particularly another animal, will form the object of attack. An example is frustrative non-reward which involves attack of an otherwise irrelevant bystander animal.

Size of target is an important factor involved in the attack decision. In the case of mice, a small conspecific is more likely to be the object of attack than a larger one. As the target animal increases in size fear becomes more probable and, in parallel, attack becomes less probable. Target movement inclines the animal to attack in many species. This is not surprising; movement usually involves an animate object.

The model of course requires considerable refinement before it can in any sense give a conclusive explanation of attack and fear behaviour. At present it claims no more than to represent the kind of basic underlying processes involved. What it clearly achieves very successfully even at this stage is to give a unifying structure to much of the experimental data, and it serves as a very economical classification system. Archer's review indicates that rather similar environmental events evoke both fear and attack, and in so doing points to the evolutionary roots of these behaviours. It perhaps gives what is at present the most parsimonious account of how the organism programmes its behaviour. In all probability it will ultimately give way to a better model, as is the case for all models of behaviour. Later we will discuss what catastrophe theory has to say about the fear/attack decision and compare this with Archer's approach. Before then we will look more closely at each of the behaviours, aggression and fear.

9.3 THE CHARACTERISTICS OF AGGRESSIVE BEHAVIOUR

9.3.1 Disparity or deprivation?

The model of aggression which Archer developed for vertebrates shows behaviour only being aroused when disparity is present. In the absence of disparity then the behavioural controller is inactive, a view which is in accordance with the arguments of Scott (1958). A long period of time may elapse without a display of aggression, but the tendency to aggression neither increases nor decreases. This is radically different from the behaviour model of Konrad Lorenz which involves an increase in tendency to aggression as a function of time elapsed since the last opportunity for aggression (e.g. Lorenz, 1966).

In terms of the Lorenz model, aggression may be said to be spontaneous. The longer the time that has elapsed since the last aggressive response the lower the threshold for aggressive behaviour. In the end virtually any stimulus is adequate. In this way aggression is seen as being almost analogous to the homeostatic interpretation of feeding and drinking. However, Toates and Archer (1978) argued that such a mechanism would be counterproductive. If aggression is to restore equilibrium in the animal's immediate environment then it should only be aroused following an intrusion. It is difficult to see the purpose of an aggressive drive increasing in intensity with the passage of time. However, one must still critically evaluate the evidence to see if it fits with the concept of an aggressive drive resulting from deprivation.

It forms the subject of much discussion as to whether or not it is unhelpful anthropomorphizing to say that animals actively seek food and water. What is beyond dispute is that the water/food-deprived organism shows an increased probability of visiting a location or emitting a response which has in the past been associated with water/food. In the present context the appropriate question is, as expressed by Hinde (1970): 'do animals go out and actively seek for fights?' Evidence from some species suggests that this is the case under certain circumstances. Domestic cocks and Siamese fighting fish can be taught operant responses for the reward of stimuli previously associated with attack. Beach (unpublished study cited by Hinde, 1970) found that C-57 mice would learn an instrumental response which was rewarded by bringing the animal into a situation where another mouse could be attacked. Rats learn an operant response for the reward of a mouse which they then kill (see Johnson, 1972, p. 141).

However, some of the commonly cited evidence is rather fragile (Hinde, 1974). Under some special conditions, Siamese fighting fish will perform an operant response for the reward of a suitable target. An example is a mirror in which the fish can attack its image. In other experiments it has been found that these fish will learn a response to *terminate* the presence of the mirror. This applied to naive fish and those which had lived for a while in harmony with conspecifics (see Hinde, 1974, p. 261). After consideration of the evidence Hinde argues against the interpretation that aggression is inevitable. The data which Lorenz interprets in terms of aggression-specific energy needing periodic discharge are explained otherwise by Hinde.

An experiment reported by Johnson (1972) is particularly interesting in this context. He found that both fighting fish and paradise fish would work for the reward of the visual stimulus of a conspecific male, but they would also work for such visual stimuli as a marble or even an empty chamber. Visual exploration may be more correctly given as the underlying motive rather than aggression, though this has been challenged by Bols (1977) who supports the idea that it is specifically the opportunity for attack which is reinforcing. Birds can be observed to return to an area where an intruder has been attacked on a previous occasion. Again, though, it may be that this behaviour is best described as exploration of a particularly significant part of the environment.

In a study performed by Azrin, Hutchinson, and McLaughlin (1965), and employing squirrel monkeys, it was found that following the pain of electric shock they would perform a learned response to obtain access to an inanimate object which was attacked. The object was a ball which was lowered into the cage after the operant response. If reinforcement was omitted the operant response to the shock extinguished. If the ball were continuously present in the cage it formed the object of attack following delivery of shock. Azrin *et al.* argue that on the basis of this result it is useful to consider aggression as a motivational state having features in common with, for example, hunger. Whereas the opportunity to eat is reinforcing for the food-deprived organism the opportunity for attack is reinforcing for the animal in an aggressive state. Although some similarity is apparent in the two situations, it must be pointed out that the tendency to attack did not arise naturally in this experiment but rather was 'injected' into the organism from outside. It may 'make sense' to the animal to attack something if the causal factors have been introduced, though this thesis would be hard to defend. Archer (1976) argues that it cannot be unambiguously assumed in this, and some similar, studies, that the opportunity for attack *per se* was the reinforcer rather than the appearance of the target object. In some other cases, though, it is difficult to escape the interpretation that opportunity for attack constitutes the reinforcer (see Archer, 1976).

In this context it is important to note that a tendency for aggression can sometimes outlast the initiating stimulus. Thus a monkey which is under attack from a superior may proceed to attack an inferior without apparent provocation. An animal which has been engaged in combat sometimes shows a higher than normal tendency to fight after the rival has gone (see Hinde, 1974, p. 255). However, even if a tendency for aggression outlasts the initiating stimulus and attack is directed at innocuous objects, this is not the same as saying that aggression arises spontaneously and must have a periodic discharge.

If aggression is homeostatic, in the sense of a tendency to aggression building up during the period since the last aggressive act, and then being lowered by attack, one might expect intermittent bursts of fighting between two animals housed together. There is some evidence that this is the case (Hinde, 1970), but it is not a well-investigated area. Even if it were the case that aggression appeared at spaced intervals it would be open to more than one possible theoretical interpretation. The homeostatic notion of motivational energy seeking discharge would require that a period of isolation be followed by a particularly strong tendency to attack. As Hinde (1970, p. 347) argues, the fact that animals reared in isolation are often exceptionally aggressive may not be relevant to this particular question. Such aggressiveness may reflect a permanent or semi-permanent developmental influence. It could possibly be viewed in Archer's terms, i.e. that virtually any intrusion in the environment of such an animal represents a significant disparity. Animals reared in the presence of conspecifics or other animals would have, to put it metaphorically, somewhat broader horizons in life. In some species, though, there is evidence that temporary isolation of the adult animal increases its aggressive tendency. Hinde (1970, p. 347) reports a study involving Siamese fighting fish which were living together in a communal tank with little sign of aggression. After one fish was removed, and held in isolation, it was more likely to show attack on meeting a conspecific. Again, though, more than one interpretation would seem possible. It is difficult to design an experiment that would unambiguously allow us to put Lorenz's ideas to the test.

9.3.2 Hormones and aggression

Archer's model shows hormones exerting an influence on the attack decision. There appears to be a quite close parallel between aggression and sexual motivation (see Brain, 1979 for a review). Castration of male animals immediately after birth normally leads to less fighting in the adult animal. If male mice are castrated before day 6 of life then less fighting is observed in the adult animal in response to testosterone injections than in controls castrated later and similarly injected (Peters, Bronson, and Whitsett, 1972). This difference in behaviour can be prevented from appearing by androgen or oestrogen injection given at around the time of birth. In the case of female mice neonatal testosterone injections increase later potential for aggression. Thus in this context, as in that of sexual behaviour, we arrive at a concept of 'androgenization' or 'masculinization' (Brain, 1979), and a proposed structural or sensitivity change caused by an early androgen presence. The extent to which the influence of early androgenization on subsequent aggressive behaviour is mediated directly via motivational (neural) mechanisms, is not absolutely clear. Muscular strength, body weight and emitted odours are all possible confounding influences in a situation which involves reciprocal interactions between at least two animals. It would certainly appear that a central motivational factor is involved (see Brain, 1979).

Adult castration has a tendency to lower the aggression of male animals (Beach, 1948; Hutchinson, Ulrich, and Azrin, 1965; Scott, 1958). However, this is not true for all species, and the effects of experience may override the hormonal effect. Some animals which have had a successful history of fighting are affected little by castration (see Archer, 1976; Brain 1979) and there is a parallel here with sexual behaviour (Beach, 1948). In those cases where fighting tendency is reduced by castration it is almost always restored by testosterone replacement (unless the animal experiences many defeats in the meantime).

In some species fighting tendency seems faithfully to reflect androgen levels. For example, seasonal breeders show maximum aggressive tendency at the time when androgen levels are maximal. The females of some species which are normally aggressive show little or no aggression at oestrus, which suggests that oestrogen may be able to inhibit fighting (Beach, 1948). Of course, this would only be true if certain developmental conditions had also been met.

Exactly how do androgens affect the adult nervous system in the context of attack behaviour? We do not know, but Archer (1976) mentions two possibilities. They could simply bias the decision mechanism in the model towards attack. They might alternatively increase the animal's persistence. In other situations, such as searching for food testosterone-injected animals show high persistence.

9.4 THE CHARACTERISTICS OF FEAR

9.4.1 General

Archer (1979) discusses the concept of a unitary scale of fear. According to such a concept intense fear and mild fear are quantitative differences in a single motivational state. The same question arises in other contexts; it may be argued that 12 h of water deprivation is qualitatively similar to 48 h of water deprivation. The animal makes the same kind of responses to 48 as to 12 h water deprivation, e.g. bar-pressing for water, reduction in food intake, etc. However, drive itself is a poorly understood concept, discussion of which usually ends up generating more heat than light, and therefore the concept of a unitary drive inevitably builds upon a quicksand. Provided this caution is held close at hand then the unitary drive concept raises some interesting questions concerned with fear. For instance, do fears which on one measure appear quantitatively similar lead to similar responses when examined by another measure? Do fear responses reliably change qualitatively when the central motivational state reaches a particular magnitude?

How do we measure fear and attempt to quantify it? In 1934 the so-called open-field test was devised by Hall. In this test the rat was placed in a closed arena for a specified period of time and the amount of urination and defaecation noted. More recent investigators simply employ the number of faeces dropped as the measure. The open-field test is often defended as providing a measure of fear (Gray, 1971). After repeated exposure to the test an animal's defaecation score decreases, and it is presumed that the novel stimulus becomes less fear-evoking. Further, there is a negative correlation between the faecal score and the amount of food eaten in the test by a hungry rat, and it may be argued that the reciprocal of food eaten is an index of fear. If a loud noise is present in the test, faecal score is increased (see Gray, 1971).

Archer (1979) doubts whether the unitary drive concept may usefully be applied to fear. He notes that the correlations between various measures of fear are sometimes poor, and that vastly different fear responses are available to the animal. The differences between individual animals are enormous, while different test situations can elicit widely different responses. Some rats freeze when placed in the open field while others run around the edge of the arena. On a common sense view both animals are showing strong fear reactions, and yet if we measure fear by lack of movement we would get two widely different measures.

Archer (1979) writes:

'We are, therefore, not dealing with one specific behavioural response characteristic of fear or escape motivation but with a complex of reflexes and co-ordinated actions, which are capable of being organized in many different ways, for the many different fear evoking circumstances which animals face.' (Reproduced by permission of Van Nostrand Reinhold.)

All of this raises the question of whether or not widely different responses such as withdrawal and immobility may be usefully classified together as fear responses.

The fact that they are both evoked by similar stimulus situations argues for such a classification. Furthermore, viewed in terms of a functional classification both serve to protect the organism from harmful stimuli.

Gray (1971) finds it helpful to categorize fear stimuli into four classes: (1) intense stimuli, (2) novelty, (3) special evolutionary danger stimuli, and (4) social interaction with conspecifics. This draws attention to the fact that evolution would favour animals which developed a particularly sensitive fear response to the most commonly encountered predator. Although one is reluctant to mention the innate or learning dichotomy, without several paragraphs of reservation and qualification, it seems almost inescapable to evoke the concept of a specific innate fear in some species. Disparity may not be the most useful model in this context. In monkeys a fear of snakes would be one such example.

Some problems bedevil the scientific study of fear which are not so apparent in other behaviours (Archer, 1979). Fear behaviour cannot be defined simply in terms of form. Fast locomotion and even immobility are forms of behaviour which occupy the animal in situations which clearly have nothing to do with fear. High intensity fear may usually be fairly unambiguously recognized, but there is no clear dividing line between low-intensity fear and negative preference. Is the avoidance of a novel food by a rat a case of fear, as is suggested by the label *neophobia*, or is it more usefully viewed as a strong negative preference?

Archer (1979) gives the following conditions which enable an unambiguous case of fear to be identified (immobility is discussed later):

(a) a normal method of locomotion would be employed but at high intensity;
(b) the direction of movement would be away from a source of potential danger, or something having a characteristic in common with a source of danger, for instance a large moving object;
(c) escape would be towards a safer location if one were present;
(d) autonomic changes characteristic of exertion would be seen, for instance heart rate would increase;
(e) expressive vocalization or odour secretion may be seen.

Immobility is also associated with autonomic changes such as high heart rate (it is worth drawing a distinction between immobility and death feigning, which is also an anti-predator device (Archer, 1979)).

Viewed in an evolutionary context, various factors may be seen to favour either immobility or escape. Predators may be particularly sensitive (or perhaps only sensitive) to moving objects. The prey may have camouflage, which would favour immobility as an anti-predator device. Conversely, if the most commonly encountered predator were slower in locomotion than the prey this might favour active escape. Within a given organism situational factors would be expected to bias behaviour towards either escape or immobility. On first exposure to a cat the laboratory rat escapes, if this is possible, but freezes if escape is blocked (Blanchard and Blanchard, 1971). As Archer (1979) points out, if the animal is already moving it might be best to continue to do so since there is a high probability that it will have been detected. Conversely, the stationary animal might find immobility the better strategy. Some experimental evidence suggests that these respective responses do occur. Furthermore, the individual history of each animal in terms of what has proved a successful strategy could also bias the decision (Hogan, 1965). There is some evidence to show that immobility is associated with diffuse and unlocalizable fear stimuli, whereas escape is more closely associated with specific and localizable stimuli (Hogan, 1965). It is not difficult to appreciate the adaptive significance of this. Gray (1971, p. 26) discusses conditioned freezing. In this study, whereas shock elicited active escape or attack in rats, a conditioned stimulus which had been associated with shock came to elicit freezing. Gray suggested that it may have adaptive value to run from stimuli but freeze in response to the warning of impending harm.

Thus it may be unhelpful always to view immobility as the response to fear more intense than that which stimulates active avoidance, though that is not to deny that high-intensity stimuli do cause immobility. The animal has at least two responses which are available, and the decision would depend upon the evolutionary experience of the species, the immediate situation, and the individual animal's past history. As Archer (1979) reminds us, to propose that the fear response is the result of a complex integration of information relevant to the animal's immediate situation is to involve a rather sophisticated control system.

Species differences in the reaction to aversive stimuli have played an important role in the development of the theory of avoidance learning proposed by Bolles (1970, 1975). In the discussion which follows it should be noted that escape refers to the animal terminating an excessive stimulus such as electric shock, either by fleeing or pecking a key, etc. Avoidance refers to avoiding it altogether by taking anticipatory action. Bolles argues that:

'. . . an animal must have a repertoire of innate species-specific defence reactions that occur when it is frightened.'

Freezing in the rat in response to fear is an example of a species-specific defence reaction (SSDR). When the animal is observed to perform fluently a single response in the laboratory it is because other SSDRs have been suppressed. Bolles (1970) reminds us of the effects of shock on normally placid domestic and laboratory animals. It evokes SSDRs such as avoidance and attack and in the presence of the experimenter the animal ultimately restricts its response repertoire to these. In the laboratory, rats may be readily taught to perform certain responses when a warning signal is given in order to avoid a shock. Other responses are learned with the greatest difficulty, if at all. For example, it is very easy to train a rat to jump or run from a location where shock occurs to a safe location. It is difficult to train them to avoid shock in a shuttlebox and some never learn (Bolles, 1970, 1975; Garcia, Clarke, and Hankins, 1973). In this apparatus the animal must constantly move from one side of a box to another in order to avoid shock; what is safe now is dangerous on the next trial. Considering the natural environment the most important feature of escape behaviour is the *direction* of the threat with respect to the threatened animal. In the laboratory, animals appear to be very capable of learning an escape or avoidance response provided it transports them from the location of danger to one of perceived safety. This is precisely what the shuttlebox does not do. The response of running in a wheel is also learned with some difficulty, presumably for the same reason, whereas running away from the location of shock is easily learned.

Bar-pressing to terminate shock (i.e. an escape response) is clearly not an SSDR and bears no similarity to any. However, it is occasionally learned by a rat, and Bolles (1970) proposes that the explanation in such cases is that the rat freezes while holding the bar. Freezing anywhere else in the box is ineffective.

This subject is discussed very lucidly in an excellent book by Rachlin (1976, p. 357). He starts by noting that it was once believed that pigeons could not be trained in the laboratory to escape or avoid aversive stimuli. The task the bird was set, but failed to learn, consisted of pecking at a key in order either to terminate or avoid shock. In association with positive reinforcement, such as food, the pigeon can, of course, readily be taught to peck a key. The SSDR of the pigeon to a sudden and intense shock is to fly away. Indeed, pigeons could be taught to raise their heads, presumably an initial component of the flight response, in order to escape or avoid intense shock. However, it was found later that pigeons could after all be trained to press a key if the shock was first applied at low intensity. If shock intensity is high they flee, whereas if it is low they attack (though, again, defence against an unusual 'predator' rather than attack may be a better description), which accords with the model proposed by Archer (1976). If the key is illuminated at the time shock is applied pigeons attack the key, either by pecking it or by hitting it with their wings. Given that these reactions are effective the pigeon becomes increasingly quick at emitting them, and full attack is moderated to a relatively gentle brushing of the key with the bird's wings. In the case where violent pecking was the initial reaction a more calm pecking response gradually emerged. Rachlin wrote:

'The response undergoes a metamorphosis from a natural species-specific defence reaction to an instrumental response that resembles the initial reaction only superficially. . . .'

Otis and Cerf (1963) studied fish in a shuttlebox. The task required the fish to swim towards the illuminated side of the division in order to avoid shock. Thus there was an unambiguous direction cue present, in distinction to the rat studies. Goldfish learned the avoidance task more easily than Siamese fighting fish. The explanation given was that withdrawal is the natural response of the goldfish to a sudden intrusion whereas the Siamese fighting fish has the specific reactions of either immobility or fighting.

Such results as these remind us of the arguments in the literature on preference and aversion in food intake, discussed earlier. In each situation animals readily perform behaviour which is biologically appropriate, and fail to form associations which would be inappropriate to the animal in the natural state. Associating the taste of food with a subsequent discomfort is biologically appropriate for a rodent, as is running away from a warning of danger. By contrast, it would hardly have been useful in their evolutionary history for rodents to associate a flashing light at the time of feeding with subsequent ill effects; neither would any kind of response other than fleeing or freezing have been appropriate in the fear context. Although bar-pressing for food is obviously not a natural response for rats, some kind of manipulation with the paws inevitably accompanies the gaining of food, but is far removed from fear-motivated behaviour.

The pain of electric shock is a very peculiar stimulus by any standards. It is, as Garcia, Clarke, and Hankins (1973), so rightly remind us, '... pain disembodied from stimulus features naturally accompanying attack'. At low intensities of shock applied through a floor grid, rats have a tendency to lift one paw, examine, and lick it, which is obviously appropriate in a natural setting to the presence of a thorn in the paw (Garcia *et al.*, 1973). If a rat has mastered an avoidance response to electric shock it will abandon this successful strategy for the unsuccessful, but biologically appropriate, response of attacking another rat.

9.4.2 Interaction between fear and other behaviours

The presence of stimuli associated with fear may either facilitate or compete with other behaviours. In this respect, according to the evidence marshalled by Archer (1979) mild fear is facilitatory and intense fear inhibitory. Electric shock can facilitate sexual behaviour in male rats (Barfield and Sachs, 1968), though whether this is best described as an interaction with fear (Archer, 1979) seems open to question. Almost any kind of behaviour can be facilitated in rats by pinching the tail (Robbins, 1978).

Intense fear can result in total suppression of other activities such as feeding and sexual behaviour. The phenomenon of *conditioned suppression* refers to the reduction in operant behaviour for food or water reward which is brought about by presentation of a stimulus previously associated with electric shock. Feeding is inhibited if food is placed in an unfamiliar container at a different location to normal.

It appears possible to place an approach tendency which we call exploration in conflict with a withdrawal tendency which we will call fear. This has been observed in a variety of animals such as Jackdaws, chimpanzees, chicks, etc. when confronted with a novel object (see Hogan, 1965 for a review). Fear usually habituates first, and the animal ends up exploring the novel object. In Archer's terms one might wish to say that the novel object becomes incorporated to some extent into the animal's expectation. Another experimental design consists of placing fear (the avoidance tendency) in opposition to hunger (the approach tendency) by putting food at a location which the animal associates with shock. The animal appears to find an equilibrium position at a certain distance from the food-shock, where approach is equal and opposite to avoidance.

9.5 A CATASTROPHE THEORY OF FEAR AND AGGRESSION

Mention of the expression 'catastrophe theory' to students usually either elicits fear or laughter, a behavioural ambivalance which itself is perhaps explicable by catastrophe theory. Lying behind catastrophe theory are some very sophisticated mathematics developed by René Thom. We are merely going to employ a very simple example of a catastrophe in a descriptive sense. Without going deeply into mathematics, we can gain illumination of the possible mechanisms behind fear and attack behaviour by means of catastrophe theory. A popular account of this theory was produced by Zeeman (1976), using fear and attack as an example, and in the construction of this section I have made extensive use of Zeeman's work. Catastrophe theory was specifically developed to deal with discontinuous behaviour, and may be applied to such diverse subject matter as animals, plants, earthquakes, and the Wall-Street crash. Traditional mathematics evolved to handle smooth and continuous phenomena.

The analysis of fear and attack which Zeeman (1976) presented is based upon somewhat different premises to those of the arguments we have up to now pursued. However, it involves some essential features in common with the present analysis. To be precise it incorporates a certain ambivalence in the animal's behaviour with regard to fear and attack. Whereas in a given situation Archer's model allows one, in theory at least, unambiguously to predict either fear or attack according to the magnitude of disparity, Zeeman's model does not allow this. For a given magnitude of causal factors either fear or attack can result. What then determines the response? It is here that we need catastrophe theory.

Zeeman discusses fear and aggression in dogs in the following manner. If a dog is frightened but anger is not aroused fleeing is very likely. If it is angered but not frightened attack is very probable. What happens if it is both frightened and angered? Other theories might predict that the two tendencies would cancel each

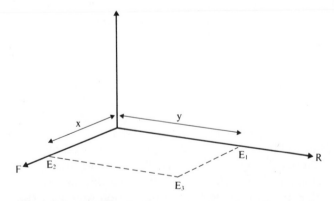

Figure 9.2 Catastrophe theory representation of causal factors. R—rage dimension, F—fear dimension. E_3 therefore has both a rage component E_1 (magnitude y) and a fear component E_2 (magnitude x)

other out, leaving no behaviour at all, but according to catastrophe theory this is very unlikely to occur. Let us refer to Figure 9.2 which shows the causal factors plotted on a two-dimensional diagram. Dimension F represents fear and dimension R represents rage. Point E_2 therefore represents a causal factor of pure fear having a magnitude x. Point E_1 represents a causal factor for pure rage having magnitude y. Causal factor E_3 has both fear (of magnitude x) and rage (of magnitude y) components. In summary, causal factors are shown in the horizontal plane as vectors, the length of the vector showing the strength of the causal factor and the direction its quality. Behaviour is shown in the vertical dimension. Figure 9.3 shows the convention which Zeeman uses. Short-length vectors indicate full retreat, while intermediate lengths indicate withdrawal and threat, and the longest length indicates full attack.

In Figure 9.4 a behaviour surface and a causal factors surface are shown. For much of the animals behaviour if we know the point on the causal factors surface we can then project upwards and we arrive at a point on the behaviour surface. The distance between the two surfaces is the length of the vector which tells us the predicted behaviour. But beware of the pleat, for it is here that the subtle message of catastrophe theory appears. The causal factor surface is a flat sheet, but the behaviour surface is a pleated sheet. If we project upwards from under the pleat we hit the surface in two places. Which of these do we take to be the behaviour the animal executes? When we project upwards from a point at the far end of vector R, one of extreme rage, we unambiguously arrive at a point corresponding to attack, indicated by the length of the vertical distance. When we project upwards from F, extreme fear causal factors, we hit the behaviour surface at a location indicating, by its short vertical distance from the causal factors plane, retreat.

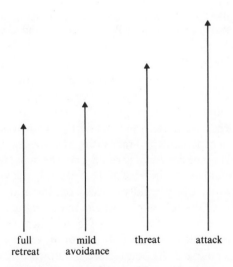

full   mild   threat   attack
retreat   avoidance

Figure 9.3 Convention adopted in the representation of fear and attack in terms of
catastrophe theory

216

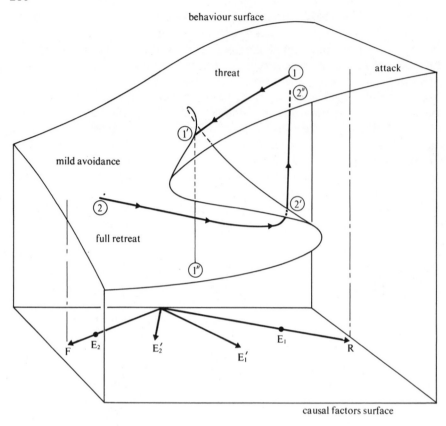

behaviour surface

threat ① attack

①'

mild avoidance ②"

② ②'

full retreat

①"

F E₂ E₂' E₁' E₁ R

causal factors surface

Figure 9.4 Catastrophe theory model of fear and attack (Based upon Zeeman, 1976 Reproduced by permission of W. H. Freeman and Company)

Imagine now causal factor E_1 being present. The animal has a rage component y (see Figure 9.2) but no fear component. Projecting upwards we hit the behaviour surface at a point (point 1) indicating threat or mild attack behaviour, such as growling. Suppose now the situation, for whatever reason, changes slightly to fear. In other words, we have the vector E_1' with both rage and fear components. Fear then increases and rage decreases. We can follow the predicted behaviour by following the course of arrow 1 over the behaviour surface. It may be seen that the point falls over the edge of the pleat at 1′ and lands after the fall at point 1″. In other words it very rapidly moves from threat to retreat. Let us now follow behaviour 2 over the behaviour surface. In the beginning causal factor E_2 is present, indicating a purely fear causal factor. Then a component of rage appears and increases in intensity so that the vector moves to E_2'. When projected upwards, this traces course 2. At point 2′ we fall upwards over the edge of the lower pleat and land at 2″. This catastrophe takes us in one instant from mild avoidance to attack.

Outside of the region which falls under the pleat the magnitude of the causal factors uniquely defines the resultant behaviour. Under the pleat region we must, in addition, know the point from which causal factors have started.

9.6 SUMMARY, GENERAL CONCLUSIONS, AND DISCUSSION

It can only be hoped that we will soon be in a position to amalgamate features of the Archer and Zeeman models. Whereas Zeeman is rather unspecific about what constitutes the causal factors, basing the model upon the anthropomorphic expressions rage and fear, Archer is more specific and objective. Given that certain constant conditions prevail, such as an escape route being present, Archer sees the causal factor for escape as being essentially only qualitatively different from the causal factor for attack. That is to say, the cause of escape is a larger disparity of the kind which causes attack. By the vector notation which he employs Zeeman demonstrates a belief in qualitatively different causal factors for fear and attack.

The fundamental problem as seen by the present writer is as follows. Until we can build simulations using actual numbers it is exceedingly difficult to make intelligent statements at any level above the rather superficially descriptive. But by their very nature attack and escape are not behaviours which can be easily quantified. The actual behaviour often involves a complex interaction of two animals with uncertain feedback consequences. By contrast, when dealing with feeding and drinking the animal's behaviour can be measured in grams of substance ingested. Sexual behaviour might at first sight be thought to be difficult to quantify. But at least in rats it can be analysed in terms of rather stereotyped intromissions and ejaculations, and the intervals between. Concerning causal factors, in the case of feeding and drinking we can begin to build quantitative models using such factors as energy and fluid state, content of diet and physical distance of food/water. To make an analogy with fear/attack, how would one begin to quantify as a causal factor the behaviour of an opponent animal? This is not to argue against model building since important conceptual issues only come into focus by constructing models. However, it is as well to state limitations in contexts such as this.

Even a superficial analysis would reveal that the model proposed by Archer is vulnerable to instability and dithering unless additional features are proposed. Consider an animal at the point of transition between avoidance and attack. One might imagine that disparity would fluctuate with the moves of an opponent. The animal might therefore dither between attack and avoidance, hardly a profitable strategy. This problem was taken up by Toates and Archer (1978), who argued that in the case of fear the system must involve a mechanism that gives momentum. In other words an initial fear-eliciting stimulus goads the appropriate escape/freezing response for a period of time which extends well beyond the time at which detectable disparity is present. For example, to end freezing and move off immediately a predator passes would invite detection. In the case of active escape the animal does not switch to attack when disparity falls to a value which would otherwise be associated with attack. There appears perhaps less need to provide momentum to attack than to fear. If the opponent flees or withdraws, the reduction in magnitude of disparity could perhaps profitably be associated with an immediate reduction in attack behaviour. It might be maladaptive if this were not the case, since the object of attack is to restore the *status quo*, not to get locked in a deadly struggle. It probably would not be wise to pursue a retreating intruder. However the results suggest that sometimes the attack tendency slightly outlives the stimulus.

As we have seen in the discussion earlier, the catastrophe model of Zeeman very clearly accounts for the problem of persistence. If we start from causal factors of unambiguous fear then escape persists even though the causal factors for escape are reduced to a minimum and causal factors for attack have arisen. Conversely, if we start with causal factors unambiguously associated with attack then a considerable fear component can enter the causal factors but still attack persists. Of course a point may be reached where if fear is very great the animal reaches an escape catastrophe. In other words this describes a very stable behavioural system, not prone to dithering. If and when a transition does occur it is decisive for the animal's behaviour. It would be interesting to simulate catastrophes with some kind of disparity signals as inputs, and hence see the effect of feedback via the animal's behaviour.

It may be thought by some to have been an omission not to have introduced at an early stage the classification system of aggression devised by Moyer (1968). He divided aggression into the following seven categories—predatory, inter-male, fear-induced, irritable, territorial, maternal, and instrumental. Such categorization was not merely for the purpose of providing an economical summary; Moyer argued that different physiological bases underlie these aggressive behaviours. By contrast Archer (1976) argues for a common mechanism being involved in a wide variety of aggressive behaviour. However, he considers predation to be quite different from other forms of aggression with its distinct motivational basis and neurophysiological machinery. With the exception of instrumental aggression, Archer argues that the other five of Moyer's classes of stimuli which provoke attack all have in common the feature that an intrusion of some kind gives rise to what Archer calls disparity. He summarizes his position as follows (Archer, 1976):

'The present model reflects both the diversity of "aggression" (emphasized in Moyer's neurophysiological approach), in that different external, internal, motivational, and hormonal factors can be involved, and also its unity, in that mechanisms for attack and fear behaviour are regarded as being activated by a common type of variable (discrepancy).'

Moyer's classification system has been criticized by Huntingford (1976). She notes that, even if we accept this scheme, in many species protection of young is performed by both sexes and therefore 'parental' should be the description rather than 'maternal'. More fundamental criticisms are also levelled, in that the criterion of classification seems to vary from situation to situation. Inter-male and predatory aggression are distinguished by exclusive categories of the target of attack. However, fear-induced aggression is in terms of an underlying emotion, whereas instrumental aggression is defined in terms of a consequence of aggression. Huntingford acknowledges that Moyer did not intend these categories to be entirely mutually exclusive, but argues that the cases where aggression falls neatly into one of the seven categories are exceptional. Inter-male and territorial aggression are usually not distinct, and she claims that one could often feel justified in attaching the labels fear-induced, sex-related, maternal (parental?) or instrumental to such aggression. Furthermore, Rodgers and Brown (1976) argue that common brain mechanisms lie at the roots of various forms of aggressive behaviour. Galef (1970a) claims that timidity and aggression share a common neural basis, and uses this to argue that both are a response to certain kinds of novelty.

Nonetheless, problems are apparent in the approach of Archer, as we have seen in the course of evaluting the model. As an explanatory device 'disparity' sometimes appears to be a blanket description which then demands *ad hoc* qualifications of the kind that the expectation model is resistant to incorporating certain stimuli. Moyer's scheme appears to overstate the position in so far as it involves distinct and endocrinological processes for seven classes of aggression. But it is a flexible model in which, by means of evolutionary selection, certain classes of aggression could be sensitized where others can be allowed to assume a secondary role. For example, as Huntingford (1976) notes, in some species predation against the young is more common from conspecifics than heterospecifics while in other species the reverse is the case. We have also had cause to consider a distinction between attack of the kind which a dominant animal makes against an intruder and self-defence 'attack'.

Finally, we must return to the model of aggression proposed by Lorenz. This is primarily a model of fighting between conspecifics. Lorenz gives examples, particularly of fish, where aggression is selectively provoked by the particular colours of a conspecific. In such a case one might say it is lack of disparity with specific internal representation which, together with other factors, provokes attack. Lorenz considers intraspecific aggression to serve an important function in the survival of the species. Thus it means not only that the fittest members reproduce, but by species-specific territoriality animals distribute themselves widely. However, this introduces the much-debated issue of whether selection can ever operate in the interests of species-survival. Dawkins (1976) argues that it does not, and this view is reflected in the title of his book *The Selfish Gene*. The other aspect of aggression which Lorenz discusses is its spontaneity. Aggression can never be truly spontaneous since it always involves a target. However, if it could be shown that animals do in general visit locations of past fights, perform operants, etc., then this would make the theory more realistic. There is some evidence that this does occur occasionally. However, the evidence is derived from very few species under particular conditions. Lorenz claims that the threshold for aggression is lowered as a function of deprivation. The evidence is not convincing on this point. Were it to be otherwise, then, as a metaphor, motivational energy-accumulating and seeking discharge might give a reasonable account, even though we don't believe it has physical reality. Although aggression may be more apparent between conspecifics usually a component of territoriality and novelty is involved as well. Intruder conspecifics are particularly likely to evoke attack. However, the model of Lorenz would have much difficulty explaining how in established rat colonies aggression is almost never shown (Galef, 1970a).

CHAPTER 10

Exploration

10.1 INTRODUCTION

Animals commonly explore a new environment into which they are placed. First, a brief summary of motivational systems will help to place such exploratory behaviour in context. Feeding and drinking depend upon the interaction of information derived from, on the one hand, biochemical and physiological events within the organism, and, on the other, perception of water and food in the environment. Behavioural temperature control may be understood in terms of the interaction of central and peripheral measures of temperature. In each of these cases the animal's physiological state and its detection of environmental events influence behaviour. In turn, behaviour is purposeful in that it has consequences for physiological state. Sexual behaviour and fear/attack can be understood in terms of the processing of sensory information by neural mechanisms which are strongly influenced by hormones.

In so far as physiologically based explanations of exploration have been made they involve only the nervous system. No extraneural influences have been postulated and there seems no need to do so. Exploration is provoked by the interaction of the state of the nervous system and the stimuli which are impinging upon its sense organs. In order to understand exploration we therefore need to consider the stimulation (or lack of it) arising in the environment and the processing of this information by the nervous system. We also need to consider the animal's memory of environments explored in the past. Because biochemical factors are not discussed in connection with exploration (except in the competition which fear offers) this is not to say they are irrelevant. Memory has an obvious biochemical aspect. Hormones may influence exploration. However, the effects are probably indirect and of little relevance to explaining the processes immediately underlying exploration.

In this chapter we will return to the subject of fear, this time in terms of its close relationship to exploration. Researchers have frequently placed animals in ambivalent situations that evoke exploration and/or fear, and have discussed theoretical models in these terms.

Shillito (1963) defines exploratory behaviour as 'behaviour which serves to acquaint the animal with the topography of the surroundings included in the range'. Exploration involves the maximization of information impinging upon all the animal's sense organs, e.g. sniffing, head rotation, ears pricking up, extension and retraction of the body, vibrissae contact, etc. Exploration is terminated once the animal is somewhat acquainted with a formerly novel object/environment or after a periodic 'patrol' of familiar territory. How do we measure exploration? Given one single novel object then the amount of time spent in close contact with it, sniffing, touching and gazing, would seem an appropriate measure. If an environment contains many complex objects then it is usual to measure the extent to which the animal moves around from object to object. If the environment consists of several maze arms or tunnels then the extent to which the animal visits all or most of these would need to be reflected in the exploration measure. This is not without its problems though. As Berlyne (1960) notes, locomotion even when accompanied by exploration is of dubious value; is an animal which passes rapidly from object to object with a cursory sniff showing more exploration than one that concentrates on a particular feature of the environment?

When allowed to, hungry rats more quickly leave a familiar home cage for a new location than food-sated rats (Fehrer, 1956). Using latency to explore as an index, it appears then that hunger potentiates exploration. However, if rats are introduced to a new complex environment and the extent to which they explore the variety of objects within it used as an index, hunger does not increase exploration (see also Einon and Morgan, 1976). As Fehrer argues, if exploration is already very high (i.e. being placed in a novel environment) it is difficult for hunger to potentiate it. Again we must pose problems of biological relevance. It would seem beneficial for hunger to motivate an animal to leave a familiar but foodless location to explore. It may not be helpful to explore a completely novel environment any faster when hungry.

Animals spend time exploring their environment even though apparently sated in all identifiable respects, e.g. ample food and water. In order to account for this, researchers have been forced to abandon or radically modify theories which were applicable to behaviour associated with specific tissue (e.g. thirst) or physiological (e.g. temperature) need. Instead only events within the nervous system could be employed in theory construction. It was only relatively recently that exploratory behaviour took its place in the literature alongside hunger, thirst and sex, etc., as a worthy subject in discussions of motivation.

In exploratory behaviour, known also as curiosity, the animal's sense organs are (Berlyne, 1966):

'... brought into contact with biologically neutral or "indifferent" stimulus patterns—that is, with objects or events that do not seem to be inherently beneficial or noxious.'

The information derived from such curiosity may, of course, subsequently prove useful to the animal.

In the West we speak of 'exploratory' behaviour, while in Eastern Europe 'orientational–investigatory activity' (Berlyne, 1966) is investigated. However, rather different experimental procedures and traditions underlie the subject matter of the two schools. Working with dogs, the Russian investigator Pavlov had observed and scientifically documented what he called the 'orientational' or 'investigatory' reflex (see Berlyne, 1966), a reflex with which we are all familiar. When a novel stimulus is suddenly presented to a dog it halts its ongoing activity, while turning its eyes and head towards the stimulus. Although this was termed an unconditional reflex, it was, in common with conditional reflexes, subject to extinction and disinhibition. If the stimulus was repeated, but with little or no consequence to the animal, then it gradually lost its power to evoke orientation.

In the West the study of exploratory behaviour developed in association with mazes and Skinner-boxes. For example, the animal would be introduced into a novel maze or chamber and its investigation of the objects present recorded. In contrast to Russian research, the animal was expected to move its whole body in the investigation, and stimuli did not appear suddenly. Both aspects of behaviour share some common properties, and sometimes both may be seen in response to a stimulus. It is often argued that they are aroused by novelty and extinguish with repeated presentation. However, O'Keefe and Nadel (1978) argue that it is misleading to classify the orientation reaction and exploratory behaviour together as responses of the organism to novelty. Orientation is an initial reaction which may be followed by either escape, approach (attack or exploration) or indifference. The orientation reaction (sometimes accompanied by defensive startle) is elicited by intense stimuli, whether novel or not. If, after orientation, the animal neither attacks nor flees, and the stimulus is novel, exploration is the likely outcome. In other words, the orientation reaction is not necessarily aroused by novelty. Neither is it necessarily followed by exploration.

Exploratory behaviour and orientation serve as cautionary tales; they were at first seen as something of a nuisance which detracted from the object of the research. Apart from the early work of Tolman and a few others, exploratory behaviour really only became respectable in Western psychology after the second World War.

Taking both Soviet and Western research into account, Berlyne (1960) finds it useful to classify exploratory behaviours in terms of three categories. *Orienting responses* consist of changing the orientation of the sense organs or of the head relative to the stimulus. *Locomotor exploration* involves locomotion of the body relative to the stimulus concerned, for example entering a tunnel or approaching and sniffing a novel object. *Investigatory responses* are those in which the animal effects a change in the environment, for example picking up an object and shaking it. Other categorizations, which would cut across these boundaries, might be needed for some purposes. Berlyne (1960) speaks of *extrinsic* and *intrinsic* exploration. Extrinsic exploration is closely associated with a biological reinforcer such as food or water. In the context of exploration, a hungry animal which selectively enters and explores an alley where food has been found on previous occasions may be said to be extrinsically motivated. By contrast, an animal which is sated for food and water but which repeatedly presses a bar for the opportunity to view a novel environment may be said to be intrinsically motivated. Two further expressions which are commonly found in the literature are *inspective* and *inquisitive* exploration. The former refers to behaviour which enhances the contact with stimuli already impinging upon the organism, for example, moving nearer an object sighted in the cage. The latter refers to behaviour which brings the animal into contact with stimuli not yet impinging on the sense organs, for example regularly taking the choice in a T-maze that leads to a complex room.

Weisler and McCall (1976) observe, within and across species, a certain similarity, one could even say *stereotype*, in the pattern of investigation. The sequence is usually as follows:

'(a) *alerting* or the recruiting of attention to the new stimulus situation,
(b) *distance-receptor scanning or examination* of the new situation,
(c) *motor-aided perceptual examination* of the situation in which the organism may perambulate the environment or manipulate the object to explore it more completely, and (d) *active physical interaction* apparently for the purpose of discerning what will happen as a consequence of the organism's interactions with the object or situation.'

Weisler and McCall note that components (b) and (c) may alternate, and as the organism matures some of the phases may be reduced or eliminated.

It serves a rather clear adaptive role for an animal to familiarize itself with its environment. For example, potential food sources and escape routes are two vital pieces of information. The kind of investigatory behaviour shown in the laboratory can also be seen to have relevance in the animal's natural environment. In monkeys, manipulation of objects, lifting and inspecting them, could be valuable in connection with food seeking (Berlyne, 1960). In some species moving vegetation aside so as to have a clear view could reveal a predator or

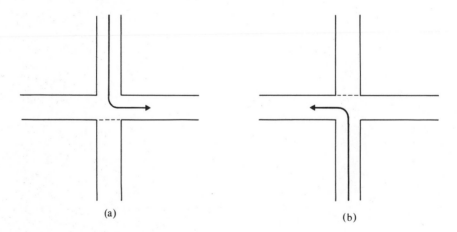

(a)   (b)

Figure 10.1 Alternation: (a) the animal turns east on the first trial; and (b) on the second turns west

quarry. Darwin noticed that chimpanzees show terror on being presented with a snake in a box, but nonetheless lift the lid occasionally and peep inside.

Although it has adaptive value for the unfamiliar to become familiar, the process of investigating the unfamiliar often involves danger. Confronted with novel stimuli some ambivalence in the animal's behaviour may be apparent. Laboratory strains of *Rattus norvegicus* show considerable inquisitiveness when confronted with novel objects, though signs of fear may also at times be detected. Newly opened-up areas of their environment are quickly explored. Although their wild conspecifics show little hesitation in exploring new areas, novel objects are approached only very reluctantly.

In this chapter we will first review the laboratory evidence on exploratory behaviour and then attempt a modest theoretical interpretation. We will then examine the arousal of both fear and exploration by novel objects.

10.2 THE EXPERIMENTAL STUDY OF EXPLORATION

10.2.1 Spontaneous alternation

The phenomenon of *spontaneous alternation* was described by Tolman in 1925 (see discussion in Glanzer, 1953). A rat is given the choice of turning left or right in a T-maze. Both routes lead to a goal-box, so no advantage is attached to either response. Sometimes rats are made hungry or thirsty and rewarded with food and water for either response. At other times no reward is provided, and sometimes rats are tested when satiated. Essentially the same result is obtained in each case (Halliday, 1967); that is that rats show a tendency to alternate between turning left and right. Some 65–70% of choices reflect alternation. Species differences need investigating; the effect is not shown by 2–6-day-old White Rock chicks (Hayes and Warren, 1963) or by hamsters (Sinclair and Bender, 1978).

There has been some controversy surrounding the interpretation of spontaneous alternation (see Glanzer, 1953). However, evidence was soon gathered to favour the interpretation that it is stimuli associated with the two alternative routes that are responsible for alternation. Another possible interpretation might have been that alternating responses, patterns of muscular effort, underlie the phenomenon. Anthropomorphically speaking, it is as though the rat is bored with the stimuli associated with the immediately previous turn and curious about the alternative (even though this alternative was traversed only slightly before the now boring last choice). It is possible to devise experiments which distinguish stimulus-based from response-based explanations of alternation. Figure 10.1 shows a cross-shaped maze in which the animal runs from, say, the north end, with the entrance to the south end blocked by the experimenter. Suppose the animal turns east. On the next trial it is released from the south, with the entrance to the north arm blocked. Now if it were to alternate locomotor responses *per se* it would tend to turn east again. But if it were alternating stimuli to which it attended it would tend to turn west, which is what tends to happen in practice.

The animal is responding on the basis of stimuli experienced on the *previous* trial. Controlled experiments rule out the possibility that the animal is simply avoiding its own odour (see Dennis, 1939), but this is one possible cue amongst several.

As the time interval between trials increases then so the strength of the alternation tendency decreases, until at about 15 min or more the animals show only chance performance, i.e. alternating on 50% of the trials (Halliday, 1967). Some experiments show alternation despite very long inter-trial intervals (Still, 1966). If the animal is detained within the arm of its choice for a period of time there is a relatively strong tendency for it to alternate on the next trial. Even if a rat is placed in one arm or goal-box by the experimenter it will tend to make the alternative choice on the next trial. Glanzer (1953) uses the expression *stimulus satiation* to describe the reaction to stimuli which have recently impinged upon the organism.

The precise cues that a rat employs in alternation have been investigated by Douglas and associates. Although rats alternate maze arms rather than responses it appears that both stimulus and response-based information is used in making the decision. Rats have a weak tendency to avoid the odour trail from the previous run (Douglas, 1966). They have a tendency also to use extra-maze environmental cues to steer them to the novel arm. These stimuli would of course normally complement each other. In addition a rat seems to have a very good orientation sense in so far as its own body is concerned (Douglas, Mitchell and Del Valle, 1974), employing the vestibular system. In other words it uses inertial navigation to construct a spatial map of the maze. Douglas *et al* obtained the interesting result shown in Figure 10.2 when mazes of different angles between the arms were employed. Rats appear to make vicarious responses at the choice point. They move their heads through a certain angle to orient first one then the next arm, and this is one cue as to spatial orientation (see also Still and Macmillan, 1975). If the arms are almost parallel this source of information is not available. As Douglas *et al.* note, response-based alternation comes back in a guise, that of responses helping the rat to select which arm was previously visited.

If a rat is allowed to take one of two paths to food, one having a fixed number of left and right turns, or one whose pattern of turns changes from trial to trial, it tends to choose the latter more often. This is complicated by the fact that there is a significant tendency to take the opposite path to that taken on the last trial (see Berlyne, 1960). In a Dashiell maze, where twenty possible routes of equal length between start and goal-box are available, rats vary their route between these possibilities. Such results lead to the conclusion that the rat is:

'. . . seeking novelty and shunning monotony' (Berlyne, 1960)

In some experiments it may be clearly demonstrated that the alternation tendency is determined by stimuli which are present only after the response has been made. Thus the strength of alternation is greater if the arms lead to two different goal-boxes than if they both lead to the same box (Sutherland, 1957), though there is a significant tendency to alternate even in the latter case (see

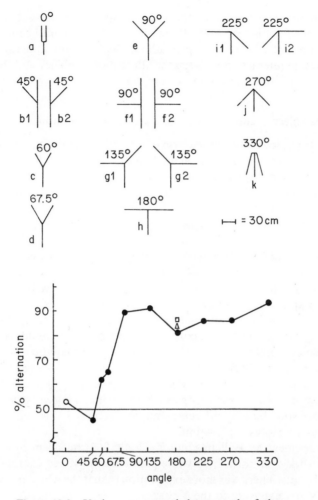

Figure 10.2 Various mazes and the strength of alternation: (a) mazes having various angular separations between the two choices of arm; (b) alternation can be seen to increase as a function of angular separation (Source: Douglas, Mitchell, and Del Valle, 1974. Reproduced by permission of the Psychonomic Society, Inc.)

Berlyne, 1960). Where stimulus objects which become available only after the response has been made serve to influence the response, there is the implication that (Berlyne, 1960, p. 134):

'. . . cues at the entrance to the path must interact with traces left in the nervous system by recent experiences.'

Similarly, Halliday (1967) points out, at a choice point an animal uses information impinging upon its sense organs (the 'inspective' case) as well as information which is stored in the nervous system as a result of previous experience (the 'inquisitive' case). It uses it so as to come into contact with the maximum amount of novel/complex stimulation that the apparatus permits. This involves an interaction between the alternation tendency and the tendency to bias the choice towards the more complex.

10.2.2 The object qualities which evoke inspective exploration

If, between trials, the colour of one arm of a T-maze is changed, rats show a high probability of exploring that arm on the next trial. Two possible explanations for this effect have been offered. If both arms were previously white and then one is changed to black (or the reverse design, since it is important to balance the colours) it could be that the rats are in some way simply 'satiated' for white. Another possible way of viewing this result is not in terms of satiation for white but the absolute attraction of the novel black choice. Dember (1956) designed an experiment to distinguish between these alternatives. He allowed rats to explore the entrance to a T-maze in which one arm was painted black and one white. They were then exposed to the choice point after both arms were made the same colour. The arm which had changed in colour between trials was chosen in preference to that which remained the same colour. This is incompatible with the explanation of satiation to one colour —except possibly colour satiation in a certain position context, which seems an unlikely explanation. These studies collectively indicate that rats are responding positively to change. They are not simply avoiding the alternative choice or the repetition of a response due to fatigue (Dember, 1965). Disparity exists between incoming sensory information and the corresponding stored representation acquired through previous exposure. This promotes exploration.

In an experiment by Williams and Kuchta (1957) replacement of a familiar black-painted Y-maze arm by a white-painted arm evoked exploration of the novel arm by rats. There was more exploration than if the white arm was present on the rat's first exposure to the Y-maze.

In a further experiment of Williams and Kuchta, rats were given daily tests of exposure to plain arms (one painted black and one white) and to an arm containing a set of novel objects. At first rats made a roughly equal number of investigations of each arm, but on subsequent days the number of visits to the complex arm increased. The number of visits to the plain arms remained roughly constant over 8 days of tests. After the first exposure, time in the object arm (spent mainly in investigation of the objects) increased at the expense of the empty arms. Williams and Kuchta suggest that the increase in percentage of time spent in the object arm over several days could be due to habituation of the fear that it initially evoked. Such fear would be in conflict with exploratory tendency due to novelty. An alternative explanation is that in the beginning even the plain arms are novel and hence evoke exploration. However, the animal rather quickly acquires an internal representation of their form, and more time is made available for the relatively demanding task of incorporating the object arm into an internal representation. Williams and Kuchta examined each 2 min interval within the daily 10 min exposure period, using the measure of number of units entered. On day 1 and 2 rats explored at a relatively low uniform rate throughout the 10 min. On subsequent days they showed a very high rate of movement between the arms in the first 2 min, but this fell sharply towards the end of the session. Any model of exploration must account for this intra-session decline, which has been widely observed by other investigators. Could it be that as trials proceed and the animal builds up an effective internal representation of its environment, on each new exposure only a cursory inspection is sufficient to confirm the internal representation? In chimpanzees, Welker (1956) found an intra-session and inter-session decline in the magnitude of exploration of objects introduced into the cage. However, the first minute of a trial showed increased exploration when compared with the last minute of the previous session.

Berlyne (1960) attempted to manipulate the novelty of objects. Rats were given 5 min in which to explore three identical objects, either wooden cubes or cardboard cylinders. Ten minutes later the rats were returned, to find one of the cylinders replaced by a cube or vice versa. The novel objects evoked more exploration than the objects that were familiar (to the extent of 5 min exposure).

Traditionally both novelty and complexity have been seen as determinants of exploration, though as Berlyne (1960) acknowledges, it is not always possible to separate them. When a rat is placed in a new environment it explores its complex components more than the simple one. For example, an object in a corner of a new cage is explored more than an empty corner (Berlyne, 1955). Access to the Dashiell maze, the reward for maze-running in some early studies, presumably offers both novelty and complexity.

One can attempt to design experiments that bring the complexity component into focus. For instance, in an experiment of Berlyne (1960) rats were given exposure to a compartment containing either a black or unpainted cube. Subsequently they were exposed to the same compartment, either when it was empty or containing a grey cube. The more complex environment, i.e. that containing the grey cube, was explored most even though, it could be argued, the empty compartment was the most novel in terms of being different from the animal's immediately previous experience. Taylor (1974) found in rats that a change in stimulus from mid-complexity to either high or low complexity effectively aroused exploratory behaviour (i.e. the novelty factor). However, the incremental change in complexity was more effective than the decremental change (i.e. complexity factor).

In some experiments it is possible to introduce what humans would call surprise, and obtain strong exploratory behaviour (Berlyne, 1960). That is to say, a new object is put in the place of a familiar object. Pure novelty is simply where a new object appears out of context.

As Halliday (1967) points out, novelty can only be defined in terms of the history of the individual organism, i.e. the stimulation it has received in the past. By contrast, complexity is a feature definable purely in terms of the stimuli in the environment. Halliday presents the rudiments of a theoretical base for exploratory behaviour, and this makes explicit some of the conclusions that have been tentatively drawn in this chapter.

Halliday notes that a change in a familiar environment is a powerful stimulus to exploration; for example, the sudden appearance of a white arm in a previously black maze. This suggests:

> '. . . presumably more exploration was elicited in the experimental group by the white arm because they were matching their previous and present experiences of the maze and finding that they did not correspond.'

Halliday argued that:

> 'An animal would therefore explore a situation with which it was unfamiliar until its internal representation of the situation matched the external environment. . . .' (Reproduced by permission of Harper and Row, Publishers, Inc.)

The intra-trial decline in exploration may be explained in terms of increased accuracy of matching. Presumably inter-trial decline occurs since it becomes progressively easier to establish a match. The relatively high rate of exploration in the first minute or so, even after several sessions, would seem to be because the animal must establish that things have not changed since the last exposure. With a good match and unchanged conditions exploration can afford to decline rapidly. Complex environments would require a large amount of exploration in order to establish a match. In Halliday's terms *surprisingness* is a case of where the incoming sensory information does not match an established internal representation, e.g. a black floor where the internal representation was of a white floor. Pure *novelty* means no internal representation exists, e.g. the contents of a goal-box on the animal's first exposure to it.

Objects are explored less, the longer they are available to the animal. If object and animal are parted for a length of time the object acquires the ability to evoke exploration on reunion, and exploratory tendency increases as a function of time apart (see Eisenberger, 1972). It could be argued that, without regular checks, objects lose their internal representation, that is to say they become novel again.

The point is sometimes raised (e.g. Halliday, 1967) that if in the course of exploration an animal moves from A to B we cannot be sure if it is approaching B (i.e. arousal of curiosity) or avoiding A (anthropomorphically speaking 'boredom'). This may not be an entirely valid dichotomy to make; all that we can really conclude is that B arouses a stronger approach than A.

It is possible to raise the objection that what appears to be exploratory behaviour caused by novel stimulus objects is really nothing more than the animal using an opportunity to perform locomotion. There are a number of grounds for rejecting this suggestion (Halliday, 1967; Myers and Miller, 1954), though of course that is not to deny that locomotion may have an additive effect at times. Returning a rat to the same maze after a period of locomotion/exploration is followed by decreasing locomotion/exploration, but placing it in a similar maze of a different colour halts the decline. Replacing an elevated Y-maze by an enclosed T-maze is also effective at arresting a decline in exploration, though there is the possibility of confounding exploration with a changing demand upon mechanisms of sensory-motor coordination and locomotion. Rats which have been denied the opportunity for physical activity show no more active exploration than those not given prior deprivation (see Berlyne, 1960), which forms evidence against the 'locomotion drive' view.

Simmel and Eleftheriou (1977), however, argue that the results of many studies that purport to observe exploratory behaviour are confounded by locomotion *per se*. For example, in the open field there may be considerable amount of locomotion, none of which could be described as exploratory. These researchers described an experiment that permitted separation of exploratory and locomotor components. Mice were introduced into a compartment having both a rich side painted with black and white stripes and a plain black-painted side. An opening in the wall allowed the animal to pass between the two sides. A factor analysis showed that locomotor activity could be measured independently from exploration of the novel side. In an experiment conducted by Leyland, Robbins and Iversen (1976), an unambiguous dissociation of exploration and locomotor activity emerged. Thus novelty and complexity increased exploration but did not affect total locomotor activity. Dissociation may also be shown under the condition of d-amphetamine injection; exploration decreases and locomotor activity increases.

In the case of the Canyon mouse, housed in a complex environment with many tunnels, provision of a running wheel decreased the amount of exploration of the environment (Brant and Kavanau, 1965). The authors concluded that in this species locomotion through the environment is determined by causal factors for both general locomotion and exploration. This result is compatible with such an interpretation, but alternatively could simply mean that time spent in the running wheel must be subtracted from time allowed in other activities and exploration suffers most.

Experiments designed to distinguish locomotion and exploration serve to raise the question of whether there is a deprivation effect in so far as inspective exploration is concerned. Does extended sensory restriction potentiate subsequent exploration? This was a particularly relevant question in the context of attempts to build general theories of drive involving features in common between thirst, hunger, exploration, etc. According to evidence reviewed by Berlyne (1960), rats deprived of the opportunity for exploration by being kept for up to 9 days in boxes with opaque walls show no stronger exploration than animals not deprived of exploratory opportunity. However, in a slightly later review of the evidence (Halliday, 1967) it was reported that such experiments yielded conflicting results; some find no effect while others find a deprivation-induced effect. Monkeys denied exploratory opportunity show operant activity rewarded by presentation of stimuli which is an increasing function of deprivation length (Butler, 1957). Evidence reviewed by Fowler (1965) shows that deprivation does increase subsequent exploration. In a study by Woods (1962) it was found that rats housed in a dull environment for 24 h prior to testing showed little decrement in exploration during a 24 min test. Those housed in normal, and particularly in rich, environments, showed a decline. We will return to this subject and to theories of exploration later.

10.2.3 Species differences in exploration

Glickman and Hartz (1964) found wide differences in exploratory behaviour when comparing rodent species. Chinchillas show a relatively high level of exploration and guinea-pigs a low level, while rats are intermediate. Over a 10min period of exposure to a new environment rodents differ very much in habituation as measured by the decline in exploration. The guinea-pig remains at a fairly constant low level, whereas the chinchilla and gerbil show a sharp decline. Qualitative differences are also evident; frequency of sniffing at novel objects is high in chinchillas and gerbils but very low in guinea-pigs. The racoon uses the forepaws for exploration, while the coati mundi uses its snout (Welker, 1961). Predominantly visual cues serve to elicit exploration in grey squirrels, tactile cues in moles and auditory cues in bats. Such differences reflect the different sensory channels normally relied upon by these species. In so far as their physical movement is an index, goldfish show systematic exploration of a new environment (Kleerekoper, Matis, Gensler and Maynard, 1974).

Thompson and Lippman (1972) observed rats and gerbils during exploration of a novel environment, a Greek cross maze having four arms. One arm was painted black, another white, while the other two were painted with black and white patterns. Gerbils showed more exploration than rats. Rats demonstrated a preference for the compartment painted black. When tested in a light phase, gerbils show very much stronger exploratory tendencies than rats (Osborne, 1977). For example, rats in this particular experiment were reluctant to leave the home cage to enter a new chamber. Gerbils showed no reluctance. It would be interesting to repeat this study in the dark phase.

Glickman and Sroges (1966) compared a very wide variety of animals in their exploration. A set of different shaped objects were placed in their home cages in Lincoln Park and Bronx Zoo. Exploration was recorded as either physical contact with, or orientation to, the novel objects. Animals were grouped into one of five categories, and showed decreasing exploration in the order, primates, carnivores, rodents, 'primitive mammals' and reptiles. There were, however, very wide differences between species within any one of these groups.

It has been suggested (see Glickman and Sroges, 1966, for references) that the degree of curiosity shown by various species correlates with the extent of their intellectual development and nervous system complexity. For example, mammals show more curiosity than reptiles. However, within a mammalian order enormous species differences with regard to exploration are evident. Glickman and Sroges argue that habitat is a more reliable predictor of exploratory tendency than any crude measure of brain sophistication (such as the ratio brain-weight/body-weight).

They suggest that feeding patterns which involve relatively complicated manipulation of the environment and discrimination should be associated with extensive exploratory tendencies. A simple stereotyped feeding response might be associated with a low exploratory tendency. During Glickman and Sroges' observations carnivores explored novel objects with little or no hesitation. Glickman and Sroges remind us that the food-seeking of such animals involves an immediate fearless response to a variety of forms. However, it could be argued that the food-seeking of some reptiles requires a similar reaction. By contrast, grazing animals would seem to have rather little need to show exploratory behaviour. Unfortunately there is little evidence available here.

10.2.4 Maze-running for the reward of exploratory opportunity

Early studies (reviewed by Berlyne, 1960) showed that a rat will cross an electrified grid for the opportunity to explore a complex maze. There are other examples of where an animal's locomotor behaviour can be influenced by stimulus objects which were perceived on earlier trials. For example, one design is for a rat to run a straight runway and enter an empty compartment. It cannot see into it until it has entered. On trial 10 the plain colouring of the floor and roof of the compartment is replaced by black and white stripes. On subsequent trials the rat runs faster. It is claimed that a novel and complex stimulus acts as reward in a similar way to water for a thirsty organism (Chapman and Levy, 1957).

Berlyne (1960) claims:

'. . . running is thus a clear example of inquisitive locomotor exploration.'

In a T-maze, rats have a tendency to choose the arm which leads to a goal-box containing complex objects rather than an empty goal-box (Berlyne, 1960).

10.2.5 Bar-pressing rewarded by a change in sensory stimulation

Rats held in darkness bar-press for the reward of turning on a light for short periods of time. Such results, which appeared first in the early 1950s, came as something of a shock (Berlyne, 1960). Rats were assumed to find light aversive; indeed, they choose black rather than white alleys. Controlled studies rule out the possibility that light is a secondary reinforcer connected with, for example, food reward.

Rats kept in light will bar-press to terminate the illumination for short periods (Glow, 1970). There is, as Glow admits, no adequate theoretical model to account for this. The lights-on and lights-off results could lead one to argue that change *per se* in the stimuli impinging upon the animal's sense organs is reinforcing since in a dull environment this maintains some kind of arousal level. However, whether this holds much explanatory value is doubtful.

Examination of Glow's results shows that initially the animal works very hard for lights-on in a dark environment, but subsequently, over each day's test, responding falls to a low asymptote. Rats earning periods of lights-out show a steady increase in responding over the days of the test. Glow suggests that at least two interpretations may be placed upon the decrease in responding shown by light-rewarded subjects (the two need not necessarily be incompatible). It could be a case of habituation to what is at first a novel stimulus. Alternatively, it could be that there is a decreasing need to scan the environment as it becomes more familiar. The latter would correspond to the explanation we have offered for the inter-trial decline in inspective exploration. The steady increase of responding by the lights-out groups could also be a reflection of a decreasing need to scan the environment visually. Let us assume that for any given time of day rats have an optimal level of illumination. This may or may not coincide with the level that best facilitates exploration in a novel environment. Some changes in rate of responding over days may therefore be expected.

In a review of the literature up to 1967, Halliday (1967) concluded that, in rats, light onset in a dark environment is reinforcing partly because it facilitates visual exploration. It is accompanied by scanning head movements. However, within limits, change *per se* may also be rewarding. Increase in light intensity in a chamber already illuminated is not so rewarding as light-onset in a dark environment. Light reduction is more rewarding than light-offset, presumably because it does not preclude visual exploration. Reduction in illumination has not always been found to serve as a reinforcer; it may do so mainly in those cases where initial illumination level is intense (Berlyne, 1960).

Opportunity for visual exploration of an accessible environment cannot explain some other operant results in this area. Animals will sometimes bar-press for the reward of a noise or the chance to see into another chamber (Berlyne, 1966), or simply for the movement of the bar (Berlyne, 1960). Although such findings have traditionally been explained in terms of optimal arousal level, I would suggest that another possible clue is to view the animal's behaviour as an extension of the kind of exploration already described in this chapter. That is to say, the animal is maximizing its 'knowledge' of the environment by manipulating it and thereby maximizing the environmental events influencing its sense organs. The laboratory testing cage is a highly unusual environment and the animal's behaviour may simply be a reflection of the same underlying processes which would, under more natural conditions, be reflected in patrolling the home range or sniffing novel objects. Working for the reward of an extra compartment to explore decreases with massed trials, and is most evident with spaced trials (Myers and Miller, 1954), a result which would conform to the theory tentatively being advanced here.

In the case of monkeys, opening a door to reveal another chamber is a powerful reinforcer. This is particularly so if the exposed room contains some activity such as a toy train or another monkey (see Berlyne, 1960 for discussion).

10.3 PATROLLING

By this expression I mean the regular movements which certain animals have been observed to make through their familiar home range. It is difficult not to attach a label of the kind *patrolling* to such behaviour, which seems to serve the purpose of either confirming the stability of the environment or drawing attention to any intrusions or disparity.

In the case of laboratory rats housed in a relatively complex environment patrolling commonly occurs after a meal has been eaten (see Cowan, 1977a). Extensive exploration of an environment immediately following a meal may mean that the causal factors for exploration were being inhibited by those for hunger (Barnett, Dickson, Marples, and Radha, 1978). They may then oust feeding by competition or be disinhibited. Cowen observed rats deprived of either food or water for 24 h. When the commodity concerned was returned it was consumed, and following this the rat explored its environment. Cowan also examined rats housed in a so-called plus maze, which consists of a central compartment with arms radiating outwards in four directions. Entrance to these four arms was blocked for 21 out of 24 h. One arm contained food, one water, while the other two were empty. The movements of rats in the four arms were observed over the 3h access period, during which rats explored extensively. Not only were the empty arms investigated, but visits to the food and water arms were not always accompanied by consummatory behaviour, and constitute what we would call 'patrols'. Such patrols sometimes did not occur if the period of access to the arms was in the light phase. Exploration was systematic in that a visit to one arm meant a low probability of an immediately following return visit and a high probability of exploring the other arms. Cowan remarked on the persistence of such exploratory behaviour; rats show little decline even after 30 days of continuous access to the whole plus-maze:

'The "novelty" of the information obtained in such circumstances is small, yet the responses do not diminish.'

Under some conditions exploration is particularly high on the first day in the plus maze (Barnett and Cowan, 1976).

When placed in a highly complex environment having many tunnels, Canyon mice gradually extend the size of the range that they traverse. Regular explorations are made throughout the range (Brant and Kavanau, 1965).

Except for work on foraging, exploration and the phenomenon of patrolling are not well integrated into the ethological literature, the weight of evidence resting upon the laboratory-based work of experimental psychologists. However, with reference to the natural environment, Jewell (1966) discusses the concept of 'home-range' which applies to many mammalian species. This is defined as:

' . . . home range is the area over which an animal normally travels in pursuit of its routine activities.'

By normal activities is meant such things as food and water-seeking, caring for the young and mating. The animal may make occasional exploratory sallies outside of the home range, and it may in some cases migrate from one home range to another. Jewell uses the alternative expression *lifetime range* to mean:

'. . . the total area with which an animal has become familiar, including seasonal home ranges, excursions for mating, and routes of movement.'

Whether or not animals in the wild make regular exploratory excursions throughout the home range unconnected with seeking essential commodities is a subject that has not, as far as I know, received much attention. Excursions for food and water, etc., would of course serve to maintain the animal's internal representation of its environment.

10.4 COGNITIVE MAPS—NEURAL MODELS

Tolman (1948), in his classic study entitled 'Cognitive maps in rats and men', discusses the theoretical interpretations that may be applied to the mastery of a complex maze by a rat. He argues that rats construct cognitive maps of their environment. Such a statement may appear to be appealing to mentalistic terms to explain rat behaviour; it certainly sounds 'softer' than the behaviourism of Watson and Skinner. However, Tolman was really only saying that somewhere in the nervous system there are formed representations or neural models of the environment. There is isomorphism between *information* within the neural structures and the environment. Furthermore, such internal maps are causal agents in the animal's locomotor behaviour. This should be placed in contrast to the view that animals make fixed locomotor responses to particular stimuli; an internal map would allow greater flexibility in behaviour. For example, if a diversion in the animal's route is imposed it can still quickly arrive at the goal. Similarly if the starting point is moved the animal can still find the goal, provided certain signs are available. For a detailed recent account of cognitive maps in primates the reader should consult the fascinating review by Menzel (1978).

A variety of experimental results lies behind Tolman's arguments. *Latent learning* refers to the fact that rats which are satiated for food can learn the route to food in a maze, this learning being suddenly revealed to the experimenter when the rats are made hungry. As Tolman expresses it:

'They had been building up a "map", and could utilize the latter as soon as they were motivated to do so.'

A similar conclusion follows if we examine animals sated for food and water, and running in a Y-maze with water at one arm and food at the other. There is no consumption of either commodity and animals are returned to the home cage when reaching either goal. However, on being made hungry or thirsty the probability of choosing the arm appropriate to the motivational state is greater than chance.

In an experiment by Petrinovich and Bolles (1954) rats were made either hungry or thirsty and obtained reward of food or water, respectively, in a T-maze. In one test the commodity was at a fixed location, while in a second test the rat was required to alternate its responses in order to be rewarded. Thirsty, water-rewarded rats were superior to hungry, food-rewarded rats when reward location was held constant, but inferior when they were required to alternate. The authors suggest that the rat's ecology is such that it would have evolved so as to 'expect' water to be in a fixed location. By contrast, the site of food would be likely to vary (see Olton, 1978, for a formal treatment of such behaviour).

Rats appear to use several sense modalities in forming maps of their environment; information from smell, vision, and touch through the vibrissae is integrated (Shorten, 1954). Shillito (1963) reported an experiment in which a brick was placed in the cage of a vole, and the animal was observed to move back and forth between the brick and a partition, as if it were measuring the distance to the new object. When placed in a new cage voles first follow its edges, and after-wards explore specific novel objects within the new environment.

Shillito (1963) believes that it is useful to distinguish two phases of exploratory behaviour that occur when a vole is in a familiar environment. There is a reconnaisance phase, where it quickly patrols the familiar area, and an investigatory phase where any novel objects are more closely examined. Shillito adapts the term 'consummatory' to exploratory behaviour:

'... the consummatory situation for the exploratory drive is the reception of adequate stimuli, which may match those received on previous occasions. ...'

Shillito suggests that fading of sensory stimuli serves to elicit a new recon-naisance. In other words, exploratory behaviour occurs when 'remembered topography has lost sharpness' and before it is forgotten. Shillito makes the important point that studies on exploratory behaviour have tended to over-emphasize its novelty-seeking aspect. Although novel stimuli are explored particularly vigorously this is only a natural amplification of the exploration of the animal's familiar environment.

In a study based upon primates, but having implications for exploratory behaviour in general, Menzel (1978) regrets that so many studies report only the quantitative aspects of exploratory behaviour. They typically report the number of maze units entered in a given time. It was found in chimpanzees that after being placed in a novel environment exploration is systematic, and from day to day they expand outwards the limits of the range of exploration. They appear to establish certain distinct frames of reference or landmarks from which to conduct further exploration. Menzel convincingly argues in favour of cognitive maps. Satiety of exploration is seen in terms of successful assimilation of information about what is 'out there'. Further, Menzel argues that:

'Older animals engage in less overt locomotion and manipulation because they are more skilled in extracting the necessary information from the environment. . . .'

Nadel and O'Keefe (1974) claim that a region of the brain known as the hippocampus is involved in the construction of cognitive maps of the environment. They write:

'. . . the theory we constructed states that the hippocampus forms a map, that is, a pre-existent framework within which all the stimuli to which an animal attends are represented. The representations drawn from a particular environment are systematically related to each other, forming what we have called a *cartoon* of that environment. As an animal moved through a familiar environment the cartoon provides continual predictions concerning the stimuli consequent upon its movements, and the confirmation or lack of confirmation of these predictions is registered. When an unexpected stimulus occurs (or an expected one fails to occur), the expectation generated by the cartoon is not confirmed, and a signal is produced which arrests on-going behaviour, directs attention towards the incongruity, and allows for its exploration. Such exploration, according to the model, is behaviour designed to incorporate the changed conditions into the cartoon for that environment. Termination of exploration occurs when there is no longer a mismatch between environment and cartoon.' (Reproduced from L. Nadel and J. O'Keefe, *Essays on the Nervous System—A Festschrift for Professor J. Z. Young* (eds. R. Bellairs and E. G. Gray) by permission of Oxford University Press.)

O'Keefe and Nadel (1978) write:

'When the animal first enters a novel situation all the misplaced detectors will be activated and exploration will continue until sufficient information is incorporated into the map of that environment. Thus our theory accords curiosity the status of a major motivation, the driving of information incorporation into cognitive maps.' (Reproduced from J. O'Keefe and L. Nadel, *The Hippocampus as a Cognitive Map* by permission of Oxford University Press.)

They proceed to argue that one does not need to postulate underlying drives for exploration; it is assumed to be aroused by unpredicted external stimuli.

Exploration is defined as activity that incorporates new information into the cognitive map. Novelty is defined as a quality of the immediate environment which does not have a representation within the cognitive map. Nadel and O'Keefe argue that one specific prediction of their theory is that animals with hippocampal damage should show no exploratory behaviour. Indeed, the evidence shows that although animals with hippocampal damage are *hyperactive*, they lack *investigative* responses to novelty.

In a study involving the Mongolian gerbil, Glickman, Higgins, and Isaacson (1970) found that hippocampal lesions increased the amount of locomotion occurring in a novel environment. However, the actual amount of investigation of novel objects was decreased by the hippocampal lesion. In such cases then increased locomotion is not correlated with increased curiosity.

Of possible relevance to the subject of cognitive maps is the result that on a two-way avoidance task rats with hippocampal lesions perform better than intact controls (Olton and Isaacson, 1968). This task was described in the last chapter and consists in moving, in response to a cue, back and forth across a barrier in order to avoid shock. What was, on the previous trial, associated with shock becomes the now safe side. Olton and Isaacson argue that rats with hippocampal damage may have a reduced tendency to avoid places previously associated with shock since their spatial map is deficient. This might also explain why they are not so good as intact controls at performing a one-way avoidance task (moving from a permanently dangerous side to a permanently safe side). Such a conclusion would dovetail neatly with the arguments of Bolles advanced in the last chapter, i.e. that a spatial vector is involved in avoidance tasks.

10.5 AMBIVALENCE BETWEEN AVOIDANCE AND EXPLORATION

From reading this and the previous chapter the reader will appreciate that the causal factors underlying avoidance and inspective exploration share common features. This section considers the ambivalence that this may involve for the animal.

It is necessary to introduce the term *commensal* to the discussion. In the case of rats it refers to those 'wild' strains which live in proximity to man and which traditionally have been directly affected by man's habits, e.g. poisoning, trapping, and supplying food to livestock. Such rats show what is known as the 'new-object reaction', the avoidance of an unfamiliar object placed in the animal's familiar surroundings (Cowan, 1976). Ultimately commensal rats do touch and explore novel objects, but it may be two or more days before food is eaten from an unfamiliar dish (Cowan, 1977b). The same object is not avoided (or only slightly avoided) if it is present when a commensal rat is first introduced into a laboratory environment. As an example of the new-object reaction, replacing a familiar food basket in the home cage by a novel one evokes a strong avoidance reaction, as does simply moving the familiar basket to a new location. The reaction shown by commensal *Rattus norvegicus* contrasts to that of conspecific laboratory strains where such a stimulus provokes approach and exploration. According to Cowan (1976) the new-object reaction of commensal rats is caused:

'... by the contrast between (a) the unfamiliar configuration of the stimuli of the object, and (b) the familiar configuration of the background against which it is presented.'

Cowan proposes that the magnitude of contrast depends upon familiarity and stability of the environment prior to the introduction of change. In other words neophobia arises from disparity between stimuli emanating in the present environment and 'the rat's existing cognitive model of its environment'. When its environment is subject to constant change, such as at rubbish tips, *Rattus norvcgicus* shows little new-object reaction and can easily be trapped.

For commensal rats, displacing a familiar object in a familiar location to a new location causes avoidance (Shorten, 1954). Cowan and Barnett (1975) cite an experiment showing that removal of a familiar object elicits little curiosity in rats, and suggest that:

'. . . perhaps stimuli must be perceived before they can be matched with an existing model.'

However, Shillito (1963) reported that voles investigate when a familiar object is removed.

In comparing commensal and domestic strains of rat in a + shaped maze it was found that a newly opened arm was investigated equally by both groups. The strong neophobia of commensals is specific to solid physical objects; they show a very long latency to enter an arm containing an unfamiliar object (Cowan, 1977b), though there exist large individual differences. Newly opened empty arms probably neither fit not fail to fit an existing 'neural model' since they have not been previously experienced (Cowan and Barnett, 1975). A novel object in a novel arm may be incongruous in terms of past experience.

When first placed in a + shaped maze both commensal and domestic rats explore extensively. Magnitude of exploration declines over a period of days, as is also the case for a newly opened extension to an existing environment. However, even after many days the new area is regularly patrolled (Cowan and Barnett, 1975).

Cowan (1977b) compared commensal, domestic, and species of truly wild rat, finding that the latter two groups were similar in their lack of aversion to new objects. It is presumed that the avoidance behaviour of commensals arises from selection pressures operating in the course of their interaction with man. The animal avoiding a novel object in a familiar location is often at an advantage, obvious examples being avoidance of traps and poisoned food. However, striking as they are, differences in neophobia between strains and species may be quantitative rather than qualitative. As Cowan (1977b) points out, there are some signs of hesitation in exploration of new objects and places even by domestic rats. A marked reluctance to enter a new compartment attached to the home cage was reported in domestic rats by Osborne (1977), though hunger overcomes this (Fehrer, 1956). Conflicting approach and avoidance tendencies are probably present in all rats. Differences in behaviour would reflect differences in the relative thresholds of exploration and avoidance, or in the time constants of fear habituation, or both.

It is widely assumed that fear inhibits exploration and indeed that neophobia is an unambiguous demonstration of this. However, somewhat at odds with common sense, is the Halliday–Lester theory, which states that mild fear enhances exploration (see Russell, 1973, for a critical discussion). Mild electric shock can be shown to facilitate exploration of a region where the rat is located at the time of shock, and this forms the base of the theory. The Halliday–Lester theory has something in common with traditional drive-reduction theories; fear promotes exploration that then reduces fear. However, as Russell points out, it is easier to accept that exploration reduces fear than that fear *per se* promotes exploration. An alternative and perhaps more parsimonious explanation for the fact that mild shock enhances exploration is given by Russell, and is as follows. There is *greater incongruity* aroused by a stimulus associated in the past with shock but no longer having such an association.

Neophobia may underlie at least part, if not all, of a commonly observed phenomenon in the feeding of domestic rats. Rats sometimes work for food by lever-pressing even though identical free food is available in a dish. Mitchell, Scott, and Williams (1973) suggest that the familiar lever is preferred to the less familiar food dish. When the two are equally familiar the rat shows a preference for the free food (though see also Carder and Beckman, 1975). In an experiment of Zimbardo and Montgomery (1957) rats were made either hungry or thirsty and then placed in a novel environment containing food or water. For all levels of deprivation studied, latency to eat was longer than latency to drink. Hungry domestic rats in the presence of food in a novel location seem to be in some ambivalence between approach and avoidance.

Ambivalence between avoidance and exploration is shown by voles (Shillito, 1963). Shiny or strongly smelling novel stimuli tend to elicit avoidance. Context is important. When voles voluntarily enter a new environment where all the objects are novel there is little sign of avoidance. By contrast, a novel object in a familiar location can result in prolonged avoidance. Confronted with a stuffed owl or its component feature, chaffinches seem to be in an approach/avoidance conflict (Hinde, 1970). Approach gives way to avoidance as the chaffinch comes closer.

Menzel (1963) described fear in chimpanzees confronted with novel objects. After repeated exposure to such objects fear very slowly gave way to active exploration and play. However, these were animals raised in isolation and with little or no opportunity for object investigation, so it is probably an exaggerated neophobia compared to that of normal animals. Fear may be evoked by situations which are extremely novel and yet it is hardly illuminating to explain this simply in terms of extreme novelty. In chimpanzees a model of the head of a conspecific without a body evokes fear, but it is a very special case (Weisler and McCall, 1976), and hardly more novel than, say, a model railway.

10.6 SUMMARY, CONCLUSIONS, AND DISCUSSION

In this chapter we have met considerable difficulty in designing a model that embodies at least some of the essential causal factors, mediating processes, and features of behaviour. We can go only a little way beyond developing a few appropriate metaphors and tentatively proposing possible processes.

We should question then what would actually constitute an explanation or theory of exploratory behaviour as compared to other motivational systems. In the case of feeding or drinking we are dealing with chemical quantities. Therefore our theory can be anchored to changes in these quantities over time. Speaking loosely, we can begin our investigation with physiology and later refine it with psychology (Booth, 1979). Although there is still much to be done, at least we know what kind of feeding and drinking models are likely to emerge and around what kind of event they will be framed. For example, blood volume and energy flow might reasonably be expected to form essential components.

In exploratory behaviour there are no obvious extraneural physiological or biochemical events upon which to focus attention. Therefore we need to look either directly at neural mechanisms, for example to measure activity in the cortex and hippocampus, or, if we prefer, to hypothesize processes without direct reference to their underlying neural locus and embodiment. There are not even clear consequences of behaviour such as in the case of sex or fear/aggression (respectively, genital friction with consequent desensitization, and separation of organism from intrusion). Unlike aggression and fear any change caused by behaviour is an internal one. Even where the animal manipulates its environment the only important consequences are in terms of assimilation of information. Although it is somewhat tautological to say that familiarity terminates exploration it is the best we can do. The behaviour elicited by novelty is investigation, novelty being defined in terms of information content and the organism's past history of experiences. Stimulus parameters which evoke exploration possess 'information potential' (Weisler and McCall, 1976). Examples include contrasts in brightness, a large number of features, movement, and the capacity to change in response to manipulation.

It is important to emphasize that when satiation of exploration is mentioned what is implied is not the satiation of some general exploratory tendency, but rather satiation to the novelty of a particular object or environment (White, 1959).

In so far as a theory of exploratory behaviour has emerged it is that the animal contains an internal representation of its environment. Admittedly this is vague; it could mean a number of different things. Nothing specific about how neural circuits might code environmental information is implied. If we adopt this hypothesis the data suggest that the internal representation is updated or, in the absence of change, consolidated by exploratory trips. Information on new objects can only be incorporated after a period of investigation. If the animal and an object with which it is familiar are parted then, on reunion, investigatory activity is needed to allow reassimilation. Such a hypothesis does not necessarily predict that animals would work for the reward of sensory stimuli or cross an electrified grid for exploratory opportunity. However, neither is it incompatible with such findings.

In describing exploratory behaviour we are particularly prone to mix causal and functional levels of explanation. In the case of feeding, to say that energy supply triggers ingestion is a statement of causality. To say a rat eats to stay alive is at the level of function served. If, at a causal level, we postulate that animals explore when an internal representation is unable to match the environment, it is imperative that we recognize that the animal must be equipped with a mechanism for translating information content into activity.

The internal model/representation/map hypothesis (or metaphor) is perhaps not the only one to be compatible with a large amount of the behavioural evidence. Other theories can better explain that animals work for sensory reward. There is a group of loosely related theories, that do not explicitly refer to internal models, but are based upon the reward value of the impingement of sensory stimuli or change in sensory stimuli. However, theories which focus upon either stimuli or internal models need not be incompatible. For an under-aroused animal, to use the accepted expression, it may be the case that a stimulus increases arousal only when no adequate internal representation of the stimulus exists (cf. Sokolov, 1960).

In psychology, exploratory behaviour can be viewed in the historical context of drive theory; animals were said to act so as to reduce their drive level. For example, a certain route in a maze leads to water. By taking this route a thirsty organism reduces its drive and so this behaviour is repeated whenever it is made thirsty. It could be argued that if an animal is placed in a novel environment then this arouses an exploratory drive. By exploring, it reduces this drive. So far an analogy with thirst might be useful. However, animals actively seek novel environments that provoke exploration. This would imply that they seek to *increase* drive, which would make the whole drive construct worthless. Some authors have therefore been forced to postulate a *boredom drive* that arises from an organism being housed in a relatively dull environment (see Eisenberger, 1972). This drive is reduced by contact with complex novel stimuli. Why then do animals sometimes avoid extreme complexity or novelty? As Eisenberger points out, by analogy, just because very large stomach loads of food are aversive, in no way does this deny that hungry organisms find small quantities of food rewarding. A theorist in this tradition is Fowler (1965). He draws an analogy with the Hullian interpretation of feeding, where drive (hunger) and incentive (food) combine to determine, for instance, the animal's performance between start-box and goal. Fowler argues that in the case of exploration the drive is boredom, and incentive is the opportunity to explore, e.g. bar-pressing for the reward of a novel stimulus. He believes that lack of stimulation (i.e. confinement in a dull environment) for a period of time increases boredom drive. This might indeed also make sense in the natural habitat of the animal; a patrol of the home range could be initiated by staying for a relatively long period in the home base. Such theories avoid problems inherent in neural model or cognitive map theories, such as why do rats regularly patrol? A synthesis of these different approaches is needed.

Some theories postulate an optimal level of sensory stimulation for an organism at any given point in time. Others develop such a notion to involve the history of the individual organism; an animal, subjected from birth to a large average information input, when later given a choice seeks a relatively complex environment. The paper of McReynolds (1962) is addressed to this issue. This investigator starts with the assumption that animals develop a cognitive structure which represents the nature of the environment. New input is assimilated into this structure by means of exploration. McReynolds begins his discussion as follows:

'The process of cognitive restructuring, in order to assimilate new input, can be thought of as going on over time.'

Perceptualization rate refers to the rate at which information is assimilated, and McReynolds argues that animals seek an optimum perceptualization rate: they avoid both monotony and extreme novelty. McReynolds believes that animals from an impoverished environment would need to explore novel objects more extensively in order to assimilate their form, and cites supportive results. Zimbardo and Montgomery (1957) found that rats reared in an enriched environment explored a relatively simple but new environment less than controls raised in a normal cage. It may also follow that the optimum perceptualization rate varies as a function of the animal's history of environmental stimulation, it being higher for enriched upbringing conditions. Turpin (1977) compared socially reared rats and isolation-reared rats in their preference for familiarity or novelty. Rats were allowed a choice between environments having the same or different characteristics as the familiar one. Rats raised in isolation showed a clear preference for familiarity whereas rats raised with conspecifics showed an equally clear preference for novelty. Similarly, handling in infancy can be an important determinant of the magnitude of exploration in adulthood. Presented with a complex environment, handled rats explore much more than non-handled controls (Denenberg, 1967).

Einon and Morgan (1976) showed that in a novel environment socially reared rats had higher initial rates of exploration than those reared in isolation, but the rate declined much faster. Can socially reared rats more quickly assimilate information about a new environment? This sounds plausible, and is suggested by Einon and Morgan's result.

Finally, we must return to what is perhaps the most intellectually challenging aspect of this subject, that we have now attributed to disparity a causal role in three quite distinct behaviours, avoidance, aggression, and exploration. All that can really be claimed is that we have labelled in rather coarse terms one vital common causal factor. It must be admitted that 'disparity' needs considerable qualification, and a number of other factors need to be included. Some of these biasing factors have been identified, e.g. a conspecific in familiar territory gives a bias towards attack. Interestingly there are numerous examples of ambivalence between these activities.

CHAPTER 11

Sleep

11.1 INTRODUCTION

It is perhaps not obvious to the student that sleep holds anything of either psychological or ethological interest, except perhaps the possibilities of dream interpretation. Indeed, the classic ethological study by Hinde (1970) contains only two sentences on sleep. Standard reference texts on animal behaviour in which the index does not mention sleep include Marler and Hamilton (1967), Eibl-Eibesfeldt (1970) and Manning (1979). It is surprising that ethologists devote so little attention to sleep since it may occupy anything up to 80% of an animal's life and provides a very rich source of comparative data. Sleep is usually well represented in textbooks of physiological psychology but this information is often not integrated into more general psychology texts. Physiological approaches very rarely have anything to say about the ethological aspect, and little on function.

The reason that sleep attracts relatively little attention in the animal behaviour literature is perhaps clear. In some respects sleep, like fainting or pathological unconsciousness, seems to be a case of non-behaving, and is therefore of medical but not behavioural interest. This is true up to a point, but as we will see later, sleep shares characteristics in common with feeding and other psychologically interesting phenomena in that it has both appetitive and consummatory aspects (see also Tinbergen, 1951). Thus, sleep itself is analogous to feeding, and being motivated to retire to a suitable sleeping site analogous to food-seeking. As Moruzzi (1969) argues, a state of rest involving a specific posture '. . . is, in fact, a behaviour'.

Meddis (1975), building upon the work of Moruzzi (1969), lists four characteristics of sleep usually shared by all animals that unambiguously are classed as sleepers. These are: (1) it consists of a prolonged period of inactivity, (2) it occurs in a circadian rhythm at usually the same times each day, (3) it is accompanied by raised response thresholds, and (4) the animal assumes specific sleep postures at particular locations (often ones of relative safety). This is a deliberate simplification; sleep is difficult to define in unambiguous terms, but a 'sleep syndrome', as Meddis calls it, does clearly emerge when we consider these characteristics. Sleep-like behaviour, if not exactly sleep, is present in insects, molluscs, fish, amphibians, reptiles, birds, and mammals (Meddis, 1975).

The literature on physiological psychology contains detailed accounts of the neural and chemical bases of sleep. It goes beyond our brief to pursue this subject in detail here, since we are treating sleep in the same way as the other systems, and looking for principles of function and behaviour. However, ultimately, as in other areas, a synthesis of, on the one hand, the behavioural, ethological or 'black-box' approach and, on the other, the neurophysiological data is indicated. The reader wishing to obtain a more detailed discussion of the physiology of sleep should consult a text such as Murray (1965).

11.2 CLASSIFICATION OF SLEEP

Many of the data on sleep are obtained by means of electrical recording from the brain, either, in animals, by implanting electrodes, or, in humans, by scalp electrodes. The record obtained in this way is known as an electroencephalogram (EEG). I will describe results obtained from humans.

The awake subject's sleep pattern lies somewhere on a continuum between two predominant EEG forms, known as *alpha* and *beta*. When the subject is resting (particularly with the eyes closed), and not mentally active, alpha rhythms are predominantly exhibited. When a demanding task is posed, or for some other reason attention needs to be focused, the alpha rhythm is replaced by beta. Alpha activity is more regular and predictable than beta, and has a discernible dominant frequency of around 8–10 Hz. Because of its more chaotic appearance the beta phase is assumed to be one where neurons are not active in unison, and is called desynchronization.

When the subject falls asleep then a low-frequency (1–4 Hz) wave of relatively high amplitude, known as the delta wave, becomes apparent in the EEG recording. This is a case of synchronization, since in order to generate such a predictable pattern a large number of neurons are presumably firing in synchrony. After an hour or so of sleep a very different EEG pattern occurs; one resembling the beta activity of concentrated activity during waking. However, we would be wrong to assume our subject had woken up. The desynchronization represents another phase of sleep, which has acquired for itself a variety of different names that need to be remembered.

Although the subject is deeply asleep the EEG is like that of an attentive awake subject and for this reason the phase is often known as *paradoxical* sleep. It is accompanied by normal saccadic eye movements and is called *rapid eye movement* (REM) sleep. If aroused by the experimenter, during this phase subjects commonly report that they were dreaming and so it is occasionally termed *dreaming* sleep. Another designation is *desynchronized* sleep (D sleep), and finally, some authors use the expression *active* sleep (Meddis, 1975). In contrast to each of these terms the other sleep phase is called non-rapid eye movement sleep (NREM), synchronized or slow-wave sleep; thankfully 'S' covers both of the latter terms.

During the course of a night the subject shows alternation between bouts of REM and non-REM. These bouts are not randomly distributed. Immediately after a sustained period of REM sleep there is a low probability of another period appearing. The probability increases as time elapses.

11.3 APPROACHES TO THE STUDY OF SLEEP

11.3.1 Introduction

Sleep stands in contrast to the other forms of behaviour described so far in that it has an opposite state, wakefulness, that attracts more attention. Should we study sleep or wakefulness? If sleeping represents a more primitive or 'resting' level from which the activities of wakefulness arise then some writers would suggest that we need only study wakefulness. After all, a working TV set is more revealing than one that is switched off for the night.

Hartmann (1973) notes that Kleitman (1963), in his classic study *Sleep and Wakefulness,* despite citing 4000 references, makes no mention of the function of sleep. Hartmann argues that Kleitman views sleep as the 'natural background state of the organism and of the brain'. If afferent impulses maintain wakefulness then a fall of such impulses to below a threshold level is associated with sleep. Indeed, more recent views of sleep place weight upon reduction in activity of the midbrain reticular formation as a causal factor in sleep. However, this reduction is caused in part by *active inhibition* from other brain regions (Williams, Holloway, and Griffiths, 1973).

According to such a passive view of sleep it is departures from the sleep state (i.e. wakefulness) that need investigating, both in terms of mechanism and function. However, Hartmann argues that there is little reason to consider either sleep or wakefulness to be the more natural. The neurophysiology of sleep, discussed later, supports this view. From a functional point of view it is reasonable that sleep should in some way serve to optimize the behaviour of the organism in activities pursued during wakefulness. Hartmann notes that the converse argument, that activities during waking ensure optimal conditions during sleep, obviously carries no weight. We will argue that sleep optimizes overall survival chances, not just that it increases efficiency of waking activities.

Concerning the function of sleep, theories generally can be divided into one of four broad categories (see also Murray, 1965):

1. An unwanted substance accumulates in the body during periods of activity and is broken down during sleep. Sleep has evolved to serve the *function* of breaking down this substance. Alternatively, or in addition, some essential process of chemical synthesis can only be carried out, or can best be carried out, during sleep.
2. Nervous system components or pathways fatigue during times of arousal, and only sleep allows recovery.
3. Sleep serves simply to immobilize the organism at times when it would be unproductive to be active (e.g. high predation risk, poor effectiveness of behaviour).
4. Processes of information-handling such as memory consolidation can best occur during sleep (this may be associated with the synthesis aspect of (1)).

When we translate from these functional explanations to the causal level various mechanisms are suggested. If (1) is accepted then the organism may be expected to be goaded to sleep by departures from optimum in the body's level of the substance concerned. Waking might be associated with new optimum levels. On the basis of (2) the causal mechanism could possibly be directly derived from the functional explanation: sleep is a failure of fatigued neural components to sustain wakefulness. Sleep restores arousability. Explanations (1) and (2) may be termed homeostatic or negative feedback. They are not necessarily mutually exclusive; due to accumulation or depletion of a substance the nervous system may be unable to maintain arousal.

Functional explanation (3) appears to stand in clear distinction since no essential intrinsic function is postulated. However, when translated to the causal level the distinction is less obvious. In no way is the necessity for an oscillation within the control system that governs sleep/waking avoided. This may take the form of the underlying stimulus for sleep being a circadian fluctuation in a biochemical quantity, perhaps a local change at a particular brain site. Sleep may feed back and reverse the biochemical conditions of sleep onset. In these terms the organism is seen to have evolved this chemical rhythm not for its own sake but for the advantages that the consequent immobility brings.

Although functional explanation (4) is in terms of the facilitation of intrinsic processes during sleep, it does not appear to suggest any possible causal mechanisms.

11.3.2 Accumulation or depletion of substances during wakefulness

The biochemical theory of sleep accords with common sense in that activity is believed either to use up a necessary substance and/or to accumulate a harmful one. Sleep is seen to reverse these processes. One might logically argue that if recovery could occur during waking, sleep would not have needed to evolve. This explanation of function has led to experimentation on the causal mechanisms, and Webb (1968) reviews some experiments carried out in this tradition. In 1910 Pieron found that cerebrospinal fluid taken from fatigued dogs and injected into rested animals caused them to sleep. More recently, Monnier, working with rabbits, showed that sleep could be induced in a similar way.

Using much more sophisticated techniques than those available to Pieron, and causing less trauma, Pappenheimer (1976) essentially repeated the study. He reasoned that the effect would be demonstrable if the donor animal were only mildly sleep-deprived, and suggested that the chemical concerned would not be species-specific. The latter belief derives from the assumption that nature would be unlikely to evolve different substances for each species. Cats and rats infused with cerebrospinal fluid taken from sleep-deprived goats became abnormally sleepy. Increased duration of quiet sleep was observed in infused rabbits. Amplitude of the slow waves was increased by the infusions, a similar effect being seen in the sleep that follows a period of deprivation. Present research in this area is concerned with identification of the sleep substance.

Adam and Oswald (1977) argue for a fundamental biochemical periodicity associated with the cycle of sleep/wakefulness, though it does not necessarily follow that this cycle provides the causal basis for behaviour and we may have to look elsewhere for this. They claim:

'... metabolic balance alters so that degradative processes are stimulated during activity or waking, and restorative synthetic processes are inevitably favoured during inactivity and sleep.' (Reproduced by permission of the *Journal of the Royal College of Physicians of London*.)

11.3.3 Neural fatigue

According to this explanation, at a causal level certain components of the nervous system fatigue during periods of wakefulness, and sleep allows recovery. Whether one can strictly call recovery a functional explanation could be debated. If the animal has absolutely no alternative and is strictly non-behaving due to exhaustion then a functional explanation may be inappropriate. However, if other neural mechanisms actively control sleep so that certain fatigued components can recover, then sleep may be said to have the functional explanation that it allows recovery. On a causal level, if sleep is a failure of a fatigued nervous system to maintain wakefulness, i.e. reduction of input to centres controlling wakefulness, then lack of sensory stimulation should assist the onset of sleep.

It is a common belief that sleep is a period of recovery from activity.

Moruzzi (1969) (see also Williams, Holloway, and Griffiths, 1973) suggested that particular neuron populations, those involved in learning and memory, could need particularly long periods of rest. Similarly Hartmann (1973, p. 116) proposed that:

'... normal waking may "wear out" certain systems, especially those in the cortex, which depend on the ascending catecholamine pathways; and D sleep may play a part in "restoring" them.' (Reproduced by permission of Yale University Press.)

However, Williams, Holloway, and Griffiths (1973), in a review of the available literature, conclude that:

'... no neuronal populations have been found whose behaviour during sleep resembles that of the fully relaxed muscle.'

This observation is particularly relevant in the case of active sleep, where neuronal activity and cerebral blood flow are particularly high.

Sleep may allow neuronal recovery, but it would be wrong to suppose that, on a causal level, we can simply explain sleep as a failure on the part of fatigued neurons/circuits to maintain waking. Changes in activity in the brain's reticular formation are associated with onset and maintenance of sleep, but this is not simply a failure of these neurons, but rather a reduction in activity due to their interactions with other parts of the control system. Active sleep mechanisms serve to inhibit activity in brain regions that are critical for waking (e.g. the midbrain reticular formation) (Williams, Holloway, and Griffiths, 1973). As these authors express it:

'. . . the transition from waking to sleep represents a dramatic shift in patterning or organization of most somatic, autonomic and CNS processes.' (Reproduced, with permission, from the *Annual Review of Psychology*, Volume 24. © 1973 by Annual Reviews Inc.)

Some brain lesions (e.g. the raphe system) promote insomnia, whereas electrical stimulation of other brain regions (particularly at certain frequencies) is rapidly followed by sleep onset (Jouvet, 1969). This is incompatible with sleep being the outcome of a passive lifting of stimulation from the waking system. Cutting the sensory tracts does not increase sleep, unlike making cuts in the reticular formation (see Webb, 1968).

On the basis of our current understanding, perhaps the safest analogy to draw is with the autonomic nervous system. Neural structures promote heightened activity of various functions and another set *actively* inhibit these same functions. For example, slowing of the heart's activity can be due to reduction in sympathetic activity, increase in parasympathetic activity, or (as is more usual) both. Thus Webb (1968) opts for the possibility of:

'. . . the reciprocal interaction between two separate systems with the sleep mechanism serving to inhibit the waking mechanism and the waking mechanism serving to override the sleep mechanism.'

It may prove to be unnecessary to speak of two distinct controllers, and, like hunger-satiety, one system in two states might provide a more parsimonious model. At present we can't devise definitive models, but we can say that if correction of neural fatigue is the primary function of sleep then the system is anticipatory and responds not to fatigue, but responds so as to avoid it.

11.3.4 The immobilization theory

According to this theory sleep has evolved primarily, if not exclusively, as a means of keeping organisms still when in terms of survival there is no net advantage in being active. It is associated with, amongst others, Synder (1966), Moruzzi (1969), and more recently, Meddis (1975, 1977), who has developed a controversial argument. It is most appropriate to consider the many aspects of this thesis together in one section, and this occurs later.

11.3.5 Information-processing theories

An early writer on this subject, Hughlings Jackson, proposed that sleep helps to erase unimportant memories and consolidate the important ones. Fishbein and Gutwein (1977) argue that active sleep is necessary for long-term memory consolidation and for stability of memory in storage. They note that in animals there is an augmentation of active sleep following learning sessions. Long-term consolidation is impaired by deprivation of active sleep. According to this argument sleep is actively concerned with processes of memory storage, in contrast to the position that it allows passive recovery of synapses (see also Hennevin and Leconte, 1977, for a very similar conclusion). In rats, taste-aversion learning is impaired by deprivation of active sleep (Danguir and Nicolaidis, 1976).

McGrath and Cohen (1978) review evidence on the role of active sleep in learning and memory. From both animal and human experiments they conclude that under certain conditions active sleep facilitates the retention of learning. It appears to be more closely associated with storage than retrieval of information. In humans there is evidence of increased length of active sleep at times of stress and demanding conditions (Hartmann, 1973, p. 96).

11.4 THE SLEEP THEORY OF MEDDIS

11.4.1 Recuperation and sleep deprivation

Meddis (1975, 1977), in building upon and elaborating theories of earlier sleep researches, sees no necessity to postulate any other functional explanation for sleep than that immobility, combined with its low metabolic cost, outweighs the advantages of being active. Meddis first marshals evidence against a valuable recuperative process necessarily being at work.

Comparative studies of sleep (reviewed in the next section) lend little or no support to recuperation theories. For example, the bat and the shrew are both small active mammals having a high metabolic rate during wakefulness, and yet they spend vastly different lengths of time in sleep. It is difficult to accept that the bat should need such an enormous length of recuperation. Meddis (1975) cites the cases of the shrew, Dall porpoise and swift, each of which, he claims, survives without sleep. Similarly the albatross appears to spend weeks in flight without sleeping.

Any author who proposes that no recuperative function is served by sleep invites some rather predictable questions. In secure conditions can animals (and humans) survive without sleep? Isn't it true that we go mad if we are not allowed to sleep? Why do we feel so terrible if we can't go to sleep? These seem intuitively simple questions that would invite unambiguous answers, but both in theory and practice complexities arise.

Meddis (1977) can find very few cases indeed of humans who lead a normal life without sleeping or feeling the need to sleep. However, we have no reason to distrust the few such reports that are available. A number of people manage not only to survive, but to perform demanding jobs, with the help of what to most of us would be a ridiculously small amount of sleep (e.g. 1–3 h per 24). It would be misleading to classify them as insomniacs since they regard their condition as normal and do not complain about it.

Religious leaders, convinced of the value of extensive prayer sessions, as well as supporters of charity marathons (and possibly a few final-year PhD students), have attempted to stay awake for very long periods, but have been forced to abandon the task after 7 days or so. There is no reliable evidence that they suffer physical or mental harm, but the fact that the pressure to sleep is so powerful must be accommodated within any acceptable theory. Webb and Cartwright (1978) review the findings of some sleep-deprivation studies in humans which extended over several days. Perhaps not surprisingly, sleepiness is unambiguously caused, but beyond this little can be said with certainty. Decrements in performance on set tasks depend upon the nature of the task concerned, though others argue that a general performance decrement is apparent after 3-4 days sleep-deprivation (Wilkinson, 1965). Hallucinations and illusions apparently seldom occur until after 60 h of deprivation, and even then are rare. According to the review of Webb and Cartwright there are no obvious biochemical changes attributable to sleep-deprivation *per se*, though the list of parameters examined is by no means exhaustive.

In various animal species prolonged sleep-deprivation has been reported to lead to death (see Hartmann, 1973, p.40), though the results are most unclear (Webb, 1975, p. 119). The ability of animals to withstand sleep-deprivation is often considerable. Thankfully this unproductive and unpleasant line of research is no longer popular. The effect of sleep-deprivation *per se* cannot be distinguished from the effect of the procedure employed to keep the animal awake. The *'per se'* of sleep-deprivation may in reality be unattainable. Suppose some one wanted to examine the effect of food-deprivation. (S)he would simply remove the animal's food. This may introduce some problems of the kind we call frustration, but presumably it is less traumatic than if food were left in place but every time the animal took a bite the food was removed from its mouth. The latter is more closely comparable to deprivation of sleep by repeated waking. In the area of sleep there appears to be no procedure exactly comparable to food-deprivation, and certainly none as sophisticated as those used in feeding studies where ingested food is removed from the duodenum by a chronically implanted tube.

Meddis (1977) does not, of course, deny that human subjects who are kept awake for long periods (e.g. laboratory subjects, insomniacs, record-breakers) report extreme discomfort. However, he does not consider that this necessarily supports a recuperative theory of function. It is sufficient explanation that if the instinctive drive to sleep is thwarted, tension is created. The system was 'never designed to be resisted for very long periods'. The frustration and misery of insomnia is seen as the result of disparity between the perceived goad of the sleep instinct and the condition of wakefulness.

It is of course necessary for Meddis to postulate a causal mechanism that is responsible for periodically goading the organism into sleep. In fact in Meddis' study this turns out to have similarities with the processes postulated by adherents of recuperative theories. As Meddis (1977) notes, many researchers believe that during wakefulness there occurs an accumulation of the chemical serotonin in the raphe nuclei of the brain stem. Activity of these cells during sleep causes release of serotonin. Meddis argues that this type of process could account for the phenomenon of *rebound*, i.e. that sleep-deprivation is followed by greater than normal amounts of sleep. A 'pressure build-up' occurs when the organism is sleep-deprived, and this may be caused by:

'. . . accumulations of the chemical serotonin in raphe nuclei of the brain stem which can only be properly released during sleep.'

It was suggested that as the accumulation of serotonin increases then the initial rate of its discharge will be higher. In consequence discharge time will increase with quantity accumulated but it will not increase proportionately. The magnitude of rebound will reach a ceiling.

11.4.2 The functional explanation of sleep

Sleeping involves reduced energy expenditure (Zepelin and Rechtschaffen, 1974). Immobility affords security against predators if the sleeping site is secure and inaccessible. Sleep occurs at times in the circadian cycle when the animal is least competent to be active, and therefore protects against accidents and often a hostile environment as well as predation. Food-seeking and other activities can be concentrated into the part of the circadian cycle when they are likely to be most profitable. In other words, in terms of optimization, sleep literally keeps an animal out of harm when, and for as long as, it has nothing better to do (though qualifications will be made later). Most birds depend upon daylight for navigation, and therefore some authors would call night-time 'spare time'. By this is meant time left over when other activities have been done, or when conditions make other behaviour unprofitable. One could always argue that in winter a blue-tit has never really eaten an adequate amount since it is so close to the limits of survival and a few extra grubs would always be helpful. The point is that, when it retires, presumably the risks outweigh the possibility of gains. Although most birds sleep by night, for the rat it is safer to emerge by night and sleep during the day.

Webb (1971) advances arguments similar to those of Meddis. Thus, sleep does not serve a recovery function, but is:

'. . . in aid of preventing non-adaptive, ineffective, or destructive behaviour from occurring . . .'

For an animal that is only able to navigate by light:

'Night activity would simply have resulted in his falling off cliffs, drowning in bogs, being consumed by effective night predators—very poor returns relative to his energy expenditure.'

Webb makes a very important point that needs emphasizing:

'. . . almost no attention has been directed towards the adaptive qualities of non-responding.' (Reproduced with permission from *Biological Rhythms and Human Performance* (ed. W. P. Colquhoun). Copyright by Academic Press Inc. (London) Ltd.)

In so far as savings on energy are concerned, Webb is not describing a feedback system, but rather an anticipatory system; exhaustion of energy reserves, etc., is prevented by enforced inactivity.

Meddis (1977) mentions, by analogy with other biological systems, a 'sleep-stat' that programmes a certain normal length of sleep for a species. Deviations from this norm are possible according to external circumstances. For instance, in birds when the length of daylight hours increases, so the length of time spent awake increases.

According to the immobility hypothesis, and its emphasis upon the adaptive significance of staying out of danger, the site chosen for sleep is vital. Vulnerability must be minimized, and indeed, where ecologically possible, animals seek safe locations. Birds choose inaccessible branches and ledges. The rat sleeps in its burrow, while members of some species group together for sleep. Fish sleep in secure locations. They sometimes wedge themselves into a crevice, or press up from underneath against a rock (Tauber, 1974). Depending upon the species concerned, the advantages that sleep brings will vary. For the more vulnerable it may serve primarily as an anti-predator device, whereas in others most advantage is derived from energy conservation. Being helpless, the human new-born can perhaps best survive by placing not unreasonable demands upon the mother, and the inactivity of sleep provides at least some respite.

The question inevitably arises of why should the organism sleep and not just remain still and quiet (Webb, 1975, p. 158). The important point is that it must be programmed *not* to respond, except to certain danger stimuli. Although one meets formidable problems in trying to theorize here, perhaps it is meaningful to say that most incoming stimuli must be blocked before they get to behavioural decision components, and that is what we mean by sleep. It is only those of us having the benefit of reading Darwin who know immobility is beneficial; the animal must be allowed no choice! But might not evolution or an omniscient creator have been wiser to have placed the immobilizer switch nearer the response end? I leave that for the reader to answer!

11.4.3 The motivational aspect of sleep

Rather than the sudden appearance of unconsciousness, a dubious behavioural asset, sleep has both appetitive and consummatory aspects. The appetitive phase covers all of the preparations for sleep, i.e. finding a suitable location and body posture (Murray, 1965; Moruzzi, 1969; Meddis, 1977). With regard to humans, Meddis writes:

'The desire to sleep is a motive which can be just as strong as the desire to eat or drink, even though psychologists rarely include it in their list of drives.' (Reproduced by permission of Routledge and Kegan Paul Ltd.)

An analogy is then made with the appetitive and consummatory phases of hunger. The subjective feeling of drowsiness stands in comparison to the feeling of hunger. When we are hungry, pleasure consists not merely in eating, it is also the spectrum of anticipatory cues such as arriving at the restaurant, smelling food, etc. These cues should be compared with the feel of the sheets and pillow in the case of sleep. An avoidance aspect is also involved. A state of tiredness causes us to avoid intense sensory stimulation, and we retire to the seclusion of a quiet sleeping location. To adapt this to animals and to employ the language of earlier sections, it can be said that a state of drowsiness permits the cues associated with a secluded sleeping place to exert a pull on the animal. A stability factor is present here. Sleep appears to be associated with reduced activity in parts of the reticular formation. Sleep (and perhaps the motivation to sleep) arises from inhibition of reticular activity. By retiring to a quiet location afferent stimulation is minimized, which further reduces reticular activity. In turn, this may lift inhibition from the sleep controller.

It is presumably necessary for the sleep control system to exert inhibition upon potentially competitive tendencies such as to feed or seek a mate. Particularly in so far as its appetitive phase is concerned, sleep must share features in common with behaviours traditionally studied in psychology and compete in some 'currency' for control of behaviour.

According to this interpretation a circadian clock controls an internal state. This periodically goads the organism to seek a sleep site. Presumably some hours later a combination of this circadian rhythm reaching another part of its cycle and a feedback consequence of sleep arouses the animal. Meddis notes that humans experience a subjective reminder of the intrinsic rhythm. If a person has to stay awake all night then s(he) feels more sleepy at 3 a.m. than at 10 a.m. the next morning, even though 10 a.m. represents 7 h more sleep-deprivation. This would not be encouraging news for a recuperation theory. Drowsiness is seen as the outcome of a circadian sleep controller and not as the consequence of fatigue *per se*, though of course it may prevent undue fatigue. After a day of doing almost nothing we can be as sleepy as after one filled with hard work.

11.5 SPECIES DIFFERENCES IN SLEEP

Species vary enormously in the parameters of sleep, and advocates of the various theories find the natural history of sleep a source of many challenges. Unlike feeding or copulating, for some species even the objective definition of sleep presents serious difficulties (Allison and Van Twyver, 1970). Fish and snakes cannot close their eyes, and horses and cattle seldom do so. For primitive amphibians such as the bullfrog and salamander, it is claimed that sleep does not occur according to either EEG or behavioural criteria. They show alternate periods of 'quiet and active wakefulness'.

Are there any predictors of how much sleep a species would be expected to show? Usually no single dominant factor emerges, but rather a number of factors are present. It would seem that in any given case evolutionary pressures have had to 'trade off' one against the other.

In an inter-specific comparison, Zepelin and Rechtschaffen (1974) find a significant tendency for animals to sleep more as the metabolic cost of keeping awake increases. A statistic they quote is that although it takes 5000 mice to equal the weight of one human their combined food consumption is 17 times as great as that of one human. In terms of energy saved a species of this size has much to gain by not being active, not needing so much food and therefore not investing so much energy in food-seeking (Horne, 1977, develops this argument). The 'intellectual' demand of existence during wakefulness is a poor predictor of sleep length. The so-called higher animals, monkeys and apes, show neither more nor less sleep (or distribution of the phases of sleep) than less advanced species. Animals which have a low caloric density diet and therefore have to invest large amounts of time in feeding, such as herbivores (e.g. cattle and elephants) sleep relatively little.

Some authors have found it helpful to divide species into 'good' and 'poor' sleepers (see Allison and Cicchetti, 1976). A good sleeper spends 8 h or more a day in sleep, and will readily sleep in the laboratory (for example, predators such as cats). Bears sleep for long periods, and also their copulation is a leisurely activity. By contrast, antelopes sleep very little, and their drinking and sexual behaviour is very rapidly performed. One might reasonably conclude that the antelope has many enemies and no safe sleeping site (Meddis, 1975). Poor sleepers obtain relatively little sleep, and when put into a new laboratory environment need a long adaptation period before reaching stable sleeping habits (species subject to predation such as rabbits fall into this group).

Vulnerability is of course strongly dependent upon habitat. Species having a relatively safe secluded sleeping place (e.g. bats) generally have longer sleeping hours than those made more vulnerable by sleep (e.g. sheep). Mice form prey for numerous species and yet sleep for long hours; their usual sleeping place is very secure (Meddis, 1977). Bats and shrews have a similar body size and yet whereas the shrew sleeps very little the bat sleeps up to 20 out of 24 h and lowers its metabolic rate very considerably during sleep (Allison and Van Twyver, 1970). This may or may not prove to be coincidental to the fact that whereas the short-tailed shrew has a life span of about 2 years, the bat can expect to attain an age of 18 years. Although the hamster and 13-lined ground squirrel are subject to predation they are good sleepers, and this is presumably explained by the security of their burrows (Allison and Van Twyver, 1970). Baboons, though a predatory species themselves, are subject to predation by leopards and do not have safe sleeping sites available. Apparently they obtain little active sleep, and Allison and Van Twyver (1970) write:

'. . . it appears that deep sleep and high percentages of paradoxical go hand in hand with a safe sleeping arrangement.'

Webb (1968) notes that the phasic patterning of sleep, i.e. the length and distribution of bouts and its position in the circadian light–dark rhythm, depends upon the animal's structure, environment, and social interactions. Animals will be nocturnally active if they function best by night in terms of food sources, mating, low chances of being eaten, etc. Lizards sleep at night when inevitably body temperature is relatively low and they are therefore sluggish and vulnerable to predation.

11.6 SUMMARY, CONCLUSIONS, AND DISCUSSION

Sleep occupies a unique position in our discussion, in that controversy rages over the question of what function it serves. By contrast no one is in much doubt as to the functional significance of feeding, drinking, etc. Even exploration is capable of being explained in fairly uncontroversial terms. A survey of the current sleep literature reveals absolutely no convergence of thinking on the question of function. Whereas Meddis (1977) dismisses biochemical explanations of sleep function (but not causation), Hartmann (1973, p. 145) concludes:

'. . . there has been no reason to question that sleep basically has a restitutive or restorative function, in accordance with our own commonsense notions.' (Reproduced by permission of Yale University Press.)

In a review of the state of the literature in 1973, Williams, Holloway, and Griffiths (1973) wrote:

'. . . neither analysis of its physiological substrate nor specific deprivation of its stages has provided a basis either for rejection or enthusiastic acceptance of any of the theories of function in the current literature.' (Reproduced, with permission, from the *Annual Review of Psychology*, Volume 24. © 1973 by Annual Reviews Inc.)

Sleep may serve more than one function. Because it may be to an animal's advantage, in terms of predation risks and energy conservation, etc., periodically to remain inactive does not preclude biochemical and recuperative functions. Sexual behaviour primarily serves fertilization but can also be employed for transmission of information on rank and territory. Suppose sleep evolved primarily as an enforced immobility response. It would surely have been remiss of nature not to exploit this time 'off-line' if some essential repair or memory consolidation process could occur then (or perhaps take place most efficiently then). Possibly such processes are carried out rapidly, which would be compatible with the fact that sleep is very short in some species. Conversely if sleep evolved primarily as a regulatory behaviour then adaptation could have shaped long sleep duration in some species.

It may prove to be misleading to view sleep either as an essential recuperative mechanism, with dire consequences (such as poisoning) that follow its deprivation, or as an adaptive behaviour that can be harmlessly resisted if one tries hard enough. A middle position may be nearer the truth: that some function such as memory retrieval/storage is damaged by sleep deprivation and that this could ultimately goad sleep by an indirect route.

The unsuspecting student is more likely to get confused between causal and functional levels of explanation in the case of sleep than other behaviours. For example, it is clear that the functional significance of feeding is to gain energy and essential dietary components. This translates to the causal level in that something (though exactly what, is not immediately apparent) to do with energy state is likely to be associated with onset and termination of meals. In the case of sleep, biochemical homeostasis may yet be shown to lie at the core of both function and causation. However, it may be relevant only to causation, a biochemical oscillation having evolved merely to immobilize the animal (this would suggest a neurotransmitter substance). Conversely the function of sleep may be to serve some biochemical synthesis or homeostatic process in the organism but that reference to this process gives no clue to the causal explanation. The causal process could be anticipatory, a potentially harmful biochemical displacement being avoided by sleep.

We discussed sleep in terms of a balance sheet of costs and benefits. A balance sheet involving several factors is still valid even if we attach most weight to a biochemical recuperative process, since overall survival is what counts. Like most evolutionary arguments and functional explanations, the ones discussed here are fraught with difficulty. We simply don't know enough about any species to attach values to the terms in the balance sheet. If one believes recuperation theories then it may well lead to the argument that sleep is of survival value *in spite of*, rather than because of, the enforced immobility. In other words, the gain (e.g. avoidance of predation), by being still and secluded, necessary for recuperation to occur does not compensate for the loss of sensory detection and resulting defencelessness. Walter Hess (see Webb, 1975, p. 161) took this view.

Some authors view sleep as a spare-time activity, i.e. if time is left over when all essential activities have been completed then it is best spent in sleep. In some cases this may be misleading. It is true that for some species, which eat high caloric value food, life seems leisurely at times of food abundance. Time can afford to be spent in sleep. An example is the large cats, where risk from predation is minimal and so the advantages of immobility would be in terms of energy conservation, avoidance of accidents, restorative processes, and possibly facilitating digestion. For some smaller species the argument about spare time seems circular and does not consider the mutual interactions and feedback between energy, energy seeking, metabolism, and temperature; time is only 'spare' *because* the animal is sleeping. One could imagine two possible stable strategies for a small mammal such as a shrew: stay awake and eat enormous amounts, or sleep and very strongly lower metabolic costs. A compromise such as sleeping but only slightly lowering metabolic rate might prove impossible since an unreasonable demand for food would arise the following day.

If no recuperative function is served by sleep it is surprising that certain species sleep at all. A member of a vulnerable species, subject to predation, would seem to be placed at a disadvantage by taking even a short sleep, though remaining still and with normal sensory thresholds might be adaptive. Thus, with reference to grazing animals, when Webb (1975, p. 159) writes

'They sleep very little because "non-behaving" has low survival value. . .'

this invites the question—why 'non-behave'? Quiet sleep may represent the best compromise between immobility and vigilance since sensory thresholds are elevated relative to active sleep.

CHAPTER 12

Conclusions

12.1 INTRODUCTION

This final chapter tries to achieve several things. Although it does not provide a summary of all the main points of the study, it is necessary to develop certain themes that emerge from individual chapters; some conclusions depend upon integration of information from various chapters. Also the relevance of the arguments to selected areas related to the study is worth pursuing. Finally, some words are needed on the philosophy that underlies the study and the development of the models.

12.2 AN OVERVIEW OF MOTIVATIONAL SYSTEMS

12.2.1 Comparison of motivational systems

A number of characteristics of motivational systems can be compared and contrasted (see also Toates and Archer, 1978). This was done to some extent in the foregoing chapters, where it seemed appropriate for explanatory purposes. Now an overview is possible.

The functional explanation tells us *why* a behaviour is present, in other words *how* evolutionary processes are supposed to have selected that particular behaviour. In earlier parts of the study the concept of fitness was mentioned; we presume that organisms have evolved in such a way as to maximize fitness. According to such an interpretation, the functional significance of any aspect of behaviour must be seen in terms of the chances it gives to perpetuation of the organism's genes. A behaviour, such as drinking, that contributes directly to the survival chances of the individual organism, thereby also contributes to the chances of gene perpetuation. When viewed in these terms, the traditional distinction made between, on the one hand, activities concerned with individual survival, such as feeding, and on the other, sexual behaviour concerned with gene perpetuation, is of little or no consequence. In other words, all behaviour has been shaped by evolutionary processes so as to maximize fitness. Thus, for example, if a starving male rat is presented with both food and a mate, it will copulate first and eat secondly (D. McFarland, private communication). This makes little or no sense in terms of individual survival, but is presumably the outcome of evolutionary processes that maximize fitness.

However, provided that this argument about fitness is kept in mind, for other purposes a useful distinction can sometimes be made between an activity such as feeding or drinking, which contributes to gene perpetuation through the mediation of the physiological state of the organism, and sexual behaviour, which, as far as we know, does not increase fitness through any effect on the organism's physiological state. It is often assumed that whereas sexual behaviour increases the chances of gene perpetuation, it decreases the survival chances of the individual organism concerned. Clearly there are disadvantages to survival in sexual behaviour. Fights may occur over potential mates. Courtship displays may attract predators, and some species are made very vulnerable to predation by copulating. The increase in body-weight associated with pregnancy makes escape more difficult. However, it cannot be assumed that there are no advantages for individual survival. Pair-bonds may be established by courtship, and strengthened by the consequences of mating, to the mutual advantage of both animals. As another example, the female may withdraw to a safe location for incubating eggs.

Of course, the functional explanation cannot account for the causation of behaviour. Presumably a starving male rat does not 'know' intellectually that the chances of gene perpetuation are maximized by copulating first. It is necessary to postulate that, in this case, the causal factors are such that in the final common path for behaviour, the signal for sexual behaviour carries more weight than that for feeding. The functional explanation only tells us the contribution to fitness that an aspect of behaviour makes, not how the biological system actually works.

However, the functional explanation may occasionally provide clues about the nature of causal mechanisms. For instance, suppose an accumulation of fat occurs in a species of bird prior to migration (McCleery, 1979). This would serve the function of providing energy for long distances of travel. It may therefore be worth looking for a biochemical change in energy conversions that is triggered by the stimulus for migration (possibly a change in ambient temperature or daylight length).

Explanations can be developed of how thirst, hunger, temperature control, fear, and aggression serve rather directly to maintain optimal conditions for the individual organism. In the first three cases this refers to specific physiological states. Even here though the situation may be very complex. An animal may consume extra protein or increase lipogenesis in connection with reproduction. This may serve gene perpetuation by making the animal better able to lay eggs or care for the young, but the extra weight *per se* may in some respects be a hindrance to survival of the individual concerned. This is obviously off-set by the advantages in terms of overall fitness. In the case of fear and aggression an even less easily defined optimum is involved. Fear (or, to be more precise, avoidance) protects against direct physical threat. Aggression may sometimes also directly serve this end, but perhaps more often contributes in less obvious ways. For instance, the individual's future physical integrity may be served by driving away competitors for its food or territory. Again, individual survival and gene perpetuation are inextricably linked; a display of aggression may serve to attract (or keep) a mate and to remove animals competing for the mate.

In each of these examples a strong negative feedback aspect is present in the postulated causal mechanisms. Whereas we discouraged the attitude that hunger and thirst are aroused simply by deficits, nonetheless displacements in one direction do potentiate ingestion and in the other direction have a restraining effect. At least superficially, we can construct analogies between, on the one hand, regulatory systems and, on the other, fear and attack. In the latter, a sudden intrusion instigates behaviour that restores the previous equilibrium. The basis of exploration appears to be similar in some respects to that of attack and fear. It was argued that exploration depends upon a stored representation or model of the environment, and is especially aroused when mild disparity is perceived between incoming sensory information and this internal model.

Sleep remains a mystery. Possibly it evolved primarily to serve some recuperative function, but we don't know. It may be useful for no other reason than that it keeps the animal out of danger and conserves energy. Another possibility is that it orginally evolved simply as a means of enforced immobilization, but later a recuperative function became attached to it. Any functional explanation, of course, requires a causal mechanism to control behaviour. An oscillation involving a chemical transmitter within the sleep controller is suggested.

Sexual behaviour is clearly somewhat different from all the other behaviours in that the common function of gene perpetuation is usually not served via the mediation of increasing the survival chances of the individual organism but by interaction with another organism. However, negative feedback is still present in the causal mechanisms, in that ejaculation desensitizes the system to further arousal. The positive feedback aspect involving genital friction is essential for achieving the end-point of ejaculation and internal fertilization. Sleep and ejaculation appear to have in common that they change system parameters (respectively, general arousal mechanisms and sexual arousability). This change in sensitivity is caused by processes that are still unidentified, and neither involve substances obtained directly from outside nor by the behaviour concerned. By contrast, feeding and drinking alter the arousability of hunger and thirst, by obvious and relatively direct routes involving external substances gained by the relevant behaviour. Interestingly, both sleep and sexual behaviour have at various times been thought to eliminate toxic substances from the body (in the case of sex, this view was advanced by, amongst others, Martin Luther and St Thomas More).

How rapidly and how fully can each behaviour be re-aroused following satiety? Drinking can be quickly re-aroused by salt-injection and feeding by insulin injection. Feeding can also be re-aroused by giving more palatable or readily ingested food (e.g. glucose solutions). In some species, following sexual exhaustion, sexual behaviour can be fully re-aroused by changing the partner. In others, this procedure is relatively ineffective in re-arousal; often desensitization presents a powerful brake to any arousal. The locus of desensitization is assumed to lie within the central nervous system. By contrast, aggression, fear, and exploration can be fully aroused by appropriate stimuli, despite having been very recently performed. This, of course, is to be expected; general satiety for fear or aggression would be counter-productive, whereas there may be adaptive significance for sexual behaviour to be restrained. In some species seminal fluid volume needs to be built up to an optimal level to maximize fertility chances (other advantages can also be found).

12.2.2 Internal and external control of behaviour

As Hogan and Roper (1978) point out, to argue that one class of behaviour, e.g. aggression, is more strongly aroused by external factors and another, e.g. hunger, by internal factors, may have something in common with trying to establish whether a particular behaviour is more strongly dependent upon innate than learned factors (see also Moruzzi, 1969). Both aggression and hunger are internally *and* externally controlled, and in one sense it is meaningless to attempt to give weightings to these factors. However, it is valid to argue that under natural conditions the rapid appearance and subsequent 'satiation' of aggression correlates with fluctuations external to the animal whereas feeding/satiety may be closely associated with either internal or external changes. In the longer term, changes in the pattern of aggression may be associated with internal changes, e.g. a seasonal effect of testosterone.

The appetitive aspect of hunger cannot be attributed purely to internal energy state. On regular schedules, the increased activity of the hungry rat is limited to the period before feeding and is reduced if external disturbance is minimized (see Hogan and Roper, 1978, for references). It seems that hunger increases the responsivity of the animal to external stimuli, particularly those which have in the past been associated with food.

Sexual arousability, an internal state of the animal, translates potential sexual arousal stimuli that arise from the partner into a pattern of appetitive and consummatory behaviour, this being somewhat analogous to the effect of energy state on the animal's response to food. The sexually exhausted animal is not aroused by sexual stimuli and may even find them aversive. The animal sated for food no longer finds the stimuli associated with food to be rewarding. Therefore the claim of some authors (see Grossman, 1967), that behaviours such as feeding are internally aroused whereas sex is externally aroused, needs very careful qualification.

Experimental reports on the proverbial aggression of the Siamese fighting fish also need scrutiny. This species may seem to have an appetite for attack, which is aroused in the absence of external stimulation (i.e. what is often called a deprivation drive). However, close examination shows the possibility of a conditioned arousal of aggression by stimuli related to a conspecific or self-image (Hogan and Roper, 1978).

Hogan and Roper review evidence from various authors which claims to show that external and internal stimuli add to give a total motivational tendency. However, I consider that additivity models can be misleading. Take the example of sexual behaviour. With a highly arousable nervous system, i.e. long deprivation and adequate testosterone, a relatively weak external stimulus may arouse sexual behaviour in the male. Conversely, it may be possible to arouse a sexually exhausted male only by a strong external stimulus (e.g. a sexually receptive and active conspecific female). In this sense, a kind of superficial additivity may fit the data. But since the components of the proposed addition are two fundamentally different entities there is no reason to believe in physical additivity of stimuli within the animal. To be precise, testosterone primes and sensitizes a sexual arousability system, analogous to turning up the volume control on a radio set. External stimuli (and, presumably, memory in the case of humans) impinge upon this system and it may be said to be put into a state of arousal. However, in no strict physiologically meaningful sense can we argue that internal and external *stimuli* are adding, or even multiplying. An incoming signal is acted upon by a *system*. As such, a multiplicative dependence between stimulus and the system's parameters may emerge, but it is not stimuli that are being multiplied. The discussion of this subject often tacitly implies that an internal stimulus is present. However, since we no longer believe that afferent nerves detect the state of seminal vesicle pressure, it is hard to see what the internal stimulus is. Blood testosterone is not 'detected' as such, but rather it appears to sensitize certain interneurons. Sometimes testosterone is more of a catalyst for these neurons, since cutting off the supply of hormone has little effect on sexual behaviour.

In the case of hunger, afferent nerves detect some property of internal energy state, and therefore it is meaningful to speak of an internal stimulus. Indeed, at one level the same stimulus quality is detected internally and externally, i.e. energy. Therefore the question is—how does the information from two sets of afferent nerves, detectors of internal and external energy, interact to control ingestion? In Chapter 4 it was proposed that an internal state gates the signal from the external stimulus so as to cause approach, avoidance, or indifference towards food.

In Chapter 3 we discussed in some detail the notion of cue strength. This was a convenient expression for the animal's perception of the availability of food or water. The possibility was mentioned that cue strength may multiply with deficit to determine a motivational tendency. We abandoned 'deficit' in Chapters 4 and 5 in preference for more precise expressions. The reader may now feel that 'cue strength' has been abandoned at Chapter 3 in favour of some more complex mode of interaction between internal and external events. The actual position is that we are dealing in each case with different kinds of explanation and different sets of questions. The idea of cue strength arose amongst Oxford researchers, largely from ethological considerations. This term does not explicitly show the small details of the stimulus side of ingestive processes, but is a more general and collective measure of distance to food, quality of food, rate of return, etc. Similarly, deficit is a convenient short-hand for saying that the animal has been deprived for a certain time. Cue strength is appropriate when dealing with such factors as the ecology of a species and the economics of searching for food. 'Deficit', 'cue strength', and 'multiplicity' form convenient first approximations which then suggest more detailed analyses.

When we consider the more 'psychological' processes of ingestion, involving factors such as palatability, then a simple multiplicative relationship between external and internal factors appears inadequate. However, it is not that one model is necessarily better than the other; they serve different ends. In the essential feature that both internal and external events govern behaviour there is a clear covergence of ideas; both kinds of model depend upon joint control. This convergence is to be welcomed, and in the future more synthesis is bound to occur in the area between ethology and physiological psychology. Indeed, a recent contribution from the Oxford group (Pring-Mill, 1979) considers both palatability and cue strength (availability) as joint external factors which, along with internal physiological state, determine motivational tendency. Parenthetically, it is worth noting that there is a parallel development to the studies described here in the work of Ludlow (1976), which gives a place of importance to reciprocal inhibition between activities in determining behavioural stability and avoiding ambivalence. There is every reason to believe that these various approaches will shortly converge, and also serve to bring ethology and physiology closer together (Ludlow, 1979).

The motivational model which is being developed here appears rather neatly to conform to the views of Bindra (1978), who wrote:

'. . . organismic conditions are important in that they serve as "gates" or limits within which certain particularly incentive stimuli become effective. . .'.

Bindra continues:

'However, no motivation for a goal is generated by the organismic conditions *per se*; the generation of particular goal-directed motivational states is critically a matter of environmental incentive stimulation.'

Bindra's theory presents a possible solution to the problem of behavioural stability. Suppose an animal, having almost equal hunger and thirst, moves towards either food or water. Having made a decision, then according to Bindra's theory we would expect momentum to be provided by the animal's movements towards and consumption of the appropriate commodity. This expresses, in alternative language, some of the ideas about stability and cue strength advanced in Chapter 3. Bindra writes:

'. . . an environmental stimulus that generates a motivational state can also serve as the goal stimulus to which the action is directed. This provides an uncomplicated mechanism for the coordination of motivation and action.' (Reproduced by permission of Cambridge University Press.)

It is worth a short digression to note that in the commentary section which followed Bindra's paper, Gallistel (1978) pointed out a particular experimental design that poses complications for this theory. As we discussed in Chapter 5, a hyponatraemic animal will locate a salt source and ingest salt, even if its only previous encounter with this particular salt location had been when it was hypernatraemic. In this case any incentive for approach must be based upon a memory that at the time of its initial storage was accompanied by an aversive or neutral stimulus from salt.

12.2.3 Reinforcement

Reinforcement has not been explicitly mentioned much in the review, though it has never been far from our interests. What makes a reinforcer reinforcing? This kind of question occupies much space in current psychology discussions. Although it is beyond our scope to tackle such questions, nonetheless the properties of different reinforcers are of importance in any systems view of behaviour.

We say that the thirsty organism finds water reinforcing, and by this we mean that water strengthens behaviour that preceded its acquisition. The animal works for the reward of water when thirsty, and masters a maze for this reward. Hogan and Roper (1978) recently presented an extensive comparative review of different reinforcers. They considered properties of schedules each in terms of the various reinforcers such as water, food, sex, etc. This section is based largely upon their review.

One kind of reinforcement schedule is known as fixed ratio, and the animal must, for example, respond five times for one reinforcement (FR 5). Given the animal's normal speed of working and, for example, eating, then if it is allowed only a limited time in the test apparatus a FR may mean that it cannot obtain as many reinforcements as on continuous reinforcement (CR or FR 1). If the daily time allowed in the Skinner-box is more generous it may be possible for the subject to compensate completely for the harder schedule. Hogan and Roper identify four possibilities that may result when the schedule is changed from CR to FR:

1. The animal may obtain as many reinforcements as on continuous reinforcement. For example, this would necessitate five times as many responses on FR 5 as on CR.
2. It may make only as many responses as on CR, in which case for FR 5, magnitude of total reinforcement is one-fifth the size of that on CR.
3. It may make an intermediate number of responses between these two extremes.
4. It may show fewer responses than on CR.

Apparently, the unlikely possibility that the animal obtains more reinforcements on FR is not encountered in practice and so is not included. Other ways of making life harder for the animal are available. Instead of increasing the ratio, the size of reinforcement may be reduced (for such reinforcers as food, water, etc.). Following these various procedures for making reinforcement 'expensive', the extent to which the animal compensates is calculated. Complete compensation means that the total amount of reinforcement obtained is unchanged, despite it being made more expensive. Conversely, if the number of responses per session is the same as on CR, compensation is said to be zero. If the number of responses is less than on CR, compensation is negative.

In a study on feeding (Collier, Hirsch, and Hamlin, 1972), that we discussed in Chapter 4, compensation was complete up to FR 20 but then started to fall. Body-weight was maintained up to FR 80, indicating the role of physiological compensation (such as increased utilization efficiency of food, decreased metabolic rate, etc.). Complete compensation at relatively low ratios, but a fall in compensation at the higher ratios, is typical of food reinforcement (Hogan and Roper, 1978). If the ratio is made very high, then the amount of energy invested in responding would not be justified by the amount of food earned. Apparently, animals abandon responding long before this point is reached.

In the case of water reward for thirsty rats, compensation does occur at the lower ratios but, unlike feeding, is never complete. Hogan and Roper suggest that in the case of fluid regulation other forms of compensation may be more effective (means which come to mind are decreased grooming and activation of ADH-mediated compensations), though the possibility of some intrinsic difference in the nature of the reinforcers must be considered.

For heat reinforcement, compensation for changes in reinforcement magnitude is good if the animal is given control over the duration of reinforcement by, for example, a heater coming on for as long as a lever is depressed. Compensation is less complete if the duration that the heater is on following the bar-press is fixed at some particular length of time and other parameters, such as heat intensity, altered. In many cases for heat reinforcement, compensation is poor on FR schedules (see Hogan and Roper, 1978). Hogan and Roper note that, unlike the need for water and energy, either behavioural or autonomic temperature control can completely compensate for moderate temperature stresses. They suggest that autonomic compensation may be particularly effective when FR size is increased. Heat reinforcement differs in some important respects from food and water. Reward is immediate and unavoidable. Even in a very cold environment, heat reward, if brief and intense, may have an aversive component because of a local effect at the animal's skin. In general, for food and water, as reinforcement size increases then so acquisition of a task such as bar-pressing or running an alley is speeded up. Larger amounts of reinforcement allow higher FRs to be sustained, and the animal shows greater vigour in its responding. For an animal in a cold environment, as heat reinforcement magnitude or duration is increased so response rate falls. Hogan and Roper note the following differences between, on the one hand, heat and, on the other, food and water. For heat reinforcement the effect is immediate but rapidly dissipates. There is nothing analogous to satiation within a session. If we assume that the animal is working so as to maintain ambient temperature, then it would be expected that responding would fall as the value of reinforcement is increased. One hour in a cooled Skinner-box, working for heat, is like a day or so of working for food. Of course, over such a period no long-term satiety effect for food is apparent. Also, if reward size were to be doubled, then, over a day or more, responses for food would fall.

As was discussed in the case of aggression (Chapter 9), most research that has attempted to compare, say, food and opportunity for attack as reinforcers has employed Siamese fighting fish. Hogan and Roper report very poor compensation on FR schedules for reinforcement consisting of the opportunity for attack.

12.2.4 The physiology and psychology of ingestive behaviour

Clearly, in order to understand the causal factors of food and water ingestion, we need to look at the internal biochemical transactions of the organism. The best possible evidence is needed on how and where the organism detects energy and fluid states. This falls within the domain of physiology and physiological psychology. In addition, though, processes involved in ingestive behaviour, which at present fall outside of the scope of traditional physiological explanation, have been described. For instance, animals form associations between the smell and taste of food and its subsequent caloric yield. Satiety cannot be predicted solely on the basis of energy states. Making food expensive to the animal by loading the bar of the Skinner-box or imposing a high FR schedule causes it to eat less and/or to change the distribution of meals. This cannot be predicted from physiology. Therefore our theories will inevitably be hybrids, composed of physiological evidence and the more 'black-box' psychological evidence. There is nothing wrong with such 'sitting on the fence'. It is surely misguided either to shun physiology or to pretend that it can, at this stage, always provide the most appropriate explanation.

12.3   THE SYSTEMS APPROACH

12.3.1   The bias of a systems approach

The reader will need little reminder that a bias is very evident in the present study. Extraneural physiological processes within the organism, representation of the environment, feedback from the environment, and overall performance characteristics of motivational systems were discussed. However, little or no attempt was made to describe brain mechanisms, anatomical locations, lesions, or chemical stimulation techniques. This is not because the author considers the latter areas of investigation to be irrelevant or necessarily uninteresting. Ultimately explanations at all levels, and drawing from a variety of experimental and theoretical techniques, will need to be integrated. There are at least two reasons that can be given to justify the bias of the present study. First, we do not always need to consider the fine details in order to answer some of the broader questions; explanations can emerge at various levels. Secondly, the subject matter of the present study is partly a reaction to the very heavy bias in the existing psychology literature towards pursuit of the small details to the exclusion of discussion of overall features and function.

It is my belief that undue faith has been placed in the value of a certain type of investigation. The approach is typified by questions of the kind—do neurons located at point x_1, y_1, z_1, in three-dimensional space have a significant role in feeding? Does feeding employ adrenergic or cholinergic pathways? What will happen if we lesion this area? Such questions sometimes yield interesting answers, but they usually remain isolated bits of information, all too often left to gather dust on a library shelf. The implications of the particular finding are not always appreciated, since they fail to be assimilated into a model of overall biological design.

I consider it reasonable to claim that the set of questions posed here is more philosophically and intellectually demanding than those raised by piece-meal experimentation. For instance, it was asked, by what processes do internal energy and food cues govern food ingestion? Clearly experimentation, with collection and presentation of empirical observations, is demanded. But this is not enough; we need models of the *kind* of process that could conceivably be responsible for this behaviour. The nature of complex processes is most unlikely to be revealed by lesions or chemical stimulation.

To take just one example, it is a reflection of the bias in the existing body of literature, that whereas one can find sexual behaviour discussed as a drive, and in physiological psychology texts there are numerous papers on lesions and chemical stimulation of sex 'centres', it is very difficult to find a discussion that relates the physiological and psychological evidence. Why do we no longer believe in the seminal vesicle tension theory? The answer to this question is hard to find.

The problems partly arise because of the nature of the biological sciences. In engineering, man designed the systems and can easily present a design blueprint or working model. In the biological sciences we are often desperately trying to get clues by devious routes. However, such difficulties are sometimes pushed too far as excuses for not integrating the findings. Until a better vehicle for presentation can be found we will often be quite unable to appreciate the results that do exist.

Ernst Mach saw science as a minimal problem, that of:

'... the completest presentment of facts with the least possible expenditure of thought.' (see Mach, 1960)

To some that may seem a little optimistic. But if we relate it to the issues raised here, then it appears that a systems approach not only economically presents facts but also delivers a commentary on their implications. It relieves the scientist of a certain amount of thinking. Conventional behavioural science is in danger of such complete 'presentment of facts' that very few researchers will be able to remember or accommodate the facts, or even locate them in the literature.

12.3.2 The properties of complex systems

Is a complex system more than the sum of its parts? This might seem a straightforward question, but in practice it is not. According to some rather simple criteria a living organism is the sum of its parts; for instance, to take a trivial case, this is true by the criterion of weighing the organism and its parts. Some might argue that in order to understand the behaviour of a complex system, all we need do is to break it down into the behaviour of its component parts. These component processes are understood one by one, and our understanding simply accumulates until we obtain the complete picture. However, it is not difficult to show the inadequacy of such an approach, since complex performance *emerges* from a combination of the properties of components and their mode of interconnection (cf. Oatley, 1978). For instance, consider when a voltage is applied to electrical components. If the performance of either of two electrical components, a capacitor or an inductor, is examined in response to application of a voltage an oscillation of current is not seen. However, if the components are connected together, application of the same voltage will cause an oscillation. The oscillation was clearly not present in the input–output characteristics of either component. Neither was it weakly present, almost present, nor present but inhibited. It emerged because of (1) the properties of the components, *and* (2) their mode of interconnection. This is the level of explanation that we need to obtain in the behavioural sciences. As a simple example of this, in the discussion of sexual behaviour (Chapter 8), the positive and negative feedback characteristics only make sense as system properties and not as the attributes of particular components.

Even if neurophysiological studies of motivation are pursued, it is most doubtful whether the results can be understood beyond a superficial level, unless the interconnections between brain regions are taken into account. But 'taking into account' will mean more than simply drawing thick and thin arrows and plus and minus signs on a diagram relating so-called centres. Ultimately it can only mean simulation by computer.

12.3.3 The philosophy of science and the use of models

Where possible, models of behaviour were developed in the present study. Models can take a variety of forms varying from the simple statement that the sun and planets form a model of the electron and nucleus, to a computer simulation of traffic flow or economic development. All share certain common features. They are aids to our thinking processes, and place in relatively familiar or accessible terms what is otherwise either unfamiliar or of impossible complexity.

In our moments of intellectual defeat (or realistic recognition of our limitations) we accept that there is little or no chance of reducing certain aspects of the physical world to explanation at a simpler level. We must be content with accepting descriptions. Other aspects usually irritate and goad us until we can explain them as the logical outcome of the descriptive processes at a simpler level. When any problem is reduced far enough we are left with a set of descriptions; in practice the level at which this occurs will depend upon the individual investigator. For example problems, which to a biologist have been reduced far enough, could be dissected further by an atomic physicist. In my own case, I do not know why objects accelerate towards the centre of the earth at a particular rate; gravity is merely a description of a phenomenon. However, if I were to be asked why, for example, a falling object first accelerates and then reaches a ceiling velocity, I could *explain* this in terms of the inevitable outcome of the *description* of component processes (i.e. air friction and gravity) and their interdependence.

Statements on the logical outcome of connecting components together in a particular way are efficiently made by the use of models. Ultimately these models will be embodied in computer simulation. There is absolutely no way that our unaided brains can give accurate predictions on the behaviour of complex systems. But that is the next stage in the study.

References

Adachi, A., Niijima, A., and Jacobs, H. L. (1976). An hepatic osmoreceptor mechanism in the rat: electrophysiological and behavioural studies. *American Journal of Physiology*, **231**, 1043–1049.

Adair, E. R., and Wright, B. A. (1976). Behavioural thermoregulation in the Squirrel monkey when response effort is varied. *Journal of Comparative and Physiological Psychology*, **90**, 179–184.

Adam, K., and Oswald, I. (1977). Sleep is for tissue restoration. *Journal of the Royal College of Physicians of London*, **11**, 376–388.

Adler, N. T. (1969). Effect of the male's copulatory behaviour on successful pregnancy of the female rat. *Journal of Comparative and Physiological Psychology*, **69**, 613–622.

Adolph, E. F. (1967). Regulation of water intake in relation to body water content. In *Handbook of Physiology*. Sec. 6. *Alimentary Canal*, Vol. 1. *Control of food and water intake* (ed. C. F. Code), pp. 163–171. Washington, D.C.: American Physiological Society.

Adolph, E. F., Barker, J. P., and Hoy, P. A. (1954). Multiple factors in thirst. *American Journal of Physiology*, **178**, 538–562.

Allison, T., and Cicchetti, D. V. (1976). Sleep in mammals: ecological and constitutional correlates. *Science*, **194**, 732–734.

Allison, T., and Van Twyver, H. (1970). The evolution of sleep. *Natural History*, **79**, 56–65.

Almli, C. R. (1970). Hyperosmolality accompanies hypovolemia: a simple explanation of additivity of stimuli for drinking. *Physiology and Behaviour*, **5**, 1021–1028.

Andersson, B. (1953). The effect of injections of hypertonic NaCl solutions into different parts of the hypothalamus of goats. *Acta Physiologica Scandinavica*, **28**, 188–201.

Andersson, B. (1973). Osmoreceptors versus sodium receptors. In *The Neuropsychology of Thirst: New Findings and Advances in Concepts* (eds. A. N. Epstein, H. R. Kissileff and E. Stellar), pp. 113–116. Washington D.C.: V. H. Winston.

Andrew, R. J. (1978). Increased persistence of attention produced by testosterone and its implications for the study of sexual behaviour. In *Biological Determinants of Sexual Behaviour* (ed. J. B. Hutchinson), pp. 255–275. Chichester: Wiley.

Archer, J. (1976). The organization of aggression and fear in vertebrates. In *Perspectives in Ethology*, Vol. 2 (eds. P. P. G. Bateson and P. Klopfer), pp. 231–298. New York: Plenum Press.

Archer, J. (1979). Behavioural aspects of fear. In *Fear in Animals and Man* (ed. W. Sluckin). Van Nostrand Reinhold.

Aronson, L. R., and Cooper, M. L. (1968). Desensitization of the glans penis, and sexual behaviour in cats. In *Perspectives in Reproduction and Sexual Behaviour* (ed. M. Diamond), pp. 303–340. Bloomington: Indiana University Press.

Arvidson, T., and Larsson, K. (1967). Seminal discharge and mating behaviour. *Physiology and Behaviour*, **2**, 341–343.

Azrin, N. H., Hutchinson, R. R., and McLaughlin, R. (1965). The opportunity for aggression as an operant reinforcer during aversive stimulation. *Journal of the Experimental Analysis of Behaviour*, **8**, 171–180.

271

272

Barfield, R. J., and Krieger, M. S. (1977). Ejaculatory and post-ejaculatory behaviour of male and female rats: effects of sex hormones and electric shock. *Physiology and Behaviour*, **19**, 203–208.

Barfield, R. J., and Sachs, B. D. (1968). Sexual behaviour: stimulation by painful electrical shock to skin in male rats. *Science*, **161**, 392–395.

Barnett, S. A., and Cowan, P. E. (1976). Activity, exploration, curiosity and fear: an ethological study. *Interdisciplinary Science Reviews*, **1**, 43–62.

Barnett, S. A., Dickson, R. G., Marples, T. G., and Radha, E. (1978). Sequences of feeding, sampling and exploration by wild and laboratory rats. *Behavioural Processes*, **3**, 29–43.

Bartoshuk, L. M. (1977). Water taste in mammals. In *Drinking Behaviour: Oral Stimulation, Reinforcement and Preference* (eds. J. A. W. M. Weijen and J. Mendelson), pp. 317–339. New York: Plenum Press.

Bateson, P. P. G. (1964). Changes in chicks' responses to novel moving objects over the sensitive period for imprinting. *Animal Behaviour*, **12**, 479–489.

Bateson, P. P. G. (1978). Early experience and sexual preferences. In *Biological Determinants of Sexual Behaviour* (ed. J. B. Hutchison), pp. 29–53. Chichester: Wiley.

Baum, M. J., and Vreeburg, J. T. M. (1973). Copulation in castrated male rats following combined treatment with estradiol and dihydrotestosterone. *Science*, **182**, 283–285.

Beach, F. A. (1947). A review of physiological and psychological studies of sexual behaviour in mammals. *Physiological Reviews*, **27**, 240–307.

Beach, F. A. (1948). *Hormones and Behaviour*. New York: Hoeber.

Beach, F. A. (1956). Characteristics of the masculine sex drive. In *Nebraska Symposium on Motivation* (ed. M. R. Jones), pp. 1–32. Lincoln: University of Nebraska Press.

Beach, F. A. (1968). Factors involved in the control of mounting behaviour by female mammals. In *Reproduction and Sexual Behaviour* (ed. M. Diamond), pp. 83–131. Indiana University Press.

Beach, F. A. (1970). Coital behaviour in dogs. VI. Long-term effects of castration upon mating in the male. *Journal of Comparative and Physiological Psychology* (Monograph), **70**, No. 3, Pt. 2, 1–32.

Beach, F. A. (1971). Hormonal factors controlling the differentiation, development, and display of copulatory behaviour in the Ramstergig and related species. In *The Biopsychology of Development* (eds. E. Tobach, L. R. Aronson and E. Shaw), pp. 249–296. New York: Academic Press.

Beach, F. A. (1975). Hormonal modification of sexually dimorphic behaviour. *Psychoneuroendocrinology*, **1**, 3–23.

Beach, F. A. (1976). Sexual attractivity, proceptivity and receptivity in female mammals. *Hormones and Behaviour*, **7**, 105–138.

Beach, F. A., and Buehler, M. G. (1977). Male rats with inherited insensitivity to androgen show reduced sexual behaviour. *Endocrinology*, **100**, 197–200.

Beach, F. A., Goldstein, A. C., and Jacoby, G. A. (1955). Effects of electroconvulsive shock on sexual behaviour in male rats. *Journal of Comparative and Physiological Psychology*, **48**, 173–179.

Beach, F. A., and Holz, A. M. (1946). Mating behaviour in male rats castrated at various ages and injected with androgen. *Journal of Experimental Zoology*, **101**, 91–142.

Beach, F. A., and Jordan, L. (1956). Sexual exhaustion and recovery in the male rat. *Quarterly Journal of Experimental Psychology*, **8**, 121–133.

Beach, F. A., Rogers, C. M., and LeBoeuf, B. J. (1968). Coital behaviour in dogs: effects of estrogen on mounting by females. *Journal of Comparative and Physiological Psychology*, **66**, 296–307.

Beach, F. A., Westbrook, W. H., and Clemens, L. G. (1966). Comparisons of the ejaculatory response in men and animals. *Psychosomatic Medicine*, **28**, 749–763.

Beach, F. A., and Wilson, J. R. (1963). Mating behaviour in male rats after removal of the seminal vesicles. *Proceedings of the National Academy of Sciences*, **49**, 624–626.

Benzinger, T. H. (1964). The thermal homeostasis of man. In *Homeostasis and Feedback Mechanism* (ed. G. M. Hughes), pp. 49–80. Cambridge University Press.

Berlyne, D. E. (1955). The arousal and satiation of perceptual curiosity in the rat. *Journal of Comparative and Physiological Psychology*, **48**, 238–246.

Berlyne, D. E. (1960). *Conflict, Arousal and Curiosity*. New York: McGraw-Hill.

Berlyne, D. E. (1966). Curiosity and exploration. *Science*, **153**, 25–33.

Bermant, G., and Westbrook, W. H. (1966). Peripheral factors in the regulation of sexual contact by female rats. *Journal of Comparative and Physiological Psychology*, **61**, 244–250.

Bindra, D. (1978). How adaptive behaviour is produced: a perceptual–motivational alternative to response-reinforcement. *The Behavioural and Brain Sciences*, **1**, 41–92.

Blake, W. D., and Lin, K. K. (1978). Hepatic portal vein infusion of glucose and sodium solutions on the control of saline drinking in the rat. *Journal of Physiology (London)*, **274**, 129–139.

Blanchard, R. J., and Blanchard, D. C. (1971). Defensive reactions in the albino rat. *Learning and Motivation*, **2**, 351–362.

Blanchard, R. J., Blanchard, D. C., and Takahashi, L. K. (1978). Pain and aggression in the rat. *Behavioural Biology*, **23**, 291–305.

Blass, E. M. (1973). Cellular-dehydration thirst: physiological, neurological and behavioural correlates. In *The Neuropsychology of Thirst: New Findings and Advances in Concepts* (eds. A. N. Epstein, H. R. Kissileff, and E. Stellar), pp. 37–72. Washington, D. C.: V. H. Winston.

Blass, E. M. (1974). The physiological, neurological and behavioural bases of thirst. *Nebraska Symposium on Motivation*, **22**, 1–47.

Blass, E. M., and Hall, W. G. (1976). Drinking termination: interactions among hydrational, orogastric, and behavioural controls in rats. *Psychological Review*, **83**, 356–374.

Blass, E. M., Jobaris, R., and Hall, W. G. (1976). Oropharyngeal control of drinking in rats. *Journal of Comparative and Physiological Psychology*, **90**, 909–916.

Bligh, J. (1972). Neuronal models of mammalian temperature regulation. In *Essays on Temperature Regulation* (eds. J. Bligh and R. Moore), pp. 105–120. Amsterdam: North-Holland Publishing Company.

Bligh, J. (1973). *Temperature Regulation in Mammals and Other Vertebrates*. Amsterdam: North-Holland Publishing Company.

Bligh, J., and Moore, R. E. (1972). *Essays on Temperature Regulation*. Amsterdam: North-Holland Publishing Company.

Blundell, J. and Latham, C. (1979). Pharmacology of food and water intake. In *Chemical Influences on Behaviour* (eds. K. Brown and S. Cooper). London: Academic Press. (In press.)

Bogert, C. M. (1959). How reptiles regulate their body temperature. *Scientific American*, **200(4)**, 105–120.

Boice, R. (1971). Excessive water intake in captive Norway rats with scar-markings. *Physiology and Behaviour*, **7**, 723–725.

Boland, B. D., and Dewsbury, D. A. (1971). Characteristics of sleep following sexual activity and wheel running in male rats. *Physiology and Behaviour*, **6**, 145–149.

Bolles, R. C. (1970). Species-specific defence reactions and avoidance learning. *Psychological Review*, **71**, 32–48.

Bolles, R. C. (1975). *Theory of Motivation*. New York: Harper and Row.

Bols, R. J. (1977). Display reinforcement in the Siamese fighting fish, *Betta splendens*: aggressive motivation or curiosity? *Journal of Comparative and Physiological Psychology*, **91**, 233–244.

Booth, D. A. (1968). Mechanism of action of norepinephrine in eliciting an eating response on injection into the rat hypothalamus. *Journal of Pharmacology and Experimental Therapeutics*, **160**, 336–348.

Booth, D. A. (1972a). Conditioned satiety in the rat. *Journal of Comparative and Physiological Psychology*, **81**, 457–571.

Booth, D. A. (1972b). Postabsorptively induced suppression of appetite and the energostatic control of feeding. *Physiology and Behaviour*, **9**, 199–202.

Booth, D. A. (1976). Approaches to feeding control. In *Appetite and Food Intake* (ed. T. Silverstone), pp. 417–498. West Berlin: Abakon/Dahlem.·

Booth, D. A. (1977). Appetite and satiety as metabolic expectancies. In *Food Intake and Chemical Senses* (eds. Y. Katsuki, M. Sato, S. F. Takagi, and Y. Oomura), pp. 317–330. Tokyo: University of Tokyo Press.

Booth, D. A. (1978). Prediction of feeding behaviour from energy flows in the rat. In *Hunger Models—Computable Theory of Feeding Control* (ed. D. A. Booth), pp. 227–278. London: Academic Press.

Booth, D. A. (1979). Is thirst largely an acquired specific appetite? *The Behavioural and Brain Sciences*, **2**, 103–104.

Booth, D. A., and Davis, J. D. (1973). Gastrointestinal factors in the acquisition of oral sensory control of satiation. *Physiology and Behaviour*, **11**, 23–29.

Booth, D. A., and Jarman, S. P. (1976). Inhibition of food intake in the rat following complete absorption of glucose delivered into the stomach, intestine or liver. *Journal of Physiology (London)*, **259**, 501–522.

Booth, D. A., Lovett, D., and McSherry, G. M. (1972). Postingestive modulation of the sweetness preference gradient in the rat. *Journal of Comparative and Physiological Psychology*, **78**, 485–512.

Booth, D. A., and Simson, P. C. (1971). Food preferences acquired by association with variations in amino acid nutrition. *Quarterly Journal of Experimental Psychology*, **23**, 135–145.

Booth, D. A., Stoloff, R., and Nicholls, J. (1974). Dietary flavour acceptance in infant rats established by association with effects of nutrient composition. *Physiological Psychology*, **2**, 313–319.

Booth, D. A., and Toates, F. M. (1974). A physiological control theory of food intake in the rat: Mark 1. *Bulletin of the Psychonomic Society*, **3**, 442–444.

Booth, D. A., Toates, F. M., and Platt, S. V. (1976). Control system for hunger and its implications in animals and man. In *Hunger: Basic Mechanisms and Clinical Implications* (eds. D. Novin, W. Wyrwicka, and G. Bray), pp. 127–143. New York: Raven Press.

Booth, J. E. (1977a). Sexual behaviour of neonatally castrated rats injected during infancy with oestrogen and dihydrotestosterone. *Journal of Endocrinology*, **72**, 135–141.

Booth, J. E. (1977b). Sexual behaviour of male rats injected with the anti-oestrogen MER-25 during infancy. *Physiology and Behaviour*, **19**, 35–39.

Brain, P. F. (1979). Effects of hormones of the pituitary–gonadal axis on behaviour. In *Chemical Influences on Behaviour* (eds. K. Brown and S. J. Cooper). London: Academic Press. (In press.)

Brant, D. H. and Kavanau, J. L. (1965). Exploration and movement patterns in the canyon mouse *Peromyscus cranitus* in an extensive laboratory enclosure. *Ecology*, **46**, 452–461.

Brattstrom, B. H. (1970). Amphibia. In *Comparative Physiology of Thermoregulation* (ed. G. C. Whittow), pp. 135–166. New York: Academic Press.

Budgell, P. (1970). The effect of changes in ambient temperature on water intake and evaporative water loss. *Psychonomic Science*, **20**, 275–276.

Budgell, P. (1971). Behavioural thermoregulation in the Barbary dove *(Streptopelia risoria)*. *Animal Behaviour*, **19**, 524–531.

Butler, R. A. (1957). The effect of deprivation of visual incentives on visual exploration motivation in monkeys. *Journal of Comparative and Physiological Psychology*, **50**, 177–179 (reproduced in Fowler (1965)).

Cabanac, M. (1971). Physiological role of pleasure. *Science*, **173**, 1103–1107.

Cabanac, M. (1972). Thermoregulatory behaviour. In *Essays on Temperature Regulation* (eds. J. Bligh and R. E. Moore), pp. 19–36. Amsterdam: North-Holland Publishing Company.

Cannon, W. B. (1947). *The Wisdom of the Body*. London: Kegan Paul, Trench, Trubner and Co.

Carder, B. and Beckman, G. C. (1975). Limitations of 'container neophobia' as an explanation of rats' responding for food in the presence of free food. *Behavioural Biology*, **14**, 109–113.

Carlsle, H. J. (1966). Heat intake and hypothalamic temperature during behavioural temperature regulation. *Journal of Comparative and Physiological Psychology*, **61**, 388–397.

Carlsle, H. J. (1968). Initiation of behavioural responding for heat in a cold environment. *Physiology and Behaviour*, **3**, 827–830.

Carlsle, H. J. (1969). Effect of fixed-ratio thermal reinforcement on thermoregulatory behaviour. *Physiology and Behaviour*, **4**, 23–28.

Carlson, N. R. (1977). *Physiology of Behaviour*. Boston: Allyn and Bacon.

Carr, W. J., and Caul, W. F. (1962). The effect of castration in rat upon the discrimination of sex odours. *Animal Behaviour*, **10**, 20–27.

Carter, C. S., Landauer, M. R., Tierney, B. M., and Jones, T. (1976). Regulation of female sexual behaviour in the Golden hamster: Behavioural effects of mating and ovarian hormones. *Journal of Comparative and Physiological Psychology*, **90**, 839–850.

Chapman, R. M., and Levy, N. (1957). Hunger drive and reinforcing effect of novel stimuli. *Journal of Comparative and Physiological Psychology*, **50**, 233–238.

Cherney, E. F., and Bermant, G. (1970). The role of stimulus female novelty on the rearousal of copulation in male laboratory rats *(Rattus norvegicus)*. *Animal Behaviour*, **18**, 567–574.

Chew, R. M. (1965). Water metabolism of mammals. In *Physiological Mammalogy*, Vol. II (eds. W. V. Mayer and R. G. Van Gelder), pp. 43–178. London: Academic Press.

Cizek, L. J. (1961). Relationship between food and water ingestion in the rabbit. *American Journal of Physiology*, **201**, 557–566.

Cizek, L. J., and Nocenti, M. R. (1965). Relationship between water and food ingestion in the rat. *American Journal of Physiology*, **208**, 615–620.

Collier, G., Hirsch, E., and Hamlin, P. H. (1972). The ecological determinants of reinforcement in the rat. *Physiology and Behaviour*, **9**, 705–716.

Collier, G., and Knarr, F. (1966). Defence of water balance in the rat. *Journal of Comparative and Physiological Psychology*, **61**, 5–10.

Collier, G., and Levitsky, D. (1967). Defence of water balance in rats: behavioural and physiological responses to depletion. *Journal of Comparative and Physiological Psychology*, **64**, 59–67.

Corbit, J. D. (1965). Effect of intravenous sodium chloride on drinking in the rat. *Journal of Comparative and Physiological Psychology*, **60**, 397–406.

Corbit, J. D. (1969a). Osmotic thirst: theoretical and experimental analysis. *Journal of Comparative and Physiological Psychology*, **67**, 3–14.

Corbit, J. D. (1969b). Behavioural regulation of hypothalamic temperature. *Science*, **166**, 256–257.

Corbit, J. D. (1970). Behavioural regulation of body temperature. In *Physiological and Behavioural Temperature Regulation* (eds. J. D. Hardy, A. P. Gagge and J. A. J. Stolwijk), pp. 777–801. Springfield: Thomas.

Corbit, J. D., and Luschei, E. S. (1969). Invariance of the rat's rate of drinking. *Journal of Comparative and Physiological Psychology*, **69**, 119–125.

Corey, D. T., Walton, A., and Wiener, N. I. (1978). Development of carbohydrate preference during water rationing: a specific hunger? *Physiology and Behaviour*, **20**, 547–552.

Cowan, P. E. (1976). The new object reaction of *Rattus rattus L:* the relative importance of various cues. *Behavioural Biology*, **16**, 31–44.

Cowan, P. E. (1977a). Systematic patrolling and orderly behaviour of rats during recovery from deprivation. *Animal Behaviour*, **25**, 171–184.

Cowan, P. E. (1977b). Neophobia and neophilia: new-object and new-place reactions of three *Rattus* species. *Journal of Comparative and Physiological Psychology*, **91**, 63–71.

Cowan, P. E., and Barnett, S. A. (1975). The new-object and new-place reactions of *Rattus rattus* L. *Zoological Journal of the Linnaean Society*, **56**, 219–234.

Crompton, A. W., Taylor, C. R., and Jagger, J. A. (1978). Evolution of homeothermy in mammals. *Nature*, **272**, 333–336.

Danguir, J., and Nicolaidis, S. (1976). Impairments of learned aversion acquisition following paradoxical sleep deprivation in the rat. *Physiology and Behaviour*, **17**, 489–492.

Davis, W. J. (1979). Behavioural hierarchies. *Trends in Neurosciences*, **2**(1), 5–7.

Davis, J. D., and Campbell, C. S. (1973). Peripheral control of meal size in the rat: effect of sham feeding on meal size and drinking rate. *Journal of Comparative and Physiological Psychology*, **83**, 379–388.

Davis, J. D., and Wirtshafter, D. (1978). Set-points or settling-points for body weight? A reply to Mrosovsky and Powley. *Behavioural Biology*, **24**, 405-411.

Dawkins, R. (1976). *The Selfish Gene.* Oxford: Oxford University Press.

Dawson, W. R., and Hudson, J. W. (1970). Birds. In *Comparative Physiology of Thermoregulation* (ed. G. C. Whittow), pp. 223–310. New York: Academic Press.

Deaux, E., and Engstrom, R. (1973). The temperature of ingested water: preference for cold water as an associative response. *Physiological Psychology*, **1**, 257–260.

Dember, W. N. (1956). Response by the rat to environmental change. *Journal of Comparative and Physiological Psychology*, **49**, 93–95.

Dember, W. N. (1965). The new look in motivation. *American Scientist*, **53**, 409–427.

Denenberg, V. H. (1967). Stimulation in infancy, emotional reactivity and exploratory behaviour. In *Neurophysiology and Emotion* (ed. D. C. Glass), pp. 161–190. New York: Rockefeller University Press.

Dennis, W. (1939). Spontaneous alternation in rats as an indicator of the persistence of stimulus effects. *Journal of Comparative Psychology*, **28**, 305–312 (reprinted in Fowler (1965)).

Denton, D. A. (1966). Some theoretical considerations in relation to innate appetite for salt. *Conditioned Reflex*, **1**, 144–170.

Denton, D. A. (1972). Instinct, appetites and medicine. *Australian and New Zealand Journal of Medicine*, **2**, 203–212.

Deutsch, J. A. (1979). Intragastric infusion and pressure. *The Behavioural and Brain Sciences*, **2**, 105.

Devenport, L. D. (1973). Aversion to a palatable saline solution in rats: interactions of physiology and experience. *Journal of Comparative and Physiological Psychology*, **83**, 98–105.

Diakow, C. (1974). Male–female interactions and the organization of mammalian mating patterns. In *Advances in the Study of Behaviour*, Vol. 5 (eds. D. S. Lehrman, J. S. Rosenblatt, R. A. Hinde, and E. Shaw), pp. 227–268. New York: Academic Press.

Dicker, J. E., and Nunn, J. (1957). The role of anti-diuretic hormone during water deprivation in rats. *Journal of Physiology (London)*, **136**, 235–248.

Doty, R. L. (1974). A cry for the liberation of the female rodent: courtship and copulation in *Rodentia. Psychological Bulletin*, **81**, 159–172.

Douglas, R. J. (1966). Cues for spontaneous alternation. *Journal of Comparative and Physiological Psychology*, **62**, 171–183.

Douglas, R. J., Mitchell, D., and Del Valle, R. (1974). Angle between choice alleys as a critical factor in spontaneous alternation. *Animal Learning and Behaviour*, **2**, 218–220.

277

Dubos, R. (1971). In defence of biological freedom. In *The Biopsychology of Development* (eds. E. Tobach, L. R. Aronson, and E. Shaw), pp. 553–560. New York: Academic Press.

Dufort, R. H., and Wright, J. H. (1962). Food intake as a function of food deprivation. *The Journal of Psychology*, **53**, 465–468.

Eibl-Eibesfeldt, I. (1970). *Ethology: The Biology of Behaviour*. New York: Holt, Rinehart and Winston.

Einon, D., and Morgan, M. (1976). Habituation of object contact in socially reared and isolated rats (*Rattus norvegicus*). *Animal Behaviour*, **24**, 415–420.

Eisenberger, R. (1972). Explanation of rewards that do not reduce tissue needs. *Psychological Bulletin*, **77**, 319–339.

Epstein, A. N. (1978). The neuroendocrinology of thirst and sodium appetite. In *Frontiers in Neuroendocrinology*, Vol. 5 (eds. W. F. Ganong, and L. Martini), pp. 101–134. New York: Raven Press.

Falk, J. L. (1961). Production of polydipsia in normal rats by an intermittent food schedule. *Science*, **133**, 195–196.

Falk, J. L. (1969). Conditions producing psychogenic polydipsia in animals. *Annals of the New York Academy of Sciences*, **157**, 569–593.

Falk, J. L. (1971). The nature and determinants of adjunctive behaviour. *Physiology and Behaviour*, **6**, 577–588.

Falk, J. L. (1977). The origin and functions of adjunctive behaviour. *Animal Learning and Behaviour*, **5**, 325–335.

Feder, H. H. (1978). Specificity of steroid hormone activation of sexual behaviour in rodents. In *Biological Determinants of Sexual Behaviour* (ed. J. B. Hutchinson), pp. 395–424. Chichester: Wiley.

Feder, H. H., Landau, I. T., Marrone, B. L., and Walker, W. A. (1977). Interactions between estrogen and progesterone in neural tissues that mediate sexual behaviour of guinea-pigs. *Psychoneuroendocrinology*, **2**, 337–347.

Fehrer, E. (1956). The effects of hunger and familiarity of locale on exploration. *Journal of Comparative and Physiological Psychology*, **49**, 549–552.

Feider, A. (1972). Feedback control of thirst in rats. *Physiology and Behaviour*, **8**, 1005–1011.

Fishbein, W. and Gutwein, B. M. (1977). Paradoxical sleep and memory storage processes. *Behavioural Biology*, **19**, 425–464.

Fitzsimons, J. T. (1971). The physiology of thirst: a review of the extraneural aspects of the mechanisms of drinking. In *Progress in Physiological Psychology*, Vol. 4 (eds. E. Stellar and J. M. Sprague), pp. 119–201. New York: Academic Press.

Fitzsimons, J. T. (1972). Thirst. *Physiological Reviews*, **52**, 468–561.

Fitzsimons, J. T. (1973). Some historical perspectives in the physiology of thirst. In *The Neuropsychology of Thirst: New Findings and Advances in Concepts* (eds. A. N. Epstein, H. R. Kissileff, and E. Stellar), pp. 3–33. Washington, D.C.: V. H. Winston.

Fitzsimons, J. T. (1976). The physiological basis of thirst. *Kidney International*, **10**, 3–11.

Fitzsimons, J. T., and LeMagnen, J. (1969). Eating as a regulatory control of drinking in the rat. *Journal of Comparative and Physiological Psychology*, **67**, 273–283.

Fitzsimons, J. T., and Oatley, K. (1968). Additivity of stimuli for drinking in rats. *Journal of Comparative and Physiological Psychology*, **66**, 450–455.

Ford, C. S., and Beach, F. A. (1952). *Patterns of Sexual Behaviour*. London: Eyre and Spottiswoode.

Fowler, H. (1965). *Curiosity and Exploratory Behaviour*. New York: Macmillan.

Fowler, S. J., and Kellogg, C. (1975). Ontogeny of thermoregulatory mechanisms in the rat. *Journal of Comparative and Physiological Psychology*, **89**, 738–746.

Freeman, S., and McFarland, D. J. (1974). RATSEX—an exercise in simulation. In *Motivational Control Systems Analysis* (ed. D. J. McFarland), pp. 479–510. London: Academic Press.

278

Friedman, M. I., and Stricker, E. M. (1976). The physiological psychology of hunger: a physiological perspective. *Psychological Review*, **83**, 409–431.

Gale, C. C., Mathews, M., and Young, J. (1970). Behavioural thermoregulatory responses to hypothalamic cooling and warming in baboons. *Physiology and Behaviour*, **5**, 1–6.

Galef, B. G. (1970a). Aggression and timidity: responses to novelty inferal Norway rats. *Journal of Comparative and Physiological Psychology*, **70**, 370–381.

Galef, B. G. (1970b). Target novelty elicits and directs shock-associated aggression in wild rats. *Journal of Comparative and Physiological Psychology*, **71**, 87–91.

Galef, B. G., and Clark, M. M. (1972),Mother's milk and adult presence: two factors determining initial dietary selection by weanling rats. *Journal of Comparative and Physiological Psychology*, **78**, 220–225.

Galef, B. G., and Henderson, P. W. (1972). Mother's milk: a determinant of the feeding preferences of weanling rat pups. *Journal of Comparative and Physiological Psychology*, **78**, 213–219.

Gallistel, C. R. (1978). The irrelevance of past pleasure. *The Behavioural and Brain Sciences*, **1**, 59–60.

Garcia, J., Clarke, J. C., and Hankins, W. G. (1973). Natural responses to scheduled rewards. In *Perspectives in Ethology* (eds. P. P. G. Bateson and P. H. Klopfer), pp. 1–42. New York: Plenum Press.

Ghent, L. (1957). Some effects of deprivation on eating and drinking behaviour. *Journal of Comparative and Physiological Psychology*, **50**, 172–176.

Gilman, A. (1937). The relation between blood osmotic pressure, fluid distribution and voluntary water intake. *American Journal of Physiology*, **120**, 323–328.

Glanzer, M. (1953). Stimulus satiation: an explanation of spontaneous alternation and related phenomena. *Psychological Review*, **60**, 257–268.

Glickman, S. E., and Hartz, K. E. (1964). Exploratory behaviour in several species of rodents. *Journal of Comparative and Physiological Psychology*, **58**, 101–104.

Glickman, S. E., Higgins, T. J., and Isaacson, R. L. (1970). Some effects of hippocampal lesions on the behaviour of Mongolian gerbils. *Physiology and Behaviour*, **5**, 931–938.

Glickman, S. E., and Sroges, R. W. (1966). Curiosity in zoo animals. *Behaviour*, **26**, 151–188.

Glow, P. H. (1970). Some acquisition and performance characteristics of response contingent sensory reinforcement. *Australian Journal of Psychology*, **22**, 145–154.

Gold, R. M., and Laforge, R. G. (1977). Temperature of ingested fluids: preference and satiation effects (pease porridge warm, pease porridge cool). In *Drinking Behaviour: Oral Stimulation, Reinforcement and Preference* (eds. J. A. W. M. Weijen and J. Mendelson), pp. 157–196. New York: Plenum Press.

Goldfoot, D. A., and Goy, R. W. (1970). Abbreviation of behavioural estrus in guinea pigs by coital and vagino-cervical stimulation. *Journal of Comparative and Physiological Psychology*, **72**, 426–434.

Goy, R. W., and Goldfoot, D. A. (1975). Neuroendocrinology: animal models and problems of human sexuality. *Archives of Sexual Behaviour*, **4**, 405–420.

Gray, J. A. (1971). *The Psychology of Fear and Stress*. London: Weidenfeld and Nicolson.

Grossman, S. P. (1967). *A Textbook of Physiological Psychology*. New York: Wiley.

Grunt, J. A. and Young, W. C. (1953). Consistency of sexual behaviour patterns in individual male guinea pigs following castration and androgen therapy. *Journal of Comparative and Physiological Psychology*, **46**, 138–144.

Guyton, A. C. (1976). *Textbook of Medical Physiology*. Philadelphia: W. B. Saunders.

Haberich, F. J. (1968). Osmoreception in the portal circulation. *Federation Proceedings*, **27**, 1137–1141.

Hafez, E. S. E. (1964). Behavioural thermoregulation in mammals and birds. *International Journal of Biometeorology*, **7**, 231–240.

Hainsworth, F. R., and Stricker, E. M. (1970). Salivary cooling by rats in the heat. In *Physiological and Behavioural Temperature Regulation* (eds. J. D. Hardy, A. P. Gagge, and J. A. J. Stolwijk), pp. 611–626. Springfield: C. C. Thomas.

Hall, W. G., and Blass, E. M. (1977). Orogastric determinants of drinking in rats: interaction between absorptive and peripheral controls. *Journal of Comparative and Physiological Psychology*, **91**, 365–373.

Halliday, M. S. (1967). Exploratory behaviour. In *Analysis of Behavioural Change* (ed. L. Weiskrantz), pp. 107–126. New York: Harper and Row.

Hammel, H. T. (1972). The set-point in temperature regulation: analogy or reality. In *Essays on Temperature Regulation* (eds. J. Bligh and R. Moore), pp. 121–137. Amsterdam: North-Holland Publishing Company.

Hardy, D. F., and DeBold, J. F. (1971). Effects of mounts without intromission upon the behaviour of female rats during the onset of estrogen induced heat. *Physiology and Behaviour*, **7**, 643–645.

Hardy, D. F., and DeBold, J. F. (1972). Effects of coital stimulation upon behaviour of the female rat. *Journal of Comparative and Physiological Psychology*, **78**, 400–408.

Harper, A. E., and Boyle, P. C. (1976). Nutrients and food intake. In *Appetite and Food Intake* (ed. T. Silverstone), pp. 177–206. West Berlin: Abakon/Dahlem.

Hart, J. S. (1970). Rodents. In *Comparative Physiology of Thermoregulation* (ed. G. C. Whittow), pp. 1–149. New York: Academic Press.

Hart, B. L. (1978). Hormones, spinal reflexes and sexual behaviour. In *Biological Determinants of Sexual Behaviour* (ed. J. B. Hutchison), pp. 319–347. Chichester: Wiley.

Hartmann, E. L. (1973). *The Functions of Sleep*. New Haven: Yale University Press.

Hatton, G. I. and Bennett, C. T. (1970). Satiation of thirst and termination of drinking: roles of plasma osmolality and absorption. *Physiology and Behaviour*, **5**, 479–487.

Hawkins, R. C. (1977). Learning to initiate and terminate meals: Theoretical, clinical and developmental aspects. In *Learning Mechanisms in Food Selection* (eds. L. M. Barker, M. R. Best, and M. Domjan), pp. 201–253. Baylor University Press.

Hayes, W. N. and Warren, J. M. (1963). Failure to find spontaneous alternation in chicks. *Journal of Comparative and Physiological Psychology*, **56**, 575–577.

Hebb, D. O. (1966). *The Organization of Behaviour*. New York: Wiley.

Heller, H. C., Crawshaw, L. I., and Hammel, H. T. (1978). The thermostat of vertebrate animals. *Scientific American*, **239**, 88–96.

Hennevin, E., and Leconte, P. (1977). Etude des relations entre le Sommeil paradoxal et les Processus d'Acquisition. *Physiology and Behaviour*, **18**, 307–319.

Hensel, H. (1970). Temperature receptors in the skin. In *Physiological and Behavioural Temperature Regulation* (eds. J. D. Hardy, A. P. Gagge, and J. A. J. Stolwijk), pp. 442–453. Springfield: C. C. Thomas.

Herberg, J. L., and Stephens, D. N. (1977). Interaction of hunger and thirst in the motivational arousal underlying hoarding behaviour in the rat. *Journal of Comparative and Physiological Psychology*, **91**, 359–364.

Hetta, J., and Meyerson, B. J. (1978). Effects of castration and testosterone treatment on sex-specific orientation in the male rat. *Acta Physiologica Scandinavica* (Suppl. 453), 47–62.

Hinde, R. A. (1970). *Animal Behaviour – A Synthesis of Ethology and Comparative Psychology*. New York: McGraw-Hill.

Hinde, R. A. (1974). *Biological Bases of Human Social Behaviour*. New York: McGraw-Hill.

Hirsch, E., and Collier, G. (1974). The ecological determinants of reinforcement in the guinea-pig. *Physiology and Behaviour*, **12**, 239–249.

Hogan, J. A. (1965). An experimental study of conflict and fear: an analysis of behaviour of young chicks toward a mealworm. Part I. The behaviour of chicks which do not eat the mealworm. *Behaviour*, **25**, 45–97.

Hogan, J. A. (1977). The ontogeny of food preferences in chicks and other animals. In *Learning Mechanisms in Food Selection* (eds. L. M. Barker, M. R. Best, and M. Domjan), pp. 71–99. Baylor University Press.

Hogan, J. A., and Roper, T. J. (1978). A comparison of the properties of different reinforcers. In *Advances in the Study of Behaviour*, Vol. 8 (eds. J. S. Rosenblatt, R. A. Hinde, C. Beer, and M. C. Busnel), pp. 155–255. New York: Academic Press.

Horne, J. A. (1977). Factors relating to energy conservation during sleep in mammals. *Physiological Psychology*, **5**, 403–408.

Horowitz, K. A., Scott, N. R., Hillman, P. E., and Van Tienhoven, A. (1978). Effects of feathers on instrumental thermoregulatory behaviour in chickens. *Physiology and Behaviour*, **21**, 233–238.

Houston, A., and McFarland, D. J. (1976). On the measurements of motivational variables. *Animal Behaviour*, **24**, 459–475.

Hsiao, S., and Lloyd, M. A. (1969). Do rats drink water in excess of apparent need when they are given food? *Psychonomic Science*, **15**, 155–156.

Huey, R. B. (1974). Behavioural thermoregulation in lizards: importance of associated costs. *Science*, **184**, 1001–1002.

Huntingford, F. A. (1976). The relationship between inter- and intra-specific aggression. *Animal Behaviour*, **24**, 485–497.

Hutchinson, R. R., Ulrich, R. E., and Azrin, N. H. (1965). Effects of age and related factors on the pain–aggression reaction. *Journal of Comparative and Physiological Psychology*, **59**, 365–369.

Hutchison, J. B. (1978). Hypothalamic regulation of male sexual responsiveness to androgen. In *Biological Determinants of Sexual Behaviour* (ed. J. B. Hutchison), pp. 277–317. Chichester: Wiley.

Iggo, A. (1970). The mechanisms of biological temperature reception. In *Physiological and Behavioural Temperature Regulation* (eds. J. D. Hardy, A. P. Gagge, and J. A. J. Stolwijk), pp. 391–407. Springfield: C. C. Thomas.

Janowitz, H. D., and Grossman, M. I. (1949). Effect of variations in nutritive density on intake of food in dogs and rats. *American Journal of Physiology*, **158**, 184–193.

Janowitz, H. D., and Hollander, F. (1955). The time factor in the adjustment of food intake to varied caloric requirement in the dog: a study of the precision of appetite regulation. *Annals of the New York Academy of Sciences*, **63**, 56–67.

Jewell, P. A. (1966). The concept of home-range in mammals. *Symposia of the Zoological Society of London*, **18**, 85–109.

Johnsen, S. G. (1973). Nogle tanker om fedme. *Ugerskrift før Laeger*, **135**, 2751–2754.

Johnson, R. N. (1972). *Aggression in Man and Animals*. Philadelphia: W. B. Saunders.

Joslyn, W. D., Feder, H. H., and Goy, R. W. (1971). Estrogen conditioning of lordosis. *Physiology and Behaviour*, **7**, 477–482.

Jouvet, M. (1969). Neurophysiological and biochemical mechanisms of sleep. In *Sleep; Physiology and Pathology* (ed. A. Kales), pp. 89–100. Philadelphia: J. B. Lippincott.

Keverne, E. B. (1976). Sexual receptivity and attractiveness in the female rhesus monkey. In *Advances in the Study of Behaviour*, Vol. 7 (eds. J. S. Rosenblatt, R. A. Hinde, E. Shaw, and C. Beer), pp. 155–200. New York: Academic Press.

Kissileff, H. R. (1969). Food-associated drinking in the rat. *Journal of Comparative and Physiological Psychology*, **67**, 284–300.

Kissleff, H. R. (1973). Nonhomeostatic controls of drinking. In *The Neuropsychology of Thirst: New Findings and Advances in Concepts* (eds. A. N. Epstein, H. R. Kissileff, and E. Stellar), pp. 163–198. Washington: V. H. Winston.

Kleerkoper, H., Matis, J., Gensler, P., and Maynard, P. (1974). Exploratory behaviour of goldfish *Carassius auratus*. *Animal Behaviour*, **22**, 124–132.

Kleitman, N. (1963). *Sleep and Wakefulness*. Chicago: University of Chicago Press.

Kow, L. M., Malsbury, C. W., and Pfaff, D. W. (1974). Effects of progesterone on female reproductive behaviour in rats; possible modes of action and role in behavioural sex differences. In *Reproductive Behaviour* (eds. W. Montagna and W. A. Sadler), pp. 179–210. New York: Plenum Press.

Krames, L. (1970). 'The Coolidge Effect in Male and Female Rats'. PhD thesis, Temple University.

Krieckhaus, E. E. (1970). 'Innate recognition' aids rats in sodium regulation. *Journal of Comparative and Physiological Psychology*, 73, 117–122.

Kuo, Z. Y. (1960). Studies on the basic factors in animal fighting. *Journal of Genetic Psychology*, 96, 201–225.

Kutscher, C. L. (1972). Interaction of food and water deprivation on drinking: effects of body water losses and characteristics of solution offered. *Physiology and Behaviour*, 9, 753–758.

Kutscher, C. L. (1979). On recognizing homeostatic behaviours. *The Behavioural and Brain Sciences*, 2, 108–109.

Larkin, S., and McFarland, D. J. (1978). The cost of changing from one activity to another. *Animal Behaviour*, 26, 1237–1246.

Larsson, K. (1966). Individual differences in reactivity to androgen in male rats. *Physiology and Behaviour*, 1, 255–258.

Larsson, K. (1978). Experiential factors in the development of sexual behaviour. In *Biological Determinants of Sexual Behaviour* (ed. J. B. Hutchison), pp. 29–53. Chichester: Wiley.

Larsson, K., Södersten, P., and Beyer, C. (1973). Sexual behaviour in male rats treated with estrogen in combination with dihydrotestosterone. *Hormones and Behaviour*, 4, 289–299.

Laudenslager, M. L., and Hammel, H. T. (1977). Environmental temperature selection by the Chukar partridge (*Alectoris chukar*), *Physiology and Behaviour*, 19, 543–548.

Lea, S. E. G. (1978). The psychology and economics of demand. *Psychological Bulletin*, 85, 441–466.

LeMagnen, J. (1955). Sur le mecanisme d'establissement des appetits caloriques. *Comptes Rendus-Academie Des Sciences*, 240, 2436–2438.

LeMagnen, J. (1971). Advances in studies on the physiological control and regulation of food intake. In *Progress in Physiological Psychology*, Vol. 4 (eds. E. Stellar and J. M. Sprague), pp. 203–261. New York: Academic Press.

LeMagnen, J., and Devos, M. (1970). Metabolic correlates of the meal onset in the free food intake of rats. *Physiology and Behaviour*, 5, 805–814.

LeMagnen, J., Devos, M., Gaudillière, J.-P., Louis-Sylvestre, J., and Tallon, S. (1973). Role of a lipostatic mechanism in regulation by feeding of energy balance in rats. *Journal of Comparative and Physiological Psychology*, 84, 1–23.

LeMagnen, J., and Tallon, S. (1968). L 'Effet du Jeûne préable sur les caracteristiques temporelles de la prise d'aliments chez le rat. *Journal de Physiologie*, 60, 143–154.

Lepkovsky, S., Lyman, R., Fleming, D., Nagumo, M., and Dimick, M. M. (1957). Gastrointestinal regulation of water and its effects on food intake and rate of digestion. *American Journal of Physiology*, 188, 327–331.

Leshner, A. I. (1975). A model of hormones and agonistic behaviour. *Physiology and Behaviour*, 15, 225–235.

Lewis, A. C., Rubini, M. E., and Beisel, W. R. (1960). A method for rapid dehydration of rats. *Journal of Applied Physiology*, 15, 525–527.

Leyland, M., Robbins, T., and Iversen, S. D. (1976). Locomotor activity and exploration: The use of traditional manipulators to dissociate two behaviours in the rat. *Animal Learning and Behaviour*, 4, 261–265.

Logan, F. A. (1964). The free behaviour situation. In *Nebraska Symposium on Motivation* (ed. D. Levine), pp. 99–134. Lincoln: University of Nebraska Press.

Lorenz, K. (1966). *On Aggression*. New York: Harcourt, Brace, and World.

282

Ludlow, A. R. (1976). The behaviour of a model animal. *Behaviour*, **58**, 131–172.

Ludlow, A. R. (1979). The importance of temporal coupling between feeding and drinking—simulations prompted by Toates' paper. *The Behavioural and Brain Sciences*, **2**, 110–111.

McCanze, R. A. (1936). Experimental sodium chloride deficiency in man. *Proceedings of the Royal Society, London*, **B119**, 245–268.

McCleery, R. H. (1978). Optimal behaviour sequences and decision making. In *Behavioural Ecology* (eds. J. R. Krebs and N. B. Davies), pp. 377–410. Oxford: Blackwell Scientific Publications.

McCleery, R. H. (1979). Homeostatic motivation theories and function. *The Behavioural and Brain Sciences*, **2**, 111.

McEwen, B. S. (1976). Interactions between hormones and nerve tissue. *Scientific American*, **235**, 48–58.

McFarland, D. J. (1966). On the causal and functional significance of displacement activities. *Zeitschrift für Tierpsychologie*, **23**, 217–235.

McFarland, D. J. (1970). Adjunctive behaviour in feeding and drinking situations. *Revue du Comportement Animal*, **4**, 64–73.

McFarland, D. J. (1971). *Feedback Mechanisms in Animal Behaviour*. London: Academic Press.

McFarland, D. J. (1973). Stimulus relevance and homeostasis. In *Constraints on Learning* (eds. R. A. Hinde and J. Stevenson-Hinde), pp. 141–152. London: Academic Press.

McFarland, D. J. (1974). Time-sharing as a behavioural phenomenon. In *Advances in the Study of Behaviour*, Vol. 5 (eds. D. S. Lehrman, J. S. Rosenblatt, R. A. Hinde, and E. Shaw), pp. 201–225. New York: Academic Press.

McFarland, D. J. (1976). Form and function in the temporal organization of behaviour. In *Growing Points in Ethology* (eds. P. P. G. Bateson and R. A. Hinde), pp. 55–93. Cambridge University Press.

McFarland, D. J., and Budgell, P. W. (1970) Determination of a behavioural transfer function by frequency analysis. *Nature*, **226**, 966–967.

McFarland, D. J., and L' Angellier, A. B. (1966). Disinhibition of drinking during satiation of feeding behaviour in the Barbary dove. *Animal Behaviour*, **14**, 463–467.

McFarland, D. J., and Rolls, B. (1972). Suppression of feeding by intracranial injections of angiotensin. *Nature*, **236**, 172–173.

McFarland, D. J., and Sibly, R. M. (1975). The behavioural final common path. *Philosophical Transactions of the Royal Society of London, B*, **270**, 265–293.

McFarland. D. J., and Wright, P. (1969). Water conservation by inhibition of food intake. *Physiology and Behaviour*, **4**, 95–99.

McGill, T. E. (1965). Studies on the sexual behaviour of male laboratory mice: effects of genotype, recovery of sex drive, and theory. In *Sex and Behaviour* (ed. F. A. Beach), pp. 76–88. New York: Wiley.

McGill, T. E., and Manning, A. (1976). Genotype and retention of the ejaculatory reflex in castrated male mice. *Animal Behaviour*, **24**, 507–518.

McGrath, M. J., and Cohen, D. B. (1978). REM sleep facilitation of adaptive waking behaviour: a review of the literature. *Psychological Bulletin*, **85**, 24–57.

Mach, E. (1960). *The Science of Mechanics*. La Salle: Open Court.

McKinley, M. J., Denton, D. A., and Weisinger, R. S. (1978). Sensors for antidiuresis and thirst—osmoreceptors or CSF sodium detectors? *Brain Research*, **141**, 89–103.

McReynolds, P. (1962). Exploratory behaviour: a theoretical interpretation. *Psychological Reports*, **11**, 311–318.

Manning, A. (1979). *An Introduction to Animal Behaviour*. London: Edward Arnold.

Marler, P., and Hamilton, W. J. (1967). *Mechanisms of Animal Behaviour*. New York: Wiley.

Marwine, A., and Collier, G. (1979). The rat at the waterhole. *Journal of Comparative and Physiological Psychology*, **93**, 391–402.

Mayer, J. (1955). Regulation of energy intake and the body weight: the glucostatic theory and lipostatic hypothesis. *Annals of the New York Academy of Sciences*, **63**, 15–43.

Meddis, R. (1975). On the function of sleep. *Animal Behaviour*, **23**, 676–691.

Meddis, R. (1977). *The Sleep Instinct*. London: Routledge and Kegan Paul.

Mendelson, J. (1977). Airlicking and cold licking in rodents. In *Drinking Behaviour: Oral Stimulation, Reinforcement and Preference* (eds. J. A. W. M. Weijen and J. Mendelson), pp. 157–196. New York: Plenum Press.

Menzel, E. W. (1963). The effects of cumulative experience on responses to novel objects in young isolation-reared chimpanzees. *Behaviour*, **21**, 1–12.

Menzel, E. (1978). Cognitive mapping in chimpanzees. In *Cognitive Processes in Animal Behaviour* (eds. S. H. Hulse, H. Fowler, and W. K. Honig), pp. 375–422. Hillsdale: Lawrence Erlbaum.

Michael, R. P., and Zumpe, D. (1978). Potency in male rhesus monkeys: effects of continuously receptive females. *Nature*, **200**, 451–452.

Milgram, N. W. (1979). On the inadequacy of a homeostatic model: where do we go from here? *The Behavioural and Brain Sciences*, **2**, 111–112.

Milgram, N. W., Krames, L., and Alloway, T. M. (1977). *Food Aversion Learning*. New York: Plenum Press.

Milgram, N. W., Krames, L., and Thompson, R. (1974). Influence of drinking history on food deprived drinking in the rat. *Journal of Comparative and Physiological Psychology*, **87**, 126–133.

Miller, D. G., and Kutscher, C. L. (1975). Effects of nutritive and bulk intake on the suppression of food-deprivation polydipsia in the gerbil. *Physiology and Behaviour*, **14**, 791–794.

Miller, N. E. (1955). Shortcomings of food consumption as a measure of hunger; results from other behavioural techniques. *Annals of the New York Academy of Sciences*, **63**, 141–143.

Miller, N. E. (1967). Behavioural and physiological techniques: rationale and experimental designs for combining their use. In *Handbook of Physiology*, sec. 6, *Alimentary Canal*, Vol. 1, *Control of Food and Water Intake'* (ed. C. F. Code), pp. 51–61. Washington, D.C.: American Physiological Society.

Milner, P. M. (1977). Theories of reinforcement, drive, and motivation. In *Handbook of Psychopharmacology*, Vol 7, *Principles of Behavioural Pharmacology* (eds. L. L. Iversen, S. D. Iversen, and S. H. Snyder), pp. 181–200. New York: Plenum Press.

Misanin, J. R., Smith, N. F., and Campbell, B. A. (1964). Relation between blood sugar level and random activity during food and water deprivation. *Psychonomic Science*, **1**, 63–64.

Mitchell, D., Atkins, A. R., and Wyndam, C. H. (1972). Mathematical and physical models of thermoregulation. In *Essays on Temperature Regulations* (eds. J. Bligh and R. Moore), pp. 37–54. Amsterdam: North-Holland Publishing Company.

Mitchell, D., Scott, D. W., and Williams, K. D. (1973). Container neophobia and the rat's preference for earned food. *Behavioural Biology*, **9**, 613–624.

Mogenson, G. J. (1977). *The Neurobiology of Behaviour*. Hillsdale: Lawrence Erlbaum.

Mook, D. G. (1974). Saccharin preference in the rat: some unpalatable findings. *Psychological Review*, **81**, 475–490.

Mook, D. G., and Kenney, N. J. (1977). Taste modulation of fluid intake. In *Drinking Behaviour: Oral Stimulation, Reinforcement and Preference* (eds. J. A. W. M. Weijnen and J. Mendelson), pp. 275–315. New York: Plenum Press.

Morin, L. P. (1977). Progesterone: inhibition of rodent sexual behaviour. *Physiology and Behaviour*, **18**, 701–715.

Morrison, S. D. (1968). Regulation of water intake by rats deprived of food. *Physiology and Behaviour*, **3**, 75–81.

Moruzzi, G. (1969). Sleep and instinctive behaviour. *Archives Italiennes de Biologie*, **107**, 175–216.

Moyer, K. E. (1968). Kinds of aggression and their physiological basis. *Communications in Behavioural Biology, Part A*, **2**, 65–87.

Mrosovsky, N., and Powley, T. L. (1977). Set points for body weight and fat. *Behavioural Biology*, **20**, 205–225.

Mugford, R. A. (1977). External influences on the feeding of carnivores. In *The Chemical Senses and Nutrition*. (eds. M. R. Kare and O. Maller), pp. 25–50. New York: Academic Press.

Murgatroyd, D., and Hardy, J. D. (1970). Central and peripheral temperatures in behavioural thermoregulation of the rat. In *Physiological and Behavioural Temperature Regulation* (eds. J. D. Hardy, A. P. Gagge, and J. A. J. Stolwijk), pp. 894–891. Springfield: C. C. Thomas.

Murray, E. J. (1965). *Sleep, Dreams and Arousal*. New York: Appleton-Century-Crofts.

Myers, A. K., and Miller, N. E. (1954). Evidence for learning motivated by 'exploration'. *Journal of Comparative and Physiological Psychology*, **28**, 305–312 (reprinted in Fowler, 1965).

Nadel, L., and O'Keefe, J. (1974). The hippocampus in pieces and patches: an essay on modes of explanation in physiological psychology. In *Essays on the Nervous System—A Festschrift for Professor J. Z. Young* (eds. R. Bellairs and E. G. Gray), pp. 368–390. Oxford: Clarendon Press.

Newman, J. C., and Booth, D. A. (1979). Gastrointestinal and metabolic consequences of a rat's meal on maintenance diet *ad libitum*. (To be submitted.)

Nicolaidis, S., and Rowland, N. (1975). Systemic versus oral and gastrointestinal metering of fluid intake. In *Control Mechanisms of Drinking* (eds. G. Peters, J. T. Fitzsimons, and L. Peters-Haefeli), pp. 14–21. Berlin: Springer–Verlag.

Novin, D. (1976). Visceral mechanisms in the control of food intake. In *Hunger: Basic Mechanisms and Clinical Implications* (eds. D. Novin, W. Wyricka, and G. Bray), pp. 247–311. New York: Raven Press.

Novin, D., and VanderWeele, D. A. (1977). Visceral involvement in feeding: there is more to regulation than the hypothalamus. In *Progress in Psychobiological Psychology*, Vol 7 (eds. J. M. Sprague and A. N. Epstein), pp. 193–241. New York: Academic Press.

Oatley, K. (1964). Changes in blood volume and osmotic pressure in the production of thirst. *Nature*, **202**, 1341–1342.

Oatley, K. (1971). Dissociation of the circadian drinking pattern from eating. *Nature*, **299**, 494–496.

Oatley, K. (1973). Simulation and theory of thirst. In *The Neuropsychology of Thirst: New Findings and Advances in Concepts* (eds. A. N. Epstein, H. R. Kissileff, and E. Stellar), pp. 199–223. Washington, D. C.: V. H. Winston.

Oatley, K. (1978). *Perceptions and Representations*. London: Methuen.

Oatley, K., and Toates, F. M. (1969). The passage of food through the gut of rats and its uptake of fluid. *Psychonomic Science*, **16**, 225–226.

O'Keefe, J., and Nadel, L. (1978). *The Hippocampus as a Cognitive Map*. Oxford: Clarendon Press.

Olton, D. S. (1978). Characteristics of spatial memory. In *Cognitive Processes in Animal Behaviour* (eds. S. H. Hulse, H. Fowler, and W. K. Honig), pp. 341–373. Hillsdale: Lawrence Erlbaum.

Olton, D. S., and Isaacson, R. L. (1968). Hippocampal lesions and active avoidance. *Physiology and Behaviour*, **3**, 719–724.

Osborne, G. L. (1977). Differences in locomotor activity between rats and gerbils in response to novelty. *Behavioural Biology*, **19**, 548–553.

Otis, L. S., and Cerf, J. A. (1963). Conditioned avoidance learning in two fish species. *Psychological Reports*, **12**, 679–682.

Overmann, S. R., and Yang, M. G. (1973). Adaptation to water restriction through dietary selection in weanling rats. *Physiology and Behaviour*, **11**, 781–786.

Pappenheimer, J. R. (1976). The sleep factor. *Scientific American*, **235(2)**, 24–29.

Peck, J. W. (1973). Thirst(s) resulting from bodily water imbalances. In *The Neuropsychology of Thirst: New Findings and Advances in Concepts*. (eds. A. N. Epstein, H. R. Kissileff, and E. Stellar, pp. 113–116. Washington, D.C.: V. H. Winston.

Peck, J. W. (1978). Rats defend different body weights depending on palatability and accessibility of their food. *Journal of Comparative and Physiological Psychology*, 92, 555–570.

Peck, J. W. (1979). Thirst, homeostasis and bodily fluid deficits. *The Behavioural and Brain Sciences*, 2, 114–115.

Peters, P. J., Bronson, F. H., and Whitsett, J. M. (1972). Neonatal castration and intermale aggression in mice. *Physiology and Behaviour*, 8, 265–268.

Petrinovich, L., and Bolles, R. (1954). Deprivation states and behavioural attributes. *Journal of Comparative and Physiological Psychology*, 47, 450–453.

Pfaff, D. (1970). Nature of sex hormone effects on rat sex behaviour: specificity of effects and individual patterns of response. *Journal of Comparative and Physiological Psychology*, 73, 349–358.

Pierce, J. T., and Nuttall, R. L. (1961). Self-paced sexual behaviour in the female rat. *Journal of Comparative and Physiological Psychology*, 54, 310–313.

Pilcher, C. W. T., Jarman, S. P., and Booth, D. A. (1974). The route of glucose to the brain from food in the mouth of the rat. *Journal of Comparative and Physiological Psychology*, 87, 56–61.

Plapinger, L., and McEwen, B. S. (1978). Gonadal steroid–brain interactions in sexual differentiation. In *Biological Determinants of Sexual Behaviour* (ed. J. B. Hutchison), pp. 153–218. Chichester: Wiley.

Porter, J. H., and Bryant, W. E. (1978). Acquisition of schedule induced polydipsia in the Monogolian gerbil. *Physiology and Behaviour*, 21, 825–827.

Powers, J. B. (1970). Hormonal control of sexual receptivity during the estrous cycle of the rat. *Physiology and Behaviour*, 5, 831–835.

Pring-Mill, A. F. (1979). Tolerable feedback: A mechanism for behavioural change. *Animal Behaviour*, 27, 226–236.

Rabe, E. F. (1975). Relationship between absolute body-fluid deficits and fluid intake in the rat. *Journal of Comparative and Physiological Psychology*, 89, 468–477.

Rachlin, H. (1976). *Introduction to Modern Behaviourism*. San Francisco: W. H. Freeman.

Ramsey, D. J., Rolls, B. J., and Wood, R. J. (1977). Body fluid changes which influence drinking in the water deprived rat. *Journal of Physiology*, 266, 453–469.

Reid, L. S., and Finger, F. W. (1955). The rat's adjustment to 23-hour food-deprivation cycles. *Journal of Comparative and Physiological Psychology*, 48, 110–113.

Revusky, S. (1974). Retention of a learned increase in the preference for a flavoured solution. *Behavioural Biology*, 11, 121–125.

Revusky, S. (1977a). Learning as a general process with an emphasis on data from feeding experiments. In *Food Aversion Learning* (eds. N. W. Milgram, L. Krames, and T. M. Alloway), pp. 1–51. New York: Plenum Press.

Revusky, S. (1977b). Interference with progress by the scientific establishment: Examples from aversion learning. In *Food Aversion Learning* (eds. N. W. Milgram, L. Krames, and T. M. Alloway), pp. 53–71. New York: Plenum Press.

Richards, S. A. (1976). Behavioural temperature regulation in the fowl. *Journal of Physiology*, 258, 122P–123P.

Richter, C. P. (1943). Total self regulatory functions in animals and human beings. *The Harvey Lectures*, 38, 63–103.

Robbins, T. (1978). A strange scientific tail. *New Scientist*, 79, 764–766.

Roberts, W. W., and Martin, J. R. (1974). Peripheral thermoregulatory responses to the rat. *Journal of Comparative and Physiological Psychology*, 87, 1109–1118.

Roberts, W. W., and Martin, J. R. (1977). Effects of lesions in central thermosensitive

286

areas on thermoregulatory responses in rats. *Physiology and Behaviour*, **19**, 503–511.

Rodgers, R. J., and Brown, K. (1976). Amygdaloid function in the central cholinergic mediation of shock induced aggression in the rat. *Aggressive Behaviour*, **2**, 131–152.

Rodgers, W., and Rozin, P. (1966). Novel food preferences in thiamine-deficient rats. *Journal of Comparative and Physiological Psychology*, **61**, 1–4.

Rolls, B. J., and McFarland, D. J. (1973). Hydration releases inhibition of feeding produced by intracranial angiotensin. *Physiology and Behaviour*, **11**, 881–884.

Rolls, B. J., Wood, R. J., and Stevens, R. M. (1978). Palatability and body fluid homeostasis. *Physiology and Behaviour*, **20**, 15–19.

Rowland, N. (1977). Regulatory drinking: do the physiological substrates have an ecological niche? *Biobehavioural Reviews*, **1**, 261–272.

Rowland, N., and Flamm, C. (1977). Quinine drinking: more regulatory puzzles. *Physiology and Behaviour*, **18**, 1165–1170.

Rowland, N., and Nicolaidis, S. (1976). Metering of fluid intake and determinants of ad libitum drinking in rats. *American Journal of Physiology*, **231**, 1–8.

Rozin, P. (1976). Psychobiological and cultural determinants of food choice. In *Appetite and Food Intake* (ed. T. Silverstone), pp. 285–312. West Berlin: Abakon/Dahlem.

Rozin, P., and Rodgers, W. (1967). Novel diet preferences in vitamin deficient rats and rats recovered from vitamin deficiencies. *Journal of Comparative and Physiological Psychology*, **63**, 421–428.

Rozin, P., and Kalat, J. W. (1971). Specific hungers and poison avoidance as adaptive specialization of learning. *Psychological Review*, **78**, 459–486.

Russek, M. (1971). Hepatic receptors and the neurophysiological mechanisms controlling feeding behaviour. In *Neurosciences Research*, Vol. 4 (eds. S. Ehrenpreis and O. C. Solnitzky), pp. 213–282.

Russek, M. (1976). A conceptual equation of intake control. In *Hunger: Basic Mechanisms and Clinical Implications* (eds. D. Novin, W. Wyrwicka, and G. Bray), pp. 327–347. New York: Raven Press.

Russell, P. A. (1973). Relationships between exploratory behaviour and fear: a review. *British Journal of Psychology*, **64**, 417–433.

Sachs, B. D., and Barfield, R. J. (1976). Functional analysis of masculine copulatory behaviour in the rat. In *Advances in the Study of Behaviour*, Vol.7 (eds. J. S. Rosenblatt, R. A. Hinde, E. Shaw, and C. Beer), pp. 91–154. New York: Academic Press.

Satinoff, E., and Hendersen, R. (1977). Thermoregulatory behaviour. In *Handbook of Operant Behaviour* (eds. W. K. Honig and J. E. R. Staddon), pp. 153–173. Englewood Cliffs: Prentice-Hall.

Sawchenko, P. E., and Friedman, M. I. (1979). Sensory functions of the liver–a review. *American Journal of Physiology*, **236**, R5–R20.

Schein, M. W., and Hale, E. B. (1965). Stimuli eliciting sexual behaviour. In *Sex and Behaviour* (ed. F. A. Beach), pp. 416–440. New York: Wiley.

Schmidt, I., and Rautenberg, W. (1975). Instrumental thermoregulatory behaviour in pigeons. *Journal of Comparative Physiology*, **101**, 225–235.

Scott, J. P. (1958). *Aggression*. Chicago: University of Chicago Press.

Seligman, M. E. P. (1970). On the generality of the laws of learning. *Psychological Review*, **77**, 406–418.

Senseman, D. M. (1977). Gastropod mollusks as model systems for the study of integrative mechanisms controlling feeding behaviour. In *The Chemical Senses and Nutrition* (eds. M. R. Kare and O. Maller), pp. 3–23. New York: Academic Press.

Seward, J. P. (1956). Comments on Professor Beach's paper. *Nebraska Symposium on Motivation*, **4**, 32–38.

Shillito, E. E. (1963). Exploratory behaviour in the short-tailed vole *Microtus agrestis*. *Behaviour*, **21**, 145–154.

Shorten, M. (1954). The reaction of the brown rat towards changes in its environment. In

Control of Rats and Mice, Vol. 2 (ed. D. Chitty), pp. 307–334. Oxford: Clarendon Press.

Sibly, R. (1975). How incentive and deficit determine feeding tendency. *Animal Behaviour*, **23**, 437–446.

Simmel, E. C., and Eleftheriou, B. E. (1977). Multivariate and behaviour genetic analysis of avoidance of complex visual stimuli and activity in recombinant inbred strains of mice. *Behaviour Genetics*, **7**, 239–250.

Sinclair, J. D., and Bender, D. D. (1978). Compensatory behaviours: suggestion for a common basis from deficits in hamsters. *Life Sciences*, **22**, 1407–1412.

Singer, J. J. (1968). Hypothalamic control of male and female sexual behaviour in female rats. *Journal of Comparative and Physiological Psychology*, **66**, 738–742.

Singer, I. (1973). *The Goals of Human Sexuality*. London: Wildwood House.

Skinner, B. F. (1938). *The Behaviour of Organisms*. New York: Appleton.

Skinner, B. F. (1974). Beyond Freedom and Dignity. Harmondsworth: Penguin.

Smith, M. H. (1972). Evidence for a learning component of sodium hunger in rats. *Journal of Comparative and Physiological Psychology*, **78**, 242–247.

Snyder, F. (1966). Towards an evolutionary theory of dreaming. *American Journal of Psychiatry*, **123**, 121–136.

Södersten, P., and Hansen, S. (1978). Effects of castration and testosterone, dihydrotestosterone or oestradiol replacement treatment in neonatal rats on mounting behaviour in the adult. *Journal of Endocrinology*, **76**, 251–260.

Sokolov, E. N. (1960). Neuronal models and the orienting reflex. In *The Central Nervous System and Behaviour* (ed. M. A. B. Brazier), pp. 187–276. New York: Josiah Macy Jr. Foundation.

Sperelakis, N. (1970). Effects of temperature on membrane potentials of excitable cells. In *Physiological and Behavioural Temperature Regulation* (eds. J. D. Hardly, A. P. Gagge, and J. A. J. Stolwijk), pp. 408–441. Springfield: C. C. Thomas.

Staddon, J. E. R. and Simmelhag, V. L. (1971). The 'Superstition' experiment: a re-examination of its implications for the principles of adaptive behaviour. *Psychological Review*, **78**, 3–43.

Stern, J. J. (1969). Neonatal castration, androstenedione, and the mating behaviour of the male rat. *Journal of Comparative and Physiological Psychology*, **69**, 608–612.

Still, A. W. (1966). Memory and spontaneous alternation in the rat. *Nature*, **210**, 400–401.

Still, A. W., and Macmillan, S. C. (1975). Location by odour and turn selection as two stages in the spontaneous alternation of rats. *Animal Behaviour*, **23**, 447–449.

Stricker, E. M. (1973). Thirst, sodium appetite, and complementary physiological contributions to the regulation of intravascular fluid volume. In *The Neuropsychology of Thirst: New Findings and Advances in Concepts* (eds. A. N. Epstein, H. R. Kissileff, and E. Stellar), pp. 73–98. Washington, D.C.: V. H. Winston.

Stricker, E. M. (1977). The renin–angiotensin system and thirst: a re-evaluation. II. Drinking elicited in rats by Caval ligation or isoproterenol. *Journal of Comparative and Physiological Psychology*, **91**, 1220–1231.

Stricker, E. M., and Wolf, G. (1967). Hypovolemic thirst in comparison with thirst induced by hyperosmolality. *Physiology and Behaviour*, **2**, 33–37.

Stricker, E. M., and Wolf, G. (1969). Behavioural control of intravascular fluid volume: thirst and sodium appetite. *Annals of the New York Academy of Sciences*, **157**, 553–568.

Stunkard, A. (1975). Satiety as a conditioned reflex. *Psychosomatic Medicine*, **37**, 383–387.

Sutherland, N. S. (1957). Spontaneous alternation and stimulus avoidance. *Journal of Comparative and Physiological Psychology*, **50**, 358–362.

Tauber, E. S. (1974). Phylogeny of sleep. In *Advances in Sleep Research*, Vol. 1 (ed. E. D. Weitzman), pp. 133–172. New York: Spectrum.

Taylor, G. T. (1974). Stimulus change and complexity in exploratory behaviour. *Animal Learning and Behaviour*, **2**, 115–118.

Templeton, J. R. (1970). Reptiles. In *Comparative Physiology of Thermoregulation* (ed. G. C. Whittow), pp. 135–166. New York: Academic Press.

Thompson, M. L., McGill, T. E., McIntosh, S. M., and Manning, A. (1976). The effects of adrenalectomy on the sexual behaviour of castrated and intact BDF_1 mice. *Animal Behaviour*, **24**, 519–522.

Thompson, R. W., and Lippman, L. G. (1972). Exploration and activity in the gerbil and rat. *Journal of Comparative and Physiological Psychology*, **80**, 439–448.

Tiefer L. (1969). Copulatory behaviour of male *Rattus norvegicus* in a multiple female exhaustion test. *Animal Behaviour*, **17**, 718–721.

Tinbergen, N. (1951). *The Study of Instinct*. Oxford: Clarendon Press.

Toates, F. M. (1971). 'Thirst, and Body Fluid regulation in the Rat'. DPhil. thesis, University of Sussex.

Toates, F. M. (1974). Computer simulation and the homeostatic control of behaviour. In *Motivational Control System Analysis* (ed. D. J. McFarland), pp. 407–426. London: Academic Press.

Toates, F. M. (1975). *Control Theory in Biology and Experimental Psychology*. London: Hutchinson Educational.

Toates, F. M. (1978). A physiological control theory of the hunger–thirst interaction. In *Hunger Models—Computable Theory of Feeding Control* (ed. D. A. Booth), pp. 347–373. London: Academic Press.

Toates, F. M. (1979a). Water and energy in the interaction of thirst and hunger. In *Chemical Influences on Behaviour* (eds. K. Brown and S. Cooper), pp. 135–200. London: Academic Press.

Toates, F. M. (1979b). Homeostasis and drinking. *The Behavioural and Brain Sciences*, **2**, 95–139.

Toates, F. M., and Archer, J. (1978). A comparative review of motivational systems using classical control theory. *Animal Behaviour*, **26**, 368–380.

Toates, F. M., and Booth, D. A. (1974). Control of food intake by energy supply. *Nature*, **251**, 710–711.

Toates, F. M., and Bowles, M. (1979). Time-sharing or competition? *Animal Behaviour*, **27**, 959–960.

Toates, F. M., and Ewart, B. (1977). Gerbil drinking patterns. *Animal Behaviour*, **25**, 782.

Toates, F. M., and Oatley, K. (1970). Computer simulation of thirst and water balance. *Medical and Biological Engineering*, **8**, 71–87.

Toates, F. M., and O'Rourke, C. (1978). Computer simulation of male rat sexual behaviour. *Medical and Biological Engineering and Computing*, **16**, 98–104.

Tolman, E. C. (1948). Cognitive maps in rats and men. *The Psychological Review*, **55**, 189–208.

Turpin, B. (1977). Variation of early social experience and environmental preference in rats. *Journal of Comparative and Physiological Psychology*, **91**, 29–32.

Vander, A. J., Sherman, J. H., and Luciano, D. S. (1975). *Human Physiology–The Mechanisms of Body Function*. New York: McGraw-Hill.

VanderWeele, D. A. (1979). Lack of fixed set-points in fluid homeostasis does not argue for learned satiety factors in drinking. *The Behavioural and Brain Sciences*, **2**, 121–122.

VanderWeele, D. A., and Sanderson, D. (1976). Peripheral glucosensitive satiety in the rabbit and the rat. In *Hunger: Basic Mechanisms and Clinical Implications* (eds. D. Novin, W. Wyrwicka, and G. Bray), pp. 383–393. New York: Raven Press.

VanderWeele, D. A., and Tellish, J. A. (1971). Adipsic to polydipsic shift in gerbils induced by food deprivation. *Psychological Reports*, **29**, 479–486.

Van Hemel, P. E., and Myer, J. S. (1970). Control of food-motivated instrumental behaviour in water-deprived rats by prior water and saline drinking. *Learning and Motivation*, **1**, 86–94.

Van Itallie, T. B., Smith, N. S., and Quartermain, D. (1977). Short-term and long-term components in the regulation of food intake: evidence for a modulatory role of carbohydrate status. *The American Journal of Clinical Nutrition*, **30**, 742–757.

Van Zoeren, J. G., and Stricker, E. M. (1977). Effects of preoptic, lateral hypothalamic, or dopamine-depleting lesions on behavioural thermoregulation in rats exposed to the cold. *Journal of Comparative and Physiological Psychology*, **91**, 989–999.

Vaughan, E., and Fisher, A. E. (1962). Male sexual behaviour induced by intracranial electrical stimulation. *Science*, **137**, 758–760.

Verney, E. B. (1947). The anti-diuretic hormone and the factors which determine its release. *Proceedings of the Royal Society, London*, **B135**, 25–106.

Walton, A. (1950). Patterns of male sex behaviour. *Proceedings of the Society for the Study of Fertility*, **1**, 40–44.

Webb, W. B. (1968). *Sleep—An Experimental Approach*. New York: Macmillan.

Webb, W. B. (1971). Sleep behaviour as a biorhythm. In *Biological Rhythms and Human Performance* (ed. W. P. Colquhoun), pp. 149–177. London: Academic Press.

Webb, W. B. (1975). *Sleep—The Gentle Tyrant*. Englewood Cliffs: Prentice-Hall.

Webb, W. B., and Cartwright, R. D. (1978). Sleep and dreams. *Annual Review of Psychology*, **29**, 223–252.

Weisler, A., and McCall, R. B. (1976). Exploration and play. *American Psychologist*, **31**, 492–508.

Weiss, B., and Laties, V. G. (1960). Magnitude of reinforcement as a variable in thermoregulatory behaviour. *Journal of Comparative and Physiological Psychology*, **53**, 603–608.

Weiss, I. P. (1969). 'Nutritional and non-nutritional factors underlying food-deprivation polydipsia in the gerbil (*Meriones ungiculatus*). PhD thesis, Syracuse University.

Welker, W. I. (1956). Some determinants of play and exploration in chimpanzees. *Journal of Comparative and Physiological Psychology*, **49**, 84–89.

Welker, W. I. (1961). An analysis of exploratory and play behaviour in animals. In *Functions of Varied Experience* (eds. D. W. Fiske and S. R. Maddi), pp. 175–226. Homewood: The Dorsey Press.

Westerterp, K. (1977). How rats economize – energy loss in starvation. *Physiological Zoology*, **50**, 331–362.

White, R. W. (1959). Motivation reconsidered: the concept of competence. *Psychological Review*, **66**, 297–333.

Whittow, G. C. (1976). Regulation of body temperature. In *Avian Physiology* (ed. P. D. Sturkie), pp. 146–173. Berlin: Springer-Verlag.

Wiepkema, P. R. (1968). Positive feedbacks at work during feeding. *Behaviour*, **39**, 266–273.

Wilkinson, R. T. (1965). Sleep deprivation. In *Physiology of Survival* (eds. O. G. Edholm and A. L. Bacharach), pp. 399–430. London: Academic Press.

Williams, R. A. (1968). Effects of repeated food deprivations and repeated feeding tests on feeding behaviour. *Journal of Comparative and Physiological Psychology*, **65**, 222–226.

Williams, C. D., and Kuchta, J. C. (1957). Exploratory behaviour in two mazes with dissimilar alternatives. *Journal of Comparative and Physiological Psychology*, **50**, 509–513.

Williams, H. L., Holloway, F. A., and Griffiths, W. J. (1973). Physiological psychology: sleep. *Annual Review of Psychology*, **24**, 279–316.

Wilson, J. R., Kuehn, R. E., and Beach, F. A. (1963). Modification in the sexual behaviour of male rats produced by changing the stimulus female. *Journal of Comparative and Physiological Psychology*, **56**, 636–644.

Wirtshafter, D., and Davis, J. D. (1977). Set points, settling points, and the control of body weight. *Physiology and Behaviour*, **19**, 75–78.

Wolf, A. V. (1958). *Thirst: Physiology of the Urge to Drink and Problems of Water Lack*. Springfield: C. C. Thomas.

Wolf, G., McGovern, J. F., and DiCare, L. V. (1974). Sodium appetite: Some conceptual and methodologic aspects of a model drive system. *Behavioural Biology*, **10**, 27–42.

Woods, P. J. (1962). Behaviour in a novel situation as influenced by the immediately preceding environment. *Journal of the Experimental Analysis of Behaviour*, **5**, 185–190.

Wright, J. W. (1976). Effect of hunger on the drinking behaviour of rodents adapted for mesic and xeric environment. *Animal Behaviour*, **24**, 300–304.

Wyrwicka, W. (1976). The problem of motivation of feeding behaviour. In *Hunger: Basic Mechanisms and Clinical Implications* (eds. D. Novin, W. Wyrwicka, and G. Bray), pp. 203–213. New York: Raven Press.

Wyrwicka, W. (1979). Sensory regulation of water intake. *The Behavioural and Brain Sciences*, **2**, 125.

Zeeman, E. C. (1976). Catastrophe theory. *Scientific American*, **234**, 65–83.

Zepelin, H., and Rechtschaffen, A. (1974). Mammalian sleep, longevity and energy metabolism. *Brain, Behaviour, and Evolution*, **10**, 425–470.

Zimbardo, P. G., and Montgomery, K. C. (1957). Effects of 'free-environment' rearing upon exploratory behaviour. *Psychological Reports*, **3**, 589–594.

Author Index

Subject Index